CHEMICAL PROPERTY ESTIMATION

Theory and Application

Edward J. Baum

Grand Valley State University
Allendale, Michigan

LEWIS PUBLISHERS

Boca Raton Boston London New York Washington, D.C.

Learning Resources
Centre

Acquiring Editor: Jane Kinney
Project Editor: Andrea Demby
Marketing Manager: Arline Massey
Cover design: Dawn Boyd
PrePress: Carlos Esser
Manufacturing: Carol Royal

Library of Congress Cataloging-in-Publication Data

Baum, Edward J.
    Chemical property estimation: theory and practice /
Edward J. Baum.
      p.   cm.
Includes bibliographical references and index.
ISBN 0-87371-938-7 (alk. paper)
      1. Pollutants--Analysis.   I. Title.
TD193.B38 1997
628.5--dc21   97-33124
                                                                      CIP

© 1998 by CRC Press LLC
Lewis Publishers is an imprint of CRC Press LLC

No claim to original U.S. Government works
International Standard Book Number 0-87371-938-7
Library of Congress Card Number 97-33124
Printed in the United States of America 1 2 3 4 5 6 7 8 9 0
Printed on acid-free paper

Preface

This book describes the theory and methods of chemical property estimation. My intention in writing it is to provide a resource, primarily for environmental scientists, for making informed decisions about emission rates, environmental clean-ups, and other issues that affect public health and well-being. But knowledge of methods of estimating chemical properties is also vital to scientists, engineers, and others who need to screen chemicals for laboratory testing, to design industrial processes, to identify outliers in experimental measurements, or to design monitoring and analysis methods. Anyone who needs to predict the behavior of chemicals in the environment, in industrial processes, or in the laboratory may need to use the estimation methods described in this book.

Good collections of data are available on some properties of chemicals such as aqueous solubility,* octanol-water partition coefficients,** and others. Chapters 10 and 11 of this book present useful data on soil partition coefficients and bioconcentration factors. But more frequently than not, the data you need will not be available, and you will have no choice but to estimate the values of chemical properties.

Moreover, published chemical property data are not always reliable. Many of the published measurements of vapor pressure, octanol-water partition coefficients, and aqueous solubility of nonvolatile and hydrophobic chemicals are of poor quality. Often, data collections contain estimated values without identifying them as such. The literature published before 1985 is particularly suspect. Measured values of chemical properties are usually more accurate than estimated values and, where they are available, you should use them. But you may also need to assess the quality of suspicious data by comparing it with estimates obtained using the methods described in this book.

Some excellent books on chemical property estimation are in print, including Lyman, Reehl, and Rosenblatt's handbook of estimation methods*** and Reid, Prausnitz, and Poling's monograph**** on the properties of gases and liquids. But, thanks to recent advances in analytical chemistry, more accurate data on the properties of chemicals of particular interest to environmental scientists have become available. And the amount of information on chemical property estimation has grown explosively in the years since these books were published. This information explosion has resulted in the development of estimation methods that offer substantial improvements in accuracy, reliability, and applicability, even in comparison to what was available just five years ago. Many of the methods described by Lyman et al. and Reid et al. have been improved or replaced, and new approaches have been developed to deal with difficult problems that were formerly ignored.

---

* Yalkowsky, S.H. and Dannenfelser, R.M., 1990. Arizona Database. 5th ed., College of Pharmacy, University of Arizona, Tucson.
** Leo, A. and Weininger, D., 1989. CLOGP, Version 3.54, Medicinal Chemistry Project, Pomona College, Claremont, CA.
*** Lyman, W. J., Reehl, W. F., and Rosenblatt, D. H., 1990. *Handbook of Chemical Property Estimation Methods*, American Chemical Society, Washington, D.C.
**** Reid, R. C., Prausnitz, J. M., and Poling, B. E., 1987. *The Properties of Gases and Liquids*, McGraw-Hill Book Company, New York.

Another factor making sophisticated estimation methods more accessible is the widespread availability of powerful microcomputers, software, and computer databases. This has led to development of many new fragment contribution schemes, as well as to the refinement of the best of the older methods described in this book.

With a few exceptions, the estimation methods I describe are methods that have been developed or revised recently. To use these methods, you don't need a high degree of chemical sophistication, and most of the input needed is readily available to environmental specialists. Yet, these methods produce accurate estimates of the properties of a broad range of chemicals of interest in the range of values observed under normal environmental conditions.

I have included methods suited to most needs, from back-of-the-envelope estimates to elaborate computer-aided computations, collected from a broad range of published environmental, industrial, and academic research. With few exceptions, as noted, these methods have been properly validated; I have included relevant statistics, range of applicability, and computational resources, as well.

ABOUT THE ORGANIZATION OF THIS BOOK

This book is an outgrowth of my consulting work in risk assessment and my experience teaching a special-topics course in the subject. Often, junior environmental specialists I have met haven't had the training or gained the experience to use chemical property estimation methods with confidence. Sometimes they use them improperly. In order to explain why some methods can be properly applied to a specific problem and others cannot, I've often found it necessary to explain the theoretical basis of chemical property estimation.

This is why I open with chapters describing the theory of chemical property estimation and its uses. Next are several chapters on estimating the partition coefficients that control the way chemicals are distributed between air, water, soil, and biota. The book concludes with several chapters on the properties of chemicals that control mass transport within the environment.

The chapters begin with parameter definitions and general background on commonly used units, data sources, and so on. Next comes the theoretical information you need to appreciate the strengths and limitations of the methods discussed in the chapter. I have designed example calculations to illustrate each method, as well as including practical advice on when it is most appropriate and what computer software, databases, or Internet resources might be helpful to you. Footnotes and appendices contain relevant background on the calculus, statistics, and scientific theory. Everything you need to know to apply estimation methods properly is included.

By presenting the physical and chemical principles and mathematical methods that underlie chemical property estimation in what I hope is a clear and understandable way, I hope to promote wider use of this quantitative environmental science. Most people (including some experts) currently understand environmental systems only in a qualitative way: we can name toxic contaminants, identify pathways of contaminant transport, and list potential adverse effects or endangered species. But our understanding of the behavior and impact of chemicals in the environment will be greatly enhanced as we learn to deal with environmental systems in a quantitative fashion. We need to predict exposure levels and assess the quantitative risks associated with the release of contaminants into the environment, if we are to meet the goal of sustaining and improving our environment.

I wrote this book while on sabbatical from Grand Valley State University, and I acknowledge the support of my university and my colleagues. I am particularly grateful to Laurel Balkema and others on the library staff for their cheerful and generous assistance. I am also grateful for the advice and suggestions I received from many of my students and colleagues. I owe a debt of gratitude as well to my teachers and mentors: the physical chemists, James N. Pitts, Jr., and James B. Ramsay, and the geochemist, George Tunnell. Melissa Brown contributed enormously to the quality of the presentation through her expert editorial skills. Finally, I thank my publisher for help and patience along the path from concept to publication of this book.

I welcome your comments and suggestions for improving this book. Please send them to me at baume@gvsu.edu.

About the Author

Edward J. Baum, Ph.D., is Professor of Chemistry at Grand Valley State University in Grand Rapids, Michigan; he has held this appointment since 1983. He received his B.Sc., chemistry, from UCLA in 1961 and Ph.D. in physical chemistry from UC Riverside in 1965. In 1967, after postdoctoral studies at UC Riverside and the University of York, Professor Baum joined the faculty of the Oregon Graduate Center, where he organized and directed the Department of Environmental Technology and the Northwest Pollution Research Institute. He held a joint appointment as Clinical Professor of Environmental Medicine at the University of Oregon Medical School.

Professor Baum left Oregon in 1978 to serve as Chief Technical Adviser to the United Nations Educational, Scientific, and Cultural Organization (UNESCO) in São Paulo, Brazil, where he helped implement graduate level research and training programs in analytical chemistry and in environmental chemistry.

Dr. Baum has been a consultant to UNESCO, the National Research Council, the Federal Highway Administration, the Oregon Department of Environmental Quality, and to major corporations and environmental engineering firms. He is the author or co-author of 37 publications in physical and environmental chemistry and holds one patent for a pollution control device. He is a member of the American Chemical Society, the International Chemometrics Society, and the International QSAR Modeling Society.

Photo credit: Melissa Brown

To Melissa

Table of Contents

Appendices

Estimating the Properties of Chemicals: The Foundation of Environmental Research

1.1 INTRODUCTION

We live in a world that is widely contaminated with high levels of chemicals that are persistent and potentially damaging to human health and welfare. Already, we spend a significant amount of time and money attempting to remedy and control pollution. And every year, the chemical industry invents thousands of new chemical products. Many are potentially dangerous; some will find their way into the environment in appreciable quantities; some will be hazardous to human health and welfare.

If we are to predict the behavior of these chemicals and control our exposure to them, we must know their physical and chemical properties. But we know all too little about the properties of the environmental contaminants we already deal with, and we know less about the new and novel chemical products introduced each year. The properties of interest to environmental specialists have not been measured for most compounds released into the environment.

Why is it that we know so little about the chemical properties of existing environmental contaminants? There are many reasons. It may be impossible to measure a chemical's properties satisfactorily, for one thing. Perhaps an authentic sample of the compound isn't available. Perhaps the chemical decomposes during the process of measuring its properties, a problem that arises in melting-point and boiling-point measurement, for example. Perhaps there simply isn't a reliable way to measure the chemical property; we have only recently developed methods of measuring the very low vapor pressures and aqueous solubilities that some chemicals exhibit under ambient conditions.

Cost is another factor. Laboratory measurements take time and money; it's impossible for environmental specialists to keep up as thousands of new chemicals

1

are added each year to the millions already in existence. Approximately 30,000 or more commercially successful chemicals have already been released into the environment in significant amounts, and hundreds more are added to this list each year. Yet a single measurement of vapor pressure or solubility can cost thousands of dollars and take days to perform. Measuring a property such as a chemical's Henry's law constant, soil sorption coefficient, or bioconcentration factor can cost considerably more. The amount of reliable data available for any one chemical is likely to be limited. And since environmental conditions differ from one location to the next, and from time to time, the scanty information available will probably not be relevant to the problem at hand. For instance, the environmental specialist who wants to assess exposure to a certain chemical may need to know its vapor pressure in a mixture at 298 K but could discover that the only information available relates to the saturation vapor pressure of the pure compound at 273 K. More likely, only the melting point or normal boiling point of the chemical will be available.

If the environmental specialist doesn't know the properties of a chemical, he or she can't predict how it will be distributed or what its fate will be in the environment. *Determining chemical properties is a key step in determining how pollutants will affect the quality of the environment.* It is also a prerequisite for developing rational policies about emission rates, cleanups, and other decisions that affect public health and well-being.

Compared to most measurement methods, the computational methods of estimating chemical properties that are described in this book cost very little. When they are properly applied, they give results that are sufficiently accurate to be useful in most environmental applications. This is why estimation methods are so useful in studying the distribution, fate, and impact of environmental contaminants.

In the following sections of this chapter, we will examine the scope of chemical property estimation methods today. The chapter describes important physicochemical properties that control chemical behavior in environmental systems, as well as exploring the kinds of mathematical relationships used in chemical property estimation. We'll discuss some details concerning units of measure, then conclude with an overview of the organization of this book.

1.2 WHY WE ESTIMATE CHEMICAL PROPERTIES

Who Uses Chemical Property Estimates?

Research scientists and engineers use chemical property estimates to get a rough idea of the behavior of a chemical. Researchers ordinarily require higher accuracy than that offered by the usual estimation method, but a rough estimate may be adequate to select chemicals for a study, to design an instrument or a measurement technique, or to investigate a suspect experimental measurement.

Regulatory personnel and environmental specialists use the information to screen chemicals for laboratory testing and to set regulatory priorities. Estimates of chemical properties are used as input to screening and management models that predict the distribution and fate of chemicals in the environment. The models are used as aids in designing environmental cleanups and in establishing permissible emission limits.

In general, experts use various stochastic (statistical) and deterministic (non-statistical) thermodynamic models to describe the way contaminants are distributed in the environment. Similarly, experts use various stochastic and deterministic kinetic models to describe the rate of transport of contaminants in the environment. These models require input of the values of various partition coefficients, diffusion coefficients, and other chemical properties.

A major problem facing environmental specialists is the difficulty of measuring key chemical properties of important classes of chemicals and the resulting unreliability of measured values reported in the literature. The unreliability of published values of aqueous solubilities and octanol–water partition coefficients of hydrophobic chemicals, for example, is frequently mentioned as a problem of special concern. As chemical property estimation methods grow increasingly sophisticated and accurate, they grow increasingly useful for identifying questionable values of chemical properties that have been published.

1.3 PREDICTING ENVIRONMENTAL PARTITIONING AND TRANSPORT

Understanding the need to estimate certain chemical properties begins with understanding, first, the way the environment itself is modeled and second, the assumptions on which our models of chemical partitioning and transport are based.

For convenience, we imagine the environment as divided into four major compartments: soil, water, air, and biota. Each compartment contains several phases: solid, liquid, and vapor. Chemicals are spontaneously transported between and within compartments from regions of high **chemical potential*** to regions of low chemical potential. The chemical transport processes are driven by departures from thermal, mechanical, and material equilibrium. The scheme is illustrated in Figure 1.1.

While net transport of chemicals between phases is the result of departures from equilibrium, the problem of modeling chemical partitioning is much simpler if we assume that the system is at equilibrium. The partitioning of chemicals among phases can then be modeled using estimates of gas–liquid, liquid–solid, and gas–solid partition coefficients.

Chemicals are spontaneously transported within phases along concentration gradients from areas of high concentration to areas of low concentration. Transport occurs by the processes of **advection**, **dispersion**, and **molecular diffusion**.** The rate of transport may be calculated with the help of estimated molecular diffusivities. Estimates of properties such as density and viscosity may also be required in order to form a detailed picture of chemical transport in the environment.

---

* **Chemical potential** is the partial molar Gibbs free energy of a chemical. It is a measure of the chemical's tendency to undergo chemical or physical change.
** **Advection** is transport resulting from mass flow in a uniform fluid medium. **Diffusion** is migration along a concentration gradient due to molecular motion. **Dispersion** (turbulent diffusion) is spreading due to turbulence or due to the complex structure of the medium.

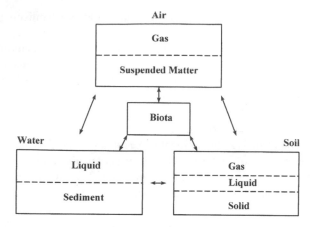

Figure 1.1 Partitioning and transport among environmental compartments.

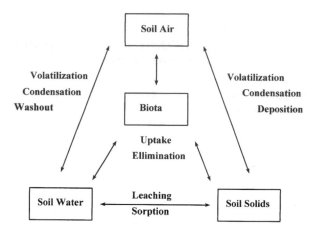

Figure 1.2 Partitioning and transport in the soil column.

Consider the partitioning and transport of chemicals deposited in a soil column as shown in Figure 1.2. The chemicals are found sorbed on soil particles and biota, dissolved in soil solution, and vaporized within soil pore spaces as well as within dislocations, worm holes, and so on. The soil sorption coefficient, the concentration ratio of chemical sorbed on soil particles to chemical dissolved in soil water at equilibrium, K_{SW}, is often estimated with the organic carbon partition coefficient, K_{OC}, or the organic matter partition coefficient, K_{OM}. The partitioning of chemical species between soil water and soil air is estimated with the air–water partition coefficient (the dimensionless Henry's law constant), K_{AW}. Estimates of bioconcentration factors, BCF, bioaccumulation transfer factors, TF, and plant uptake factors, B_V, will also be useful in determining partitioning. If the chemical is found in an organic phase in the soil column, the chemical's vapor pressure and aqueous solubility must be considered, also. (A useful review of chemical partitioning and transport in soils is given by Lyman et al., 1992.)

Table 1.1 Chemical Properties That Control Chemical Partitioning and Fate

Environmental Partitioning

Boiling point
Melting point
Gas and liquid density
Surface tension
Vapor pressure
Aqueous solubility
Air–water partition coefficient
Sorption coefficient for soil and sediment
Bioconcentration factor

Environmental Transport and Fate

Diffusivity in air and water
Phase transfer coefficient for air–water
Phase transfer coefficient for air–soil

While diffusion of sorbed species in soils is usually negligible, diffusion of chemicals in soil water and in soil air is fast enough to be important. In addition, dissolved chemicals are transported along with the soil water by wicking and percolation in the unsaturated zone and by advection and dispersion in the saturated zone. Chemicals are transported in soil air by barometric pumping resulting from sporadic changes in atmospheric pressure and by displacement with soil water. The estimated diffusivities of chemicals in air and water, along with information about the medium such as tortuosity and hydraulic conductivity, are used to evaluate chemical movement in the soil column due to the various transport processes.

Table 1.1 summarizes some important chemical properties that control the partitioning and transport of chemicals in the environment. Measured values are often available for properties such as melting point, boiling point, aqueous solubility, and octanol–water partition coefficient, although the reported values are not always reliable. Measured values of vapor pressure, air–water partition coefficient, soil sorption coefficient, bioconcentration factor, and diffusion coefficients are rarely available (Altschuh and Bruggemann, 1993). Many of these are commonly estimated from the magnitude of the chemical's octanol–water partition coefficient, K_{ow}. Methods of estimating the magnitude of the properties of chemicals and some key properties of environmental media are fundamental to modeling the behavior of chemicals in the environment and are presented in this book.

1.4 ORGANIZATION OF THE BOOK

The book opens with a discussion of the types of methods used to estimate chemical properties: quantitative structure–property relationships and quantitative property–property relationships. We review some of the fundamentals in order to alert you to the strengths and weaknesses of the various types of models. Several chapters follow that deal with properties controlling the partitioning of chemicals between air, water, soil, and biota. The book ends with several chapters dealing with properties that control chemical transport in the absence of equilibrium.

Each chapter begins with a review of background information and theory explaining the range of application and limitations of the methods to be described. Chemical property estimation methods are presented along with sample calculations and practical advice on how and when the methods are best used. The methods range from simple, back-of-the-envelope calculations to complex procedures requiring the aid of a computer. Most of those described apply to the wide range of chemicals that are of interest to environmental specialists. All of the methods described yield suitably accurate estimates in the range of values expected under normal ambient conditions. Finally, information is given concerning method accuracy and reliability and about the response of the chemical property to changing environmental conditions.

There is a large and growing body of computer software, on-line data bases, and Internet resources available for chemical property estimation. These resources are cited in the following chapters. In addition, a supplemental catalog of resources is given in Appendix 8.

Other supplementary material at the end of the book reviews fundamental constants, units of measure, estimating the uncertainty of a chemical property estimate, fundamentals of empirical modeling, describing chemical structures with the molecular connectivity index, and the Calculus.

The definitions of most technical terms used in this book may be found in any standard reference work on physical chemistry or chemical engineering. Two excellent sources are the physical chemistry textbook by Levine (1996) and the environmental engineering textbook by Valsaraj (1995). Useful general reference works on environmental chemistry have been published by Manahan (1991) and by Schwartzenbach et al. (1993). Two major reference works in chemical property estimation are the books by Lyman et al. (1990) and by Reid et al. (1987). Both are very useful for exploring the literature prior to 1980. There are, of course, many other excellent sources of information on environmental science and engineering, and these are cited in the following chapters.

Finally, a word about when *not* to estimate. You should avoid estimating chemical properties if at all possible. Measured values of chemical properties are usually (but not always) more accurate than estimated values. The common practice of stringing chemical property calculations into a chain, estimating the values of chemical properties from the estimated values of other chemical properties, is especially dangerous. Thanks to propagation of error through the sequence of calculations, the uncertainty in the final result is often so large as to make the exercise useless. You should limit an estimation procedure to one calculation if possible.

Extensive and reliable compilations of octanol–water solubility coefficients, vapor pressures, and aqueous solubilities are available. These are identified in the relevant chapters. A list of useful sources of measured chemical property values is also given in Appendix 8. Extensive lists of measured values of soil sorption coefficients and bioconcentration factors are given in Chapters 10 and 11, respectively.

1.5 UNITS OF MEASURE

The environment is of interest to specialists of all kinds, and many different systems of measure are in use. In order to promote uniformity and reduce confusion, the *Système International d'Unités*, abbreviated SI system, was proposed in 1960 at the General Conference on Weights and Measures for global scientific use. While SI units are now widely used in science and engineering, environmental specialists continue to use non-SI units such as the torr, the liter, and the hour because it is convenient to do so. For this reason, non-SI units are used in the illustrative examples given in this book. In describing the published estimation methods, the system of units used by the author is also retained. The SI system, along with useful conversion factors and numerical constants, is described in Appendix 2.

REFERENCES

Altschuh, J. and Bruggemann, R., 1993. Limitations on the Calculation of Physicochemical Properties. In *Chemical Exposure Prediction,* Calamari, D., Ed. Lewis Publishers, Boca Raton, FL, pp. 1–11.

Levine, I.N., 1996. *Physical Chemistry.* Fifth Edition. McGraw-Hill, New York.

Lyman, W.J., Reehl, W.F., and Rosenblatt, D.H., 1990. *Handbook of Chemical Property Estimation Methods.* American Chemical Society, Washington, D.C.

Lyman, W.J., Reidy, P.J., and Levy, B., 1992. *Mobility and Degradation of Organic Contaminants in Subsurface Environments.* C. K. Smoley, Chelsea, MI.

Manahan, S. E., 1991. *Environmental Chemistry.* Fifth Edition. Lewis Publishers, Boca Raton, FL.

Reid, R.C., Prausnitz, J.M., and Poling, B.E., 1987. *The Properties of Gases and Liquids.* McGraw-Hill, New York.

Schwartzenbach, R.P., Gschwend, P.M., and Imboden, D.M., 1993. *Environmental Organic Chemistry.* John Wiley & Sons, New York.

Valsaraj, K.T., 1995. *Elements of Environmental Engineering.* Lewis Publishers, Boca Raton, FL.

CHAPTER 2

Concepts and Theory of Chemical Property Estimation

2.1 INTRODUCTION

The objective of chemical property estimation is to calculate the value of a physicochemical property which may be difficult, expensive, or even impossible to measure directly. An estimation method consists of a mathematical relationship between two chemical properties or between a computable structural descriptor and a chemical property coupled with a procedure for using the relationship to obtain quantitative predictions. The mathematical models used for this purpose relate the values of physical and chemical properties of interest to the values of molecular and molar properties which are easy or inexpensive to obtain. Satisfactory models are accurate, reliable, and no more complex than necessary.

Two general approaches are commonly used for chemical property estimation. In one approach, estimates are made with **quantitative property–property relationships**, QPPRs, *mathematical models that relate the property of interest to other properties of the chemical for which measured values are available*. In the other approach, estimates are made with **quantitative structure–property relationships**, QSPRs, *mathematical models that use fragment contributions and structural descriptors to relate the property of interest to molecular composition and geometry*.

Property–property relationships usually are simple to apply and reliable when dealing with monofunctional chemicals, compounds belonging to only one class of chemical. Structure–property relationships are usually more broadly applicable than QPPRs, and, unlike many QPPRs, they are unambiguous when applied to multifunctional chemicals. Now that powerful digital computers are almost universally available, the increased computational effort needed to apply many QSPRs is no longer an impediment to their use. The theoretical foundation of QPPR and QSPR modeling is discussed in this chapter, and the advantages and limitations of the two approaches are described in detail.

2.2 QUANTITATIVE PROPERTY–PROPERTY RELATIONSHIPS

Property–property relationships can be classified as being fundamental, empirical, or semi-empirical. A **fundamental QPPR** is *a property–property relationship that is derived strictly from first principles*. It has no empirical elements, only system parameters and universal constants. Such a model is applicable to ideal systems at thermal, mechanical, and material equilibrium. By definition, intermolecular forces do not exist in ideal systems, and fundamental relationships describe the behavior of real systems insofar as intermolecular interactions are weak enough to be ignored. A **semi-empirical QPPR** is *a fundamental relationship to which empirical parameters have been added to enhance the accuracy of prediction when intermolecular forces are important and cannot be ignored*. An **empirical QPPR** usually takes the form of *a regression model that describes the correlation between two chemical properties*.

2.2.1 Fundamental Models

Some estimation methods arise directly from the fundamental relationships of thermodynamics. For instance, the **Clausius–Clapeyron equation** is a relationship between the saturation vapor pressure of a liquid, P^S (Pa), and the ambient temperature, T (K). The Clausius–Clapeyron equation is:

$$\frac{dlnP^S}{dT} = \frac{\Delta H_{VAP}}{RT^2} \tag{2.1}$$

where ΔH_{vap} (J/mol) is the **(latent) heat of vaporization** of the liquid*, and R ($8.314 \ J \ mol^{-1} \ K^{-1}$) is the ideal gas constant. Equation 2.1 arises from the proposition that the chemical potential of a pure liquid is equal to the chemical potential of the vapor in contact with the liquid at equilibrium. A number of assumptions are made in deriving the Clausius–Clapeyron equation, and it is only strictly valid for ideal systems at temperatures far from the critical temperature. The derivation is described in Chapter 6. In this section, we explore some basic concepts used to develop the fundamental and semi-empirical models described in the following chapters.

The thermodynamic state of an environmental system is described by specifying the values of state variables of the system such as temperature and pressure.** An extremely useful state variable is the **Gibbs free energy** (also called the free energy and the Gibbs function), G (J). The Gibbs free energy is a measure of the degree of disorder in the system plus its surroundings.

Most chemical and physical processes of interest to us are rapid compared to changes in environmental conditions, and it simplifies the task of estimating chemical properties to assume that the ambient pressure is fixed at 1 standard atmosphere

---

* The **(latent) heat of vaporization** is the thermal energy required to vaporize 1 mol of substance.
** Any part of the environment of interest is called "**the system**". The rest of the universe is called "**the surroundings**". In the first part of the book, we will be concerned with closed systems. A **closed system** does not exchange matter with its surroundings. In the second part of this book, we will be concerned with open systems. An **open system** exchanges matter through mass transport or chemical reaction.

(101.325 kPa) and the ambient temperature is fixed at an appropriate value between −40 and +40°C. For a finite, spontaneous process that takes place in a closed system at constant temperature, the change in Gibbs free energy is

$$\Delta G = \Delta H + T\Delta S \tag{2.2}$$

where H (J) is the system's enthalpy and S (J/K) is the system's entropy. The value of ΔG is a measure of the spontaneity of a process that takes place in a system at constant temperature and pressure. If ΔG is negative, the process is spontaneous. If ΔG is positive, the process is not spontaneous, and work must be done to force it to take place.

If $\Delta G = 0$, the system is at equilibrium. A system is in **equilibrium** if its properties do not change value when the system is isolated, or when after a slight disturbance, its properties return to their original values.* The first part of this book deals with equilibrium properties of environmental systems.

For a chemical component i present in the system with quantity n_i (mol), the change in Gibbs free energy with change in the quantity of chemical is called the component's **chemical potential**, $dG/dn_i = \mu_i$ (J/mol). You can think of the chemical potential just like you think of electric or gravitational potential. It is a measure of (electrical, gravitational, centrifugal, etc.) forces that move matter from one place to another and transform matter from one form to another. Matter moves from regions and chemical forms of high chemical potential to regions and chemical forms of low chemical potential. At equilibrium, the value of μ_i is the same throughout the system.

Later on in this book, we will deal with chemical properties related to the nonequilibrium processes that take place in environmental systems. It simplifies matters to think of nonequilibrium processes as taking place at a fixed rate. *A process that occurs at a fixed rate* is called a **steady-state** process.

Imagine that the chemical is distributed between two phases, α and β, as illustrated in Figure 2.1. No net transfer of mass occurs between phase α and phase β when the chemical potential is the same in both phases:

$$\mu_i(\alpha) = \mu_i(\beta) \tag{2.3}$$

Temperature and pressure changes are important controls of processes that occur in environmental systems. The way in which $\mu_i(\alpha)$ varies with change in T and P is described by:

$$d\mu_i(\alpha) = -S_{m,i}(\alpha)dT + V_{m,i}(\alpha)dP \tag{2.4}$$

where $S_{m,i}(\alpha)$ (J/K) is the chemical's molar entropy and $V_{m,i}(\alpha)$ is the chemical's molar volume in phase α. A similar relationship can be written for the chemical in

---

* A **metastable** system can adopt a new state after a perturbation. An **inert** system appears to be stable because its properties change values imperceptibly slowly.

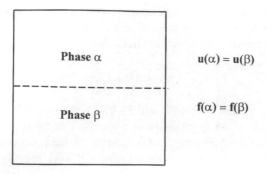

Figure 2.1   At equilibrium, the chemical potential of component i in phase α equals the component's chemical potential in phase β. The component's fugacity is the same in both phases, also.

phase β. If the temperature and pressure of the system change in such a way that the two phases stay in equilibrium, $d\mu_i(\alpha) = d\mu_i(\beta)$ and

$$-S_{m,i}(\alpha)dT + V_{m,i}(\alpha)dP = -S_{m,i}(\beta)dT + V_{m,i}(\beta)dP \qquad (2.5)$$

or

$$\frac{dp}{dT} = \frac{\Delta S_{m,i}}{\Delta V_{m,i}} \qquad (2.6)$$

where $\Delta S_{m,i} = S_{m,i}(\beta) - S_{m,i}(\alpha)$ and $\Delta V_{m,i} = V_{m,i}(\beta) - V_{m,i}(\alpha)$. Equation 2.6, the **Clapeyron equation**, is useful for estimating the properties of chemicals, and with a few simple assumptions, it is easily converted to the Clausius–Clapeyron equation, Equation 2.1.

Except for temperature, pressure, and entropy, we do not calculate the absolute value of a thermodynamic state variable. We calculate the difference between the state variable's value in the system and the variable's value in a standard state; the pure, ideal chemical at a pressure of either 1 atm or 1 bar. For an ideal gas i, the chemical potential is

$$\mu_i = \mu_i^0 + RT\ln\left(\frac{P_i}{P^0}\right) \qquad (2.7)$$

where μ_i^0 is the component's chemical potential in the standard state and P^0 is the standard pressure, 1 atm or 1 bar. For a real gas, P_i is an effective pressure, and Equation 2.7 does not accurately describe the relationship between μ_i and P_i. Equation 2.7 must be corrected by replacing P_i with the component's **fugacity**, f_i (Pa), *the thermodynamic pressure of a chemical or the pressure of an ideal gas that has chemical potential μ_i.*

$$\mu_i = \mu_i^0 + RT \ln\left(\frac{f_i}{P_0}\right) \tag{2.8}$$

You can think of the fugacity as the escaping tendency of a chemical. The reference fugacity is chosen to be the same for both the solute in solution and its vapor. It is the equilibrium vapor pressure of the pure liquid chemical at the standard pressure and ambient temperature.

For an ideal component i of solution,

$$\mu_i = \mu_i^0 + RT \ln\left(x_i\right) \tag{2.9}$$

where x_i is the component's mole fraction. For a real component of solution, x_i is an effective concentration, and Equation 2.9 does not accurately describe the relationship between μ_i and x_i. Equation 2.9 must be corrected by replacing x_i with the component's **activity** a_i, its *thermodynamic concentration or the concentration that an ideal solute that has chemical potential μ_i*:

$$\mu_i = \mu_i^0 + RT \ln a_i \tag{2.10}$$

where $\alpha_i = \gamma_i x_i$, and γ_i, the **activity coefficient** is a measure of the deviation of the chemical's behavior from ideal behavior. Here, $\mu_i^0(T,P) = \mu_i^S(T,P)$, the chemical potential of the pure substance at T and P of the solution.

It is awkward to work directly with the chemical potential. First, its absolute value cannot be calculated or measured. Only differences between chemical potentials can be determined. As a result, the standard states of vapor, liquid, solid, and solute must be carefully defined, and it is not always a straightforward matter to eliminate the standard chemical potentials from thermodynamic relationships. Second, the standard state conventions in use are sometimes confusing. Third, chemical potential is an exponential function of vapor pressure and concentration. This has awkward consequences when, for instance, the value of the chemical potential approaches $-\infty$ as the value of pressure or concentration approaches 0.

These problems do not arise when working with the fugacity. As a result, partition–coefficient relationships are almost always derived from fugacity relationships. (See, for example, Mackay, 1991.) The process of deriving useful partition–coefficient relationships begins with the basic premise that the fugacity of a chemical is the same in all phases at equilibrium:

$$f_i(\alpha) = f_i(\beta) \tag{2.11}$$

where $f_i(\alpha)$ is the fugacity of chemical i in phase α and $f_i(\beta)$ is the fugacity of chemical i in phase β.

2.2.2 Empirical Models

If no fundamental model relates the value of the chemical property to be estimated to the value of a property that has been measured, an empirical model is used. **Trouton's rule**, an empirical relationship between a chemical's heat of vaporization and its **normal boiling point**, T_b^*, is probably the first QPPR proposed. Trouton's rule is

$$\Delta H_{vap}\left(T_b\right) = 10.5 \ RT_b \tag{2.12}$$

where $\Delta H_{vap}(T_b)$ (J/mol) is the heat of vaporization at T_b. Trouton's rule is based on the empirical observation that the entropy of vaporization, $\Delta S_{vap} = \Delta H_{vap}/T$, is roughly equal to 87 J mol^{-1} K^{-1} or 10.5 R for nonpolar liquids at the boiling point. The relationship fails for polar liquids, especially hydrogen-bonded polar liquids, and for liquids that do not boil between 150 and 1000 K.

Empirical models usually take the form of a polynomial regression formula such as:

$$q = \beta_0 + \beta_1\chi + \beta_2\chi^2 + \beta_3\chi^3 + \dots \tag{2.13}$$

Here, q is the quantity of interest and χ is a measurable property or a molecular descriptor of the chemical. The regression parameters $\beta_0, \beta_1, \beta_2, \beta_3 \dots$ are computed with values of q and χ obtained for a group of chemicals called the training set. The formula is usually reported along with information on the number of chemicals in the training set, n, the standard error, s, the correlation coefficient, r, and the F ratio.

The number of terms employed in the regression formula depends upon the linearity of the relationship between chemical properties and is the minimum number needed to achieve the desired accuracy. Regardless of the underlying relationship between chemical properties, if a linear model provides the required degree of accuracy, two terms are the best choice. If more accuracy is required, more than two terms are used.

Two chemical properties can be related with an empirical model even if there is no underlying fundamental relationship between them. This is perfectly adequate for the purposes of property estimation, but such a model is useless beyond the limitations of the training set. Empirical models cannot be extrapolated beyond the range of property values exhibited by the training set.

2.2.3 Semi-Empirical Models

Semi-empirical thermodynamic relationships are used to correlate the chemical properties when intermolecular interactions significantly control chemical behavior. Often, they are derived from fundamental relationships by replacing effective parameters such as concentration and vapor pressure with thermodynamic parameters such as activity and fugacity.

---

* The **normal boiling point** is the temperature at which a liquid's vapor pressure equals 1 atm.

A simple semi-empirical model based on the Clausius–Clapeyron equation is

$$lnP^S = \beta_0 + \frac{\beta_1}{T} \qquad (2.14)$$

where β_0 and β_1 are regression parameters derived from experimental data. Experiment shows that the relationship between P^S, the saturation vapor pressure,* and T^{-1} is not linear, and so the following quadratic model might better fit the data:

$$lnP^S = \beta_0 + \frac{\beta_1}{T} + \frac{\beta_2}{T^2} + \dots \qquad (2.15)$$

In selecting a model for use, the required precision of the result must be considered. Rough estimates and worst-case calculations are made with the simplest mathematical models. Complex formulas are used only if necessary.

Another example of a semi-empirical model is the relationship between the aqueous solubility, C_W^S (mol/l), of a liquid organic compound and its octanol–water partition coefficient, K_{OW}:

$$\log K_{OW} = \beta_0 + \beta_1 \log C_W^S \qquad (2.16)$$

Equation 2.16 is based on a fundamental relationship between the solubility and partition coefficient. The fundamental relationship has an intercept of 0.8 and a slope of −1.0 (Yalkowsky et al., 1992). The regression coefficients, β_0 and β_1, vary somewhat depending on the class of compounds in the training set. Based on measurements of 111 compounds of mixed class, β_0 and β_1 were found to be 0.515 and −1.016, respectively (Valvani et al., 1981).

Departures from equilibrium or steady–state may become so great that chemical behavior cannot be modeled accurately with the usual semi-empirical relationships. Additional system parameters may be added to the equilibrium relationships in order to describe chemical behavior in such systems. This is discussed in subsequent chapters.

The regression coefficients of empirical and semi-empirical correlations are evaluated with experimental measurements made on real systems. They are valid only for the range of local conditions used to make the measurements. Equation 2.16, along with the values of β_0 and β_1 reported above, will only be accurate when applied to compounds (and environmental systems) which are similar to those of the training set. Semi-empirical models are not reliable when dealing with chemicals (and environmental systems) which are very different from those of the training set. In this sense, QPPRs are related to molecular structure.

QPPRs are usually simple, linear relationships between molar chemical properties. When the relationship has some theoretical basis, it may be broadly applicable to many different classes of chemical compound. More often, the relationship is an

---

* The **saturation vapor pressure** is the vapor pressure of the pure chemical.

empirical or semi-empirical one, in which case it can be applied only to the class of compound used to calibrate the model. If multifunctional chemicals are included in the training set, this can be ambiguous and confusing.

2.3 QUANTITATIVE STRUCTURE–PROPERTY RELATIONSHIPS

Molar properties of chemicals are determined by the molecular properties of chemicals. Molecular properties are determined by the structure of molecules, the type and geometric arrangement of their atoms. Quantitative structure–property relationships use fundamental relationships, fragment contributions, or topological indices to account for the relationship between the molar properties and the molecular structure of chemicals.

2.3.1 Fundamental Relationships

Given the geometric arrangement of atoms in an isolated molecule, the methods of molecular and quantum mechanics can be used to determine molecular properties such as equilibrium geometry, electron distribution, electronic energy levels, dipole moment, and polarizability (Clark, 1988; Cartwright, 1993; Haile, 1992). The molecular properties can be related, in turn, to molar thermodynamic properties such as molar volume (Wong et al., 1996). Several of the software packages for chemical workstations mentioned in Appendix 8 include QSAR programs for log K_{OW}, for instance.

Plenty of computer software is available for computational chemistry. While the methods are not widely applied in environmental studies, they are reasonably accurate, and the effort and expense involved are low enough that this resource is well within the reach of most providers of environmental services.

2.3.2 Fragment Contributions and Structural Factors

By molecular group or fragment, we mean an atom or group of atoms in a molecule. The value of a chemical property of interest, q, may be estimated as the sum of contributions made by each fragment. For instance,

$$\log q = \sum_i m_i \cdot \delta_i \tag{2.17}$$

where m_i is the number of fragments of type i in the molecule. The factor δ_i is an average contribution of the molecular fragment i to the total value of log q. The average is computed from information on the behavior of all types of chemical compounds which contain the fragment.

Equation 2.17 implies that the fragment contributions, δ_i, are perfectly additive, i.e., the contribution of a particular fragment is independent of the presence of other molecular fragments and of molecular topology (the geometric arrangement of fragments in a molecule). An **additive property** is one that *can be accurately evaluated by adding simple fragment contributions.*

Molecular weight is a perfect example of an additive chemical property. The weight of a molecule is exactly equal to the sum of the weights of the atoms of which it is composed. The atomic weights are determined only by atomic identity. They are not altered by the presence of other atoms in the molecule. The molecular weight does not depend on the arrangement of the atoms. While there are potentially an infinite number and variety of chemicals, the molecular weight of every one can be estimated with a list of a few (109 to be exact) atom contributions. Because of the simplicity, compactness, and convenience, molecular weight is always estimated rather than measured.

The fragment-contribution approach has been applied to most other properties of interest to scientists and engineers. Langmuir (1925) first suggested that intermolecular interactions between a solute and a solvent may be estimated by summing the interactions of solute fragments with the solvent. Experimental confirmation of Langmuir's suggestion has been obtained. For example, Fujita et al. (1964) showed that substituting –H for groups such as –CH_3 increases the partition coefficients of many chemicals by about the same amount. Group additivity is discussed in detail by Exner (1966, 1967).

The interactions that determine the magnitudes of solubility-related partition coefficients and other chemical properties appear to be highly complex. The values of properties such as aqueous solubility, C_W^S, activity coefficient, γ_i, and octanol–water partition coefficient, K_{OW}, are not accurately estimated as simple sums of contributions made by the chemical's fragments. Intramolecular interactions of the molecular fragments (proximity effects) must be accounted for, also. For instance, the value of the fragment contribution, δ_i, may depend upon the position of the fragment in the molecule. An **additive-constitutive property** is one that *can be accurately evaluated only when fragment contributions are adjusted to account for topographic features of the molecule.* Highly successful group contribution methods have been developed to estimate additive-constitutive properties such as K_{OW} and C_W^S.

One way of accounting for the influence of topology on the value of a chemical property is to adjust the value of the fragment contributions depending on the fragment–fragment interactions which occur in the molecule. Another way of accounting for topology is to add a **structural factor** term, Δ_j, to Equation 2.17, so that

$$\log q = \sum_i m_i \cdot \delta_i + \sum_j m_j \cdot \Delta_j \qquad (2.18)$$

where m_j is the number of structural features of type j in the molecule and the factor Δ_j is the average contribution of a chemical's structure feature to log q. The value of q is the product of the additive and the constitutive contributions, $q = 10^{\sum_i m_i \delta_i} \cdot 10^{\sum_j m_j \Delta_j}$, and the constitutive contribution is a multiplicative correction factor just like an activity coefficient or a fugacity coefficient. Structural features which are accounted for with structural factors include type of chemical bond, chain length, chain branching, ring structure, intramolecular hydrogen bonding, group proximity, and substitution pattern. The approach is used in estimating properties such as the octanol–water partition coefficient (Hansch and Leo, 1979). One major problem is that structural factors are often evaluated with small training sets and may not be reliable.

The difference between the value of a property q_{R-X} of compound R–X and the value of the same property q_{R-Y} of another compound R–Y is

$$q_{R-X} - q_{R-Y} = \Pi = \Sigma_i \left[m_i(R-X) - m_i(R-Y) \right] \delta_i$$
$$- \Sigma_j \left[m_j(R-X) - m_j(R-Y) \right] \cdot \Delta_j$$

(2.19)

where $m_i(R-X)$ and $m_i(R-Y)$ are the number of fragments of type i and $m_j(R-X)$ and $m_j(R-Y)$ are the number of structural features of type j in molecules R–X and R–Y, respectively. The **group exchange** factor, Π, accounts for the change in property value on replacing the group Y in the chemical with the group X. Group interchange methods avoid the inherent error in attempting to correct for topological features that enhance intramolecular interactions. They are a central element in QSAR application programs such as DESOC (Drefahl and Reinhard, 1993).

The group exchange approach is most useful in simple cases where one atom in a reference compound is replaced by a similar atom to form the compound of interest. If the chemical's structural features remain substantially unchanged on making the substitution, the value of Π will be equal to the difference between the fragment contributions of groups X and Y. In this case, chemical properties are estimated accurately by adding the value of the group exchange factor to the value of the chemical property of the reference compound:

$$q_{R-X} = q_{R-Y} + \Pi$$

(2.20)

When estimating the value of Π by taking the sum and difference of group contributions to properties like log K_{OW}, one must be careful to use the correct fragment constants. The group exchange factor was proposed originally by Fujita et al. (1964) to estimate log K_{OW} values. They define Π as the difference between log K_{OW} of a molecular substituent and log K_{OW} of the hydrogen atom it replaces in the parent compound. By definition, $\Pi(H) = 0$.

Hansch and Leo (1979) report that the contribution of an H atom to log K_{OW} is 0.23. This value is valid only for H atoms present in the hydrophobic portion of a molecule. The contributions of polar fragments to log K_{OW} include the H atom contributions, and there is no separate group contribution for H present in the polar fragments of a molecule.

The fundamental premise on which group contribution methods are based is that the Gibbs free energy change of a chemical process, ΔG (J), is an additive and constitutive function of molecular structure. The premise was first applied by Hammett (1940) to derive a model that describes the effect of molecular structure on chemical equilibria and reaction rates.* The approach was applied by Lambert (1967) to obtain a structure–activity model for the soil sorption coefficient. The following treatment is similar to Lambert's.

Imagine a molecule X–Y–Z composed of fragments X, Y, and Z. The molar Gibbs energy of X–Y–Z, G_{mXYZ}, is an additive and constitutive function. It is related to the contributions of the molecular fragments by:

$$G_{mXYZ} = g_x + g_Y + g_Z + g_{X,Y} + g_{X,Z} + g_{Y,Z} \qquad (2.21)$$

where the contributions g_X (J), g_Y (J), and g_Z (J) are independent functions of the groups X, Y, and Z, respectively, and the contributions $g_{X,Y}$ (J), $g_{X,Z}$ (J), and $g_{Y,Z}$ (J) result from group interactions of X with Y, X with Z, and Y with Z, respectively.

Suppose the molecule X–Y–Z is partitioned between liquid phase 1 and liquid phase 2; X–Y–Z (l_1) $\rightleftharpoons$ X–Y–Z (l_2). The equilibrium constant, K_{XYZ}, is

$$K_{XYZ} = \frac{C_{XYZ}(l_2)}{C_{XYZ}(l_1)} \qquad (2.22)$$

where $C_{XYZ}(l_1)$ (mol/l) is the equilibrium concentration of X–Y–Z in liquid phase 1 and $C_{XYZ}(l_2)$ (mol/l) is the equilibrium concentration of X–Y–Z in phase 2. The change in standard Gibbs energy of the process is related to K_{XYZ} by:

$$\begin{aligned}
\Delta_r G_{mXYZ} &= G_{mXYZ}(l_2) - G_{mXYZ}(l_1) \\
&= \left[g_X(l_2) + g_Y(l_2) + g_Z(l_2) + g_{X,Y}(l_2) + g_{X,Z}(l_2) + g_{Y,Z}(l_2) \right] \\
&\quad - \left[g_X(l_1) + g_Y(l_1) + g_Z(l_1) + g_{X,Y}(l_1) + g_{X,Z}(l_1) + g_{Y,Z}(l_1) \right] \\
&= -RT \ \ln K_{XYZ}
\end{aligned} \qquad (2.23)$$

Rearranging Equation 2.23 and dividing all terms by –RT, we obtain:

$$\ln K_{XYZ} = \delta_X + \delta_Y + \delta_Z + \Delta_{X,Y} + \Delta_{X,Z} + \Delta_{Y,Z} = \sum_i m_i \delta_i + \sum_j m_j \Delta_j \qquad (2.24)$$

where m_i is the number of fragments i found in the molecular structure, $\delta_X = -[g_X(l_2) - g_X(l_1)]/RT$, $\delta_Y = -[g_Y(l_2) - g_Y(l_1)]/RT$, $\delta_Z = -[g_Z(l_2) - g_Z(l_1)]/RT$, m_j is the number of proximity effects j which must be accounted for, $\Delta_{X,Y} = -[g_{X,Y}(l_2) - g_{X,Y}(l_1)]/RT$, $\Delta_{X,Z} = -[g_{X,Z}(l_2) - g_{X,Z}(l_1)]/RT$, and $\Delta_{Y,Z} = -[g_{Y,Z}(l_2) - g_{Y,Z}(l_1)]/RT$. In practice, some group contributions and structural factors may be insignificant compared to the others, and only the most important may be used to estimate the value of K_{XYZ}.

---

* Hansch (1993) states that we do not know why Hammett's method works. Perhaps it is because, for small changes in physical-chemical properties, the mathematical form of the model is not important.

Group contribution methods are used by environmental specialists because they usually apply to a wide variety of chemical types and they are not ambiguous when applied to multifunctional chemicals. All group contribution methods are limited by lack of data, however. The limitation arises in three ways: (1) a fragment constant may not be available for the functional group of interest; (2) the constant may be derived from such a small training set that it is not reliable; or (3) the method may not be applicable to the type of compound of interest.

2.3.3 Topological Indices

Group contribution methods that account for intramolecular interactions can be complicated. With the addition of structural factors, the simple relationships involving fragment contributions are converted to complex nonlinear relationships. Also, the methods tend to evolve, the contributions of larger and larger molecular fragments being evaluated, until nearly the entire molecule is treated as a fragment. Continuing adaptive growth of this sort has become a hallmark of QSAR methods such as CLOGP used to estimate the value of log K_{OW} (Leo, 1993). This problem is avoided by using multivariate regression models involving topological indices, descriptors that encode the existence of topological features in a molecule:

$$\log q = \beta_0 + \beta_1 \chi_1 + \beta_2 \chi_2 + \beta_3 \chi_3 + \dots \quad (2.25)$$

Here, the β_i are regression parameters and the χ_i are computable molecular descriptors such as the molecular connectivity indices (Kier and Hall, 1986). The descriptors are computed on the basis of important topological features of the chemical such as the spatial arrangement and chemical bonding of the structural backbone.

Multivariate regression models employing a host of structural descriptors selected with the aid of neural networks are available to estimate the values of chemical properties such as normal boiling point (Egolf et al., 1994; Wessel and Jurs, 1996), octanol–water partition coefficient (Basak et al., 1996), aqueous solubility (Sutter and Jurs, 1966), air–water partition coefficient (Katritzky and Mu, 1996), and soil–water partition coefficient (Lohninger, 1994). With the enormous computing power now offered by the omnipresent desktop computer, this approach to property estimation is becoming more widely applied even as it grows increasingly complex. The method of calculating connectivity indices is discussed in Appendix 5. Regression analysis is discussed in Appendix 4.

REFERENCES

Basak, S.C., Gute, B.D., and Grunwald, G.D., 1996. A comparative study of topological and geometrical parameters in estimating normal boiling point and octanol/water partition coefficient. *J. Chem. Inf. Comput. Sci.*, **36**, 1054–1060.

Cartwright, H.M., 1993. *Applications of Artificial Intelligence in Chemistry.* Volume 11. Oxford Chemistry Primer Series. Oxford University Press, New York.

Clark, T., 1988. Molecular Orbital and Force-Field Calculations for Structure and Energy Predictions. In *Physical Property Prediction in Organic Chemistry. Proceedings of the Beilstein Workshop, Schloss Korb, Italy.* Jochum, C., Hicks, M., and Sunkel, J., Eds. Springer-Verlag, Berlin, pp. 95–102; *A Handbook of Computational Chemistry. A Practical Guide to Chemical Structure and Energy Calculations.* Wiley-Interscience, New York.

Drefahl, A. and Reinhard, M., 1993. Similarity-based search and evaluation of environmentally relevant properties for organic compounds in combination with the group contribution approach. *J. Chem. Inf. Comput. Sci.*, **33**, 886–895.

Egolf, L.M., Wessel, M.D., and Jurs, P.C., 1994. Prediction of boiling points and critical temperatures of industrially important organic compounds from molecular structure. *J. Chem. Inf. Comput. Sci.*, **34**, 947–956.

Exner, O., 1966. Additive physical properties. I. *Collect. Czech. Chem. Commun.*, **31**, 3222–3251.

Exner, O., 1967. Additive physical properties. II. *Collect. Czech. Chem. Commun.*, **32**, 1–22; Additive physical properties. III. *Collect. Czech. Chem. Commun.*, **32**, 24–54.

Fujita, T., Iwasa, J., and Hansch, C., 1964. A new substituent constant, π, derived from partition coefficients. *J. Am. Chem. Soc.*, **86**, 5175–5180.

Haile, J.M., 1992. *Molecular Dynamics Simulation: Elementary Methods.* John Wiley & Sons, New York.

Hansch, C. and Leo, A., 1979. *Substituent Constants for Correlation Analysis in Chemistry and Biology.* John Wiley & Sons, New York.

Hansch, C., 1993. Quantitative structure-activity relationships and the unnamed science. *Acc. Chem. Res.*, **26**, 147–153.

Hammett, L.P., 1940. *Physical Organic Chemistry.* McGraw-Hill, New York.

Katritzky, A.R. and Mu, L., 1996. A QSPR study of the solubility of gases and vapors in water. *J. Chem. Inf. Comput. Sci.*, **36**, 1162–1168.

Kier, L.B. and Hall, L.H., 1986. *Molecular Connectivity in Structure–Activity Analysis.* John Wiley & Sons, New York.

Lambert, S.M., 1967. Functional relationship between sorption in soil and chemical structure. *J. Agr. Food Chem.*, **15**, 572–576.

Langmuir, I., 1925. The distribution and orientation of molecules. *Am. Chem. Soc. ACS Colloid Symp. Monogr.*, **3**, 48–75.

Leo, A.J., 1993. Calculating log P_{OCT} from structures. *Chem. Rev.*, **93**, 1281–1306.

Lohninger, H., 1994. Estimation of soil partition coefficients of pesticides from their chemical structure. *Chemosphere*, **29**, 1611–1626.

Mackay, D., 1991. *Multimedia Environmental Models. The Fugacity Approach.* Lewis Publishers, Chelsea, MI.

Sutter, J.M. and Jurs, P.C., 1966. Prediction of aqueous solubility for a diverse set of heteroatom-containing organic compounds using a quantitative structure–property relationship. *J. Chem. Inf. Comput. Sci.*, **36**, 100–107.

Valvani, S.C., Yalkowski, S.H., and Roseman, T.J., 1981. Solubility and partitioning. IV. Aqueous solubility and octanol–water partition coefficients of liquid nonelectrolytes. *J. Pharm. Sci.*, **70**, 502–507.

Yalkowsky, S.H., Pinal, R., and Bannerjee, S., 1992. *Aqueous Solubility. Methods of Estimation for Organic Compounds.* Marcel Dekker, New York.

Wessel, M.D. and Jurs, P.C., 1996. Prediction of normal boiling points of hydrocarbons from molecular structure. *J. Chem. Inf. Comput. Sci.*, **35**, 68–76; Prediction of normal boiling points for a diverse set of industrially important organic compounds from molecular structure. *J. Chem. Inf. Comput. Sci.*, **35**, 841–850.

Wong, M.W., Wiberg, K.B., and Frisch, M.J., 1996. *Ab initio* calculation of molar volumes: comparison with experiment and use in solvation models. *J. Comput. Chem.*, **16**, 385–394.

CHAPTER 3

Boiling Point and Melting Point

3.1 INTRODUCTION

Boiling point and melting point values are among the easiest information to obtain about chemicals. That is why many methods of estimating chemical properties such as vapor pressure, aqueous solubility, and K_{OW} begin with boiling point or melting point measurements. However, authentic samples of chemicals are not always available with which to make the measurements, and some chemicals sublime at temperatures below the melting point or decompose at temperatures below the boiling point. In such cases, estimation methods are the only means of obtaining needed boiling and melting point values.

The boiling point of a liquid is the temperature at which its vapor pressure equals the ambient pressure. The **normal boiling point**, T_b, is *the temperature at which the vapor pressure of the pure liquid equals 1 atm.*

The **(normal) melting point**, T_m, is *the temperature at which solid and liquid chemicals are in equilibrium when the ambient pressure is 1 atm.* The melting point of the solid and the freezing point of the liquid are identical for pure chemicals but not for contaminated chemicals or for mixtures.

Extensive tabulations of boiling and melting point measurements are published by fine-chemical manufacturers and others (for example, see Dean, 1992 and Lide, 1994). Boiling point is a sensitive function of pressure, and the pressure at which the reported value is measured, if it is not 1 atm, is always noted. Since pressure has no significant effect on the melting point, the reported value is almost always the normal melting point. If measured values cannot be found in the literature, approximate values of boiling and melting point can be estimated with the methods presented in this chapter.

3.2 METHODS OF ESTIMATING NORMAL BOILING POINT

3.2.1   Joback and Reid's Method

Joback and Reid (1987) proposed a group contribution method that gives an approximate value of the boiling point of aliphatic and aromatic hydrocarbons. Stein and Brown (1994) increased the range of applicability and accuracy of the Joback–Reid method primarily by increasing the number of structural groups from 41 to 85. The boiling point is estimated with the sum of contributions of all structural groups found in the molecule:

$$T_b = 198.2 + \sum n_i \Delta_{b,i} \tag{3.1}$$

where T_b = normal boiling point, K
$\quad\quad$ n_i = number of i groups in molecule
$\quad\quad$ $\Delta_{b,i}$ = Joback contribution of group i to T_b, K

Values of $\Delta_{b,i}$ reported by Stein and Brown are given in Table 3.1. The size of the training set used to derive each group contribution value is given, also. The size of the training set indicates the reliability of the $\Delta_{b,i}$ value; the larger the set, the more reliable the value.

Equation 3.1 tends to overestimate boiling point values in the temperature range above 500 K. The following two polynomial relationships correct for this.

$$T_b(\text{corr}) = T_b - 94.84 + 0.5577 \ T_b - 0.0007705 \ T_b^2, \quad T_b \le 700 \ \text{K}$$

$$= T_b + 282.7 - 0.5209 \ T_b, \quad\quad\quad\quad T_b > 700 \ \text{K} \tag{3.2}$$

$$n = 4426, \ \text{AAE} = 15.5 \ \text{K, and AAPE} = 3.2\%,$$

where T_b $\quad$ = normal boiling point, K
$\quad\quad$ n $\quad\quad$ = number of chemicals in training set
$\quad\quad$ AAE   = average absolute error, K
$\quad\quad$ AAPE = average absolute percent error

Using the appropriate polynomial correction, Equation 3.2, the model produces an average absolute error of 15.5 K and an average absolute percent error of 3.2% with the training set of 4426 diverse organic chemicals listed in the Aldrich (1990) chemical catalog. The chemicals in the training set have normal boiling points in the range from 250 to 900 K. The model, when validated with a test set of 6584 diverse chemicals (Graselli, 1990), produced an average absolute error of 20.4 K, and an average absolute percent error or predictive error of 4.3%.

Table 3.1 Joback Group Contributions to the Boiling Point, $\Delta_{b,i}$, and the Melting Point, $\Delta_{m,i}$

Structural Group	$\Delta_{b,i}$[a]	$\Delta_{m,i}$[b]	Structural Group	$\Delta_{b,i}$[a]	$\Delta_{m,i}$[b]
		Carbon Increments			
$-CH_3$	21.98 (2832)	−5.10	$=C_rH-$ [c]	28.03 (285)	8.13
$>CH_2$	24.22 (2200)	11.27	$=C<$	23.58 (169)	11.14
$>C_rH_2$ [c]	26.44 (757)	7.75	$=C_r<$ [c]	28.19 (257)	37.02
$>CH-$	11.86 (603)	12.64	aaCH [d]	28.53 (1727)	8.13
$>C_rH-$ [c]	21.66 (457)	19.88	aaC−	30.76 (1740)	37.02
$>C<$	4.50 (370)	46.43	aaaC	45.46 (110)	37.02
$>C_r<$ [c]	11.12 (118)	60.15	$\equiv CH$	21.71 (43)	−11.18
$=CH_2$	16.44 (284)	−4.32	$\equiv C-$	32.99 (77)	64.32
$=CH-$	27.95 (405)	8.73			
		Oxygen Increments			
$-OH$	106.27 (14)	44.45	$\phi-OH$ [f]	70.48 (179)	82.83
1-OH [e]	88.46 (370)	44.45	$-O-$	25.16 (525)	22.23
2-OH [e]	80.63 (205)	44.45	$-O_r-$ [c]	32.98 (217)	23.05
3-OH [e]	69.32 (46)	44.45	$-OOH$	72.92 (1)	[66.68]
		Carboxyl Increments			
$-CHO$	83.38 (178)	36.90	$-C(O)NH_2$	230.39 (10)	[128.09]
$>CO$	71.53 (419)	61.20	$-C(O)NH-$	225.09 (19)	[113.86]
$>C_rO$ [c]	94.76 (134)	75.97	$-C_r(O)N_rH-$ [c]	246.13 (6)	[177.48]
$-C(O)O-$	78.85 (530)	53.60	$-C(O)N<$	142.77 (19)	[110.04]
$-C_r(O)O_r-$ [c]	172.49 (59)	53.60	$-C_r(O)N_r<$ [c]	180.22 (17)	[124.81]
$-C(O)OH$	169.83 (169)	155.50			
		Nitrogen Increments			
$-NH_2$	61.98 (217)	66.89	$=N-$	31.32 (12)	—
$\phi-NH_2$	86.63 (146)	66.89	$=N_r-$ [c]	43.54 (62)	68.40
$>NH$	45.28 (116)	52.66	$=N_rN_rH-$ [c]	179.43 (6)	[169.91]
$>N_rH$ [c]	65.50 (86)	101.51	$-N_r=C_rRN_rH-$ [c]	284.16 (3)	[206.93]
$>N-$	25.78 (138)	48.84	$-N=NNH-$	257.29 (2)	—
$>N_r-$ [c]	32.77 (105)	48.84	$-N=N-$	90.87 (3)	—
$>NOH$	104.87 (1)	[93.29]	$-NO$	30.91 (4)	—
$>NNO$	184.68 (2)	—	$-NO_2$	113.99 (117)	127.24
aaN	39.88 (171)	68.40	$-CN$	119.16 (129)	59.89
$=NH$	73.40 (3)	68.91	$\phi-CN$	95.43 (33)	59.89
		Halogen Increments			
$-F$	0.13 (136)	−15.78	$\phi-Cl$	36.79 (213)	13.55
$\phi-F$	−7.81 (163)	−15.78	$-Br$	76.28 (227)	43.43
$-Cl$	34.08 (180)	13.55	$\phi-Br$	61.85 (?)	43.43
1-Cl [e]	62.63 (158)	13.55	$-I$	111.67 (38)	41.69
2-Cl [e]	49.41 (56)	13.55	$\phi-I$	99.93 (36)	41.69
3-Cl [e]	36.23 (45)	13.55			

Table 3.1 (continued) Joback Group Contributions to the Boiling Point, $\Delta_{b,i}$, and the Melting Point, $\Delta_{m,i}$

Structural Group	$\Delta_{b,i}{}^a$	$\Delta_{m,i}{}^b$	Structural Group	$\Delta_{b,i}{}^a$	$\Delta_{m,i}{}^b$
		Sulfur Increments			
–SH	81.71 (45)	20.09	>SO	154.50 (4)	—
φ–SH	77.49 (24)	20.09	>SO$_2$	171.58 (15)	—
–S–	69.42 (79)	34.40	>CS	106.20 (2)	—
–S$_r$– ^{c}	69.00 (84)	79.93	>C,S ^{c}	179.26 (2)	—
		Phosphorus Increments			
–PH$_2$	59.11 (1)	—	>P–	43.75 (10)	—
>PH	40.54 (1)	—	>PO–	107.23 (9)	—
	Silicon Increments			Miscellaneous Increments	
>SiH–	27.15 (9)	—	>B–	–27.27 (9)	—
>Si<	8.21 (70)	—	–Se–	92.06 (1)	—
>Si$_r$<	–12.16 (2)	—	>Sn<	62.89 (6)	—

^{a} The number of times the group occurs in the training set is given in parentheses.
^{b} Values in brackets are composites of values from the published increment set.
^{c} Atoms with subscript r are in ring structures.
^{d} The symbol a denotes an aromatic bond.
^{e} Numbers 1, 2, and 3 denote attachment to primary, secondary, and tertiary carbon atoms, respectively.
^{f} The symbol φ denotes an aromatic system.

From Stein, S.E. and Brown, R.L., 1994. *J. Chem. Inf. Comput. Sci.*, **34**, 581–587; Reid, R.C., Prausnitz, J.M., and Poling, B.E., 1987. *The Properties of Gases and Liquids.* Fourth Edition. McGraw-Hill, New York.

Example

Estimate the normal boiling point of phenol, C_6H_6O, using the Joback–Reid group contribution method.

1. Phenol consists of five aaCH groups, one aaC– group, and one phenolic –OH group. Given the values of $\Delta_{b,i}$ listed in Table 3.1,

$$\Sigma n_i \cdot \Delta_{b,i} \ 1 = (5)(28.53 \ K) + 30.76 \ K + 70.48 \ K$$

$$= 243.89 \ K$$

2. Estimate the boiling point using Equation 3.1.

$$T_b = 198.2 + \Sigma n_i \cdot \Delta_{m,i} \ K$$

$$= 198.2 + 243.89 \ K$$

$$= 442.1 \ K$$

3. Estimate $T_b(corr)$ with Equation 3.2. For $T_b \leq 700$ K,

$$T_b(corr) = T_b - 94.84 + 0.5577 \ T_b - 0.0007705 \ T_b^2$$

$$= 442.1 - 94.84 + (0.5577)(442.1) - (0.0007705)(442.1)^2$$

$$T_b(corr) = 443.2 \ K$$

The measured boiling point of phenol is 181.7°C (454.9 K) (Lide, 1994).

A number of general methods of estimating the normal boiling point of chemicals have been reviewed (Rechsteiner, 1990; Reid et al., 1966). The Joback–Reid method is as accurate as any and is the most broadly applicable. The method does not distinguish among geometric isomers. This can result in significant discrepancies between the estimated and actual values.

3.2.2 Other Methods

Constantinou and Gani (1994) describe a group contribution method that does distinguish among isomers. An absolute error of 5.35 K, an absolute percent error of 1.42%, was reported with a test set of 392 chemicals of various types.

When dealing with simple monofunctional chemicals, some improvement in accuracy might be gained by using a narrowly focused and refined estimation method. White (1986) developed a simple quantitative structure–property relationship for planar polycyclic hydrocarbons. The molecular connectivity model distinguishes molecular size and topology. White's model is

$$T_b = 225.71 + 76.21\,^1\chi^v$$

(3.3)

$$n = 30, \ s = 8.59 \ K, \ r = 0.994$$

where T_b = normal boiling point, K
  $^1\chi^v$ = first-order valence molecular connectivity
     = $\Sigma(\delta_i^v \delta_j^v)^{-0.5}$
  δ_i^v = valence delta value
  n = number of chemicals in the training set
  s = standard deviation, K
  r = correlation coefficient

The training set consists of 30 planar polycyclic hydrocarbons. The boiling point range of chemicals in the training set is from 491 to 792 K.

Simamora et al. (1993) report a group contribution method to estimate the boiling point of rigid aromatic chemicals. Group contributions are given for fragments such as halo, methyl, cyano, and nitro groups. Hydrogen-bonding groups included are hydroxyl, aldehyde, carboxyl, primary amino, and amido groups. A standard error of 17.3 K is obtained in estimating T_b values of the 246 rigid aromatic compounds in the training set. Hoshino et al. (1985) describe a method of estimating the boiling point of polysubstituted alkylbenzenes. An average error of 0.4% and a maximum error of 4.7% are reported in estimating the T_b values of the 48 substituted benzenes in the training set.

Balaban et al. (1992a) report a quantitative structure–boiling-point relationship for polyhaloalkanes with 1 to 4 carbon atoms. The relationship is

$$T_b = 138.62 \ {}^1\chi - 15.35 \ N_H - 76.35 \ N_F - 37.21 \ N_{Cl} - 14.29 \ N_{Br}$$

$$+ 14.45 \ N_I - 53.78 \tag{3.4}$$

$$n = 532, \ s = 11.59 \ K, r^2 = 0.97, \ F = 2620$$

where T_b = normal boiling point, K
       ${}^1\chi$ = simple first-order molecular connectivity
          = $\Sigma(\delta_i\delta_j)^{-0.5}$
       δ_i = simple delta value
       N_i = number of atoms i in the molecule
       n = number of chemicals in the training set
       s = standard deviation, K
       r = correlation coefficient
       F = F ratio

The training set consists of 532 halo- and polyhaloalkanes with 1 to 4 carbon atoms.

The Joback–Reid group contribution method overestimates the boiling point of long-chain and perfluorinated hydrocarbons. Devotta and Pendyala (1992) describe a revised method that accounts for intramolecular interactions between H and F atoms in aliphatic halocarbons. Improved estimates of the boiling points of perfluorinated hydrocarbons might be obtained with the revised Joback–Reid method as reported by Devotta and Pendyala. It produces an average absolute error of 12.9 K, an average percent error of 3.6%, when used to estimate the T_b values of the 438 halogenated aliphatic hydrocarbons in the training set.

Balaban et al. (1992b) describe a quantitative structure–boiling-point relationship for acyclic saturated ethers, peroxides, acetals, and their sulfur analogs. It is

$$T_b = -59.10 + 44.30 \ {}^1\chi + 42.88 N_S$$

$$n = 185, \ s = 9.0 \ K, r^2 = 0.964, \ F = 2390 \tag{3.5}$$

where T_b = normal boiling point, K
       ${}^1\chi$ = first-order molecular connectivity
          = $\Sigma(\delta_i\delta_j)^{-0.5}$
       δ_i = number of bonds to atom i in the hydrogen-suppressed structure
       N_S = number of sulfur atoms in the molecule
       n = number of chemicals in the training set

s = standard deviation, K
r = correlation coefficient
F = F ratio

The range of normal boiling points of the chemicals in the training set is 250 to 513 K. Balaban et al. also report a more complicated structure–boiling-point relationship for which the standard error is 8.2 K, r^2 is 0.9714, and the F ratio is 2048.

Example

Estimate the boiling point of naphthalene, $C_{10}H_8$, using White's molecular connectivity model, Equation 3.3.

1. Calculate $^1X^v$. The hydrogen-suppressed structure of naphthalene is given in Figure 3.1. The assigned δ^v values are given in parentheses for each carbon atom.

$$^1X^v = (4 \cdot 3)^{-1/2} + (3 \cdot 3)^{-1/2} + (3 \cdot 3)^{-1/2} + (3 \cdot 3)^{-1/2}$$

$$+ (3 \cdot 4)^{-1/2} + (4 \cdot 3)^{-1/2} + (3 \cdot 3)^{-1/2} + (3 \cdot 3)^{-1/2} + (3 \cdot 3)^{-1/2}$$

$$+ (3 \cdot 4)^{-1/2}$$

$$= 3.155$$

2. Estimate T_b using Equation 3.3.

$$T_b = 225.71 \text{ K} + 76.1 \ ^1X^v \ \text{K}$$

$$= 225.71 \text{ K} + 76.21 \, (3.155) \ \text{K}$$

$$= 446 \text{ K}$$

The measured boiling point of naphthalene is 218°C (491 K) (Lide, 1994).

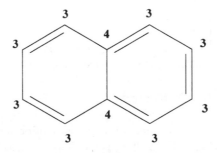

Figure 3.1   The hydrogen-suppressed graph of naphthalene. δ^v values are given alongside each atom.

The molecular connectivity concept has been extended by Hall and Kier (1995) to include an electrotopological state (E-state) index that is somewhat more complicated to compute than the molecular connectivity index. Hall and Story (1996) describe a method of estimating normal boiling-point values that employs the E-state index. The model produces a correlation coefficient (r) of 0.9975 and a mean absolute error of 3.86 K with a training set of 268 diverse organic chemicals. The training set data are compiled by Egolf et al. (1994). The model gives a correlation coefficient of 0.9967 and a mean absolute error of 4.57 K with a test set of 30 diverse chemicals.

Many chemicals decompose at temperatures below their normal boiling point, hence only their melting point is found in the literature. Walters et al. (1995) report a simple group-contribution method for estimating the boiling point of a chemical from its melting point, rotational symmetry, and structure. The model produces a correlation coefficient (r^2) of 0.913 and a standard error of 28.1 K with a training set of 1419 diverse chemicals.

The model was validated with a test set of 40 simple and multifunctional hydrocarbons including chemicals that contain oxygen, nitrogen, sulfur, and halogen groups. An average absolute error of 23.9 K was obtained with the test set. The method's accuracy was shown to be comparable to that of the original method of Joback and Reid (1987).

3.2.3 Sensitivity to Environmental Parameters

The normal boiling temperature is a sensitive function of ambient pressure and liquid composition. This chapter describes methods of estimating the boiling point of a liquid at 1 atm pressure. The boiling point of a liquid at any pressure other than 1 atm can be estimated with the vapor pressure models presented in Chapter 6. The boiling point of mixtures of miscible liquids can be estimated using Raoult's law, which is also discussed in Chapter 6.

3.3 METHODS OF ESTIMATING MELTING POINT

There are no simple, yet reliable, methods of estimating a chemical's melting point. This is because the value of T_m is determined, in part, by the geometric arrangement of molecules in the solid, and so it is controlled by molecular geometry. Chemicals with the same molecular formula can have different geometric structures and different melting points. Another complication is that, due to the existence of polymorphic crystal structures, some chemicals exhibit multiple melting points.

3.3.1 Joback and Reid's Method

Joback and Reid (1987) describe a group contribution method of estimating a "very approximate" value of the melting point. Only 40 structural groups are identified, and the range of applicability is somewhat limited. The melting-point value is estimated as the sum of contributions of all groups present in the molecule,

$$T_m = 122 + \sum n_i \Delta_{m,i}$$

(3.6)

$$n = 388, \ \text{AAE} = 24.7 \ \text{K}, \ \text{AAPE} = 11.2\%$$

where T_m = melting point, K
n_i = number of i groups in molecule
$\Delta_{m,i}$ = Joback contribution of group i to T_m, K
n   = number of chemicals in the training set
AAE   = average absolute error, K
AAPE   = average absolute percent error

Values of $\Delta_{m,i}$ are listed in Table 3.1. The training set consisted of 388 aliphatic and aromatic hydrocarbons of various types.

Example

Estimate the melting point of phenol, C_6H_6O, using the Joback–Reid method.

1. Phenol consists of five =CH– ring groups, one =C< ring group, and one phenolic –OH group. Given the values of $\Delta_{m,i}$ listed in Table 3.1:

$$\Sigma n_i \cdot \Delta_{m,i} \ 1 = (5)(8.13 \ \text{K}) + 37.02 \ \text{K} + 82.83 \ \text{K}$$

$$= 160.50 \ \text{K}$$

2. Estimate the melting point with Equation 3.6.

$$T_m = 122 + \Sigma n_i \cdot \Delta_{m,i} \ \text{K}$$

$$= 122 + 160.50 \ \text{K}$$

$$= 282.50 \ \text{K}$$

The measured melting point of phenol is 43°C (316 K) (Lide, 1994).

The Joback–Reid method does not distinguish between many types of molecular structural features. For instance, aliphatic aldehydes and carboxylic acids are not distinguished from aromatic aldehydes and carboxylic acids. Structural isomers are not distinguished, either. A more accurate estimate of the T_m value of a chemical might be made by starting with the melting point of a reference chemical that is structurally similar to the chemical of interest. Then, as we saw in Chapter 2:

$$T_{m,R-X} = T_{m,R-Y} + \Pi$$

(3.7)

where π is the group exchange factor, the difference between fragment contributions and correction factors for T_m attributable to groups x and y.

Example

Estimate the melting point of benzoic acid, C_6H_5COOH, using the measured melting point of phenol, C_6H_6O. The melting point of phenol is 43°C (316 K) (Lide, 1994).

1. The value of π is the difference between the group contributions of the phenolic OH group and the carboxylic acid group. Values of $\Delta_{m,j}$ are given in Table 3.1.

$$\Pi = \Delta_{m,COOH} - \Delta_{m,OH}$$

$$= 155.50 - 82.82 \ K$$

$$= 72.67 \ K$$

2. Estimate the melting point of benzoic acid using Equation 3.7.

$$T_{m,R-COOH} = T_{m,R-OH} + \Pi$$

$$= 316 \ K + 72.67 \ K$$

$$= 388.67 \ K$$

The measured melting point of benzoic acid is 122.4°C (396.6 K) (Lide, 1994).

Care must be taken when estimating T_m values of straight-chain aliphatic hydrocarbons. The melting points of these compounds increase irregularly with increasing chain length. The correlation between melting point and chain length is different for chemicals with odd numbers of carbon in the chain than it is for chemicals with even numbers of carbon atoms (Burrows, 1992; Somayajulu, 1990). This **odd–even effect** stems from the different stereochemical limitations that are imposed on the packing arrangements of odd and even straight-chain hydrocarbons in the solid state. Therefore, the melting point of straight-chain aliphatic hydrocarbons should not be estimated with a group exchange method using the T_m value of an adjacent member of the homologous series.

3.3.2   Other Methods

The most accurate estimate of the melting point of a monofunctional chemical might be obtained with a narrowly focused group contribution method. Abramowitz and Yalkowsky (1990) describe a structure–melting-point relationship for polychlorinated biphenyl (PCB) congeners. The relationship, involving five computable molecular descriptors, gives a correlation coefficient (r) of 0.91 and a standard deviation of 22.1 K when applied to the training set of 58 congeners having melting points between 300 and 600 K. The model was used to estimate the melting points of 208 PCB congeners.

Simamora et al. (1993) suggest a method of estimating T_m values of rigid aromatic ring systems. Group contributions are given for substituents such as halo,

methyl, cyano, and nitro groups. Hydrogen-bonding groups such as hydroxyl, aldehyde, carboxyl, primary amino, and amido groups are included. The rotational
symmetry of the molecule, intramolecular hydrogen bonding, and the presence of
biphenyl groups are accounted for. The training set consists of 1181 compounds.
The standard error of the linear least-squares regression is 36.63 K, and r^2 is 0.9910.

Constantinou and Gani (1994) propose a group contribution method for melting
point that distinguishes among isomers. An absolute error of 14.03 K and an absolute
percent error of 7.23% is reported when the method is applied to the training set of
312 diverse chemicals.

REFERENCES

Abramowitz, R. and Yalkowsky, S.H., 1990. Estimation of aqueous solubility and melting
point of PCB congeners. *Chemosphere*, **21**, 1221–1229.

Aldrich Handbook of Fine Chemicals. 1990. Aldrich Chemical Company, Milwaukee, WI.

Balaban, A.T., Joshi, N., Kier, L.B., and Hall, L.H., 1992. Correlations between chemical
structure and normal boiling points of halogenated alkanes, C_1–C_4. *J. Chem. Int. Comput.
Sci.,* **32**, 233–237.

Balaban, A.T., Kier, L.B., and Joshi, N., 1992. Correlations between chemical structure and
normal boiling points of acyclic ethers, peroxides, acetals, and their sulfur analogs.
J. Chem. Int. Comput. Sci., **32**, 237–244.

Burrows, H.D., 1992. Studying odd–even effects and solubility behavior using α,ω-dicarbox-
ylic acids. *J. Chem. Ed.,* **69**, 69–73.

Constantinou, L. and Gani, R., 1994. New group contribution method for estimating properties
of pure compounds. *AIChE J.,* **40**, 1697–1710.

Dean, J.A., 1992. *Lange's Handbook of Chemistry*, 14th Edition. McGraw-Hill, New York.

Devotta, S. and Pendyala, V.R., 1992. Modified Joback group contribution method for normal
boiling point of aliphatic halogenated compounds. *Ind. Eng. Chem. Res.,* **31**, 2042–2046.

Egolf, L.M., Wessel, M.D., and Jurs, P.C., 1994. Prediction of boiling points and critical
temperatures of industrially important organic compounds from molecular structure,
J. Chem. Inf. Comput. Sci., **34**, 947–956.

Graselli, J., Ed., 1990. *Handbook of Data of Organic Compounds*. CRC Press, Boca Raton, FL.

Hall, L.H. and Kier, L.B., 1995. Electrotopological state indices for atom types: a novel
combination of electronic, topological, and valence state information. *J. Chem. Inf.
Comput. Sci.,* **35**, 1039–1045.

Hall, L.H. and Story, C.T., 1996. Boiling point and critical temperature of a heterogeneous
data set: QSAR with atom type electrotopological state indices using artificial neural
networks. *J. Chem. Inf. Comput. Sci.,* **36**, 1004–1014.

Hoshino, D., Zhu, X.R., Nagahama, K., and Hirata, M., 1985. Prediction of vapor pressures
for substituted benzenes by a group contribution method. *Ind. Eng. Chem. Fundam.,* **24**,
112–114.

Joback, K.G. and Reid, R.C., 1987. Estimation of pure-component properties from group-
contributions. *Chem. Eng. Comm.,* **57**, 233–243.

Lide, D.R., 1994. *CRC Handbook of Chemistry and Physics*. 74th Edition. CRC Press, Boca
Raton, FL.

Rechsteiner, C.E., 1990. Boiling Point. Chapter 12. In *Handbook of Chemical Property
Estimation Methods. Environmental Behavior of Organic Compounds,* Lyman, W.J.,
Reehl, W.F., and Rosenblatt, D.H., Eds. American Chemical Society, Washington, D.C.

Reid, R.C., Prausnitz, J.M., and Poling, B.E., 1966. *The Properties of Gases and Liquids*. Second Edition. McGraw-Hill, New York.

Reid, R.C., Prausnitz, J.M., and Poling, B.E., 1987. *The Properties of Gases and Liquids*. Fourth Edition. McGraw-Hill, New York.

Simamora, P., Miller, A.H., and Yalkowsky, S.H., 1993. Melting point and normal boiling point correlations: application to rigid aromatic compounds. *J. Chem. Int. Comput. Sci.,* **33**, 437–440.

Somayajulu, G.R., 1990. The melting points of ultralong paraffins and their homologues. *Int. J. Thermophys.,* **11**, 555–572.

Stein, S.E. and Brown, R.L., 1994. Estimation of normal boiling points from group contributions. *J. Chem. Inf. Comput. Sci.,* **34**, 581–587.

Walters, A.E., Myrdal, P.B., and Yalkowsky, S.H., 1995. A method for estimating the boiling points of organic compounds from their melting points. *Chemosphere*, **31**, 3001–3008.

White, C.M., 1986. Prediction of boiling point, heat of vaporization, and vapor pressure at various temperatures for polycyclic aromatic hydrocarbons. *J. Chem. Eng. Data,* **31**, 198–203.

CHAPTER 4

Density and Molar Volume

4.1 INTRODUCTION

Density, ρ (kg/m^3), is *the mass of a substance per unit volume*. The molar volume of a substance, V_m (m^3/mol), is a related property. The density is related to molar volume by:

$$\rho = \frac{M}{V_m} \tag{4.1}$$

where ρ = density, kg/m^3
 M = molar mass, kg/mol
 V_m = molar volume, m^3/mol

The **specific gravity**, $d^{T1/T2}$, is *the density of a substance at temperature T_1 relative to the density of water at T_2*. Usually, the reference density is that of water at 4°C, 1.0000 g/cm^3.

The relative density of a chemical with respect to the density of air or water can be an important determinant of the chemical's environmental distribution. Moreover, the value of density can be used to estimate a chemical's surface tension, its viscosity, and various of its partition coefficients.

Extensive tabulations of measured density values are published by chemical manufacturers and others. (See, for instance, Dean, 1992 and Lide, 1994.) Published values range from about 0.5 to 3 g/cm^3 for organic liquids and from about 0.9 to 4 g/cm^3 for organic solids. This chapter describes several methods of estimating the density and molar volume of organic liquids and solids that may be useful when published values are not available.* Other estimation methods for density are described by Reid et al. (1987) and by Nelken (1990).

---

* The densities of organic vapors range from 0.5 to 3 g/l. The density of a vapor is estimated with the ideal gas law:

$$\rho_g = \frac{MP}{RT}$$

where M (kg/mol) is molar mass, P (Pa) is partial pressure, R (8.3145 J/mol•K) is the ideal gas constant, and T (K) is the temperature of the vapor.

Table 4.1   Relative Atomic Volume

Element	$V_{a,i}$
H	1
Period 2 elements, Li to F	2
Period 3 elements, Na to Cl	4
Period 4 elements, K to Br	5
Period 5 elements, Rb to I	7.5
Period 6 elements, Cs to Bi	9

From Girolami, G.S., 1994. *J. Chem. Ed.*, **11**, 962–964.

4.2 ESTIMATION METHODS

4.2.1 Girolami's Method

Girolami (1994) proposes a simple atom contribution method of estimating the density of a pure liquid at room temperature, usually taken to mean $25 \pm 5°C$. The molecular volume of a solid chemical may also be estimated. The method is applicable to organic, inorganic, and metal–organic chemicals. For liquids:

$$\rho = \frac{M}{5\sum_i n_i \cdot v_{a,i}} \tag{4.2}$$

where ρ = density of liquid chemical, g/cm^3
M = molar mass of chemical, g/mol
n_i = number of atoms i in molecule
$v_{a,i}$ = relative volume of atom i

The scaling factor of 5 cm^3/mol in Equation 4.2 is derived from the relationship between the approximate volume of one molecule, $8.30\sum_i n_i \cdot v_{a,i}$ $Å^3$/molecule, and the molar volume of a liquid chemical, $[8.30\sum_i n_i \cdot v_{a,i}$ $Å^3$/molecule]$\cdot[6.022 \times 10^{23}$ molecules/mol]$\cdot[1 \times 10^{-24}$ $cm^3/Å^3] = 5$ cm^3/mol. The relative atomic volumes, $v_{a,i}$, are given in Table 4.1 tabulated by period of the periodic table. They are derived from published van der Waals radii (Pauling, 1960).

Some structural features enhance the attractive forces between molecules, increasing liquid density. The estimate of liquid density must be corrected if any of the features listed in Table 4.2 are present in the molecule structure. If several of the listed features are present, the density estimate is increased to a limit of 30%.

Extensive intermolecular association occurs in some inorganic liquids due to bridging by oxygen and halogen. This occurs in liquid SO_3 and SbF_5, for instance. Density estimates should be increased by 10% in this case.

Table 4.2   Structural Features for Second-Order
Density Correction[a]

Structural Feature

Hydroxyl, −OH
Carboxylic acid, −C(=O) OH
Inorganic acid, for each acidic −OH
Primary amine, −NH$_2$
Secondary amine, −NHR
Amide, −C(=O) NH$_2$
N-substituted amide, −C(=O) NHR
Sulfoxide, −S(=O)−
Sulfone,[b] −S(=O)$_2$−
Unfused ring
Fused ring[c]

[a] Increase density by 10% for each feature present
to a limit of 30%.
[b] Treat the sulfone group as two sulfoxide groups.
[c] Increase density by 7.5% for each fused ring.

From Girolami, G.S., 1994. *J. Chem. Ed.*, 11, 962–964.

Girolami (1994) reports the root-mean-square error of his method to be 0.049 g/cm^3 with a test set of the 166 organic liquids listed in *The Chemists Companion* (Gordon and Ford, 1972). The measured densities of chemicals in the set range from 0.6 to 2.9 g/cm^3. With the exception of the density estimates for acetonitrile and dibromochloromethane, the relative absolute error of the estimate is less than 0.1 g/cm^3 in all cases. The method does not distinguish between isomers.

Girolami suggests that the molecular volume of a crystalline solid chemical is roughly $6.4\sum_i \cdot n_i \cdot V_{a,i}$ Å^3/molecule. The molar volume of a solid is approximately $V_m = [6.4\sum_i \cdot n_i \cdot V_{a,i}$ Å^3/molecule][6.022 × 10^{23} molecules/mol]·[1 × 10^{-24} cm^3/Å^3] = 3.85 $\sum_i \cdot n_i \cdot v_{a,i}$ cm^3/mol. The relative atomic volumes, $v_{a,i}$, are listed in Table 4.1.

Girolami calls his approach a "back-of-the-envelope" method. Still, it is remarkably accurate for such a simple procedure and is more broadly applicable than most other estimation methods. The relatively good accuracy is probably due to the fact that, except for the alkali metals and alkaline earths (groups 1A and 2A of the periodic table), covalent radii vary by less than about 20% across periods 2 and 3 and less than a few percent across periods 4, 5, and 6 of the periodic table. Also, the approach accounts for important dipolar interactions and hydrogen bonding with corrections for contributing structural features.

Example

Estimate the density of α-bromonaphthalene, α-BrC$_{10}$H$_7$, at 20°C. The molar mass of α-bromonaphthalene is 207.08 g/mol (Lide, 1994).

1. Estimate the molar volume of α-bromonaphthalene by summing the atomic contributions as in Equation 4.2. The values of $v_{a,i}$ are found in Table 4.1.

$$v_m = 5 \cdot \Sigma_i n_i \cdot v_{a,i} \quad cm^3/mol$$

$$= (5) \left[(1)(5) + (10)(2) + (7)(1) \right] \quad cm^3/mol$$

$$= 160 \quad cm^3/mol$$

2. Estimate the liquid density using Equation 4.2. Correction factors for important structural features of the molecule are listed in Table 4.2. Since the naphthalene structure consists of two fused rings, the density estimate must be increased by 15 (= 2 × 7.5)%.

$$\rho = (1.15)(207 \text{ g/mol}) / (160 \text{ cm}^3/\text{mol})$$

$$= 1.49 \text{ g/cm}^3$$

The temperature for which the estimated density is made is not carefully specified. The measured value of d$^{20/4}$ is 1.4826 g/cm³ (Lide, 1994). The density estimate differs from the measured value of d$^{20/4}$ by 0.35%.

4.2.2 Grain's Method

Grain suggests a method of estimating the density of a liquid at any temperature between the melting point and the boiling point. It is described by Nelken (1990). Grain's model is

$$\rho_l = \frac{M}{V_b} \left[3 - 2 \left(\frac{T}{T_b} \right) \right]^n \tag{4.3}$$

where ρ_l = liquid density at temperature T, g/cm³
       M = molar mass, g/mol
       V_b = molar volume at normal boiling point, cm³/mol
       T = temperature, K
       T_b = normal boiling point, K

The molar volume at the normal boiling point, V_b, is estimated with Schroeder's method described below. The value of the exponent n is 0.25 for alcohols, 0.29 for hydrocarbons, and 0.31 for other organic chemicals. Nelken (1990) demonstrates the use of Grain's method with a test set of 28 chemicals that includes aliphatic and aromatic hydrocarbons, halocarbons, oxygen-containing compounds, and nitrogen-containing compounds. The measured densities of chemicals in the test set range from 0.801 to 2.1792 g/mL. The average error of an estimate is 3.1% and the maximum error is 8.6%.

Nelken (1990) also estimates the density of five chemicals — acetone, isoamyl propionate, methane, n-octane, and dimethyl ether — at two to three different temperatures using Grain's method. The predictive error usually lies within 4% but error as large as 11% is observed.

4.2.3 Schroeder's Method

This method uses group contributions to estimate the molar volume of a liquid at the normal boiling point. The value is used in Grain's method of estimating liquid density as a function of temperature as discussed above. Schroeder's method is described by both Reid et al. (1987) and Nelken (1990). The molar volume is given by:

$$V_b = \sum_i n_i \cdot v_{b,i} \qquad (4.4)$$

where V_b = molar volume at normal boiling point, cm^3/mol
   n_i = number of features i in molecule
   $v_{b,i}$ = contribution of feature i molar volume, cm^3/mol

The group contributions, $v_{b,i}$, are listed in Table 4.3.

Except for highly associated liquids, Schroeder's method produces estimates that are within 3 or 4% of experimental values. It produces a 3.0% average error with a test set of 32 organic and inorganic chemicals of various types (Reid et al., 1987). The measured molar volumes of chemicals in the test set range from 18.7 to 162 cm^3/mol.

Table 4.3 Schroeder's Group Contributions to Molar Volume

Feature	V_{bi}, cm^3/mol
Atoms	
C, H, N, O	7
F	10.5
Cl	24.5
Br	31.5
I	38.5
S	21
Rings	−7
Bonds	
Double	7
Triple	14

From Reid, R.C., Prausnitz, J.M., and Poling, B.E., 1987. *The Properties of Gases and Liquids*. Fourth Edition. McGraw-Hill, New York.

Example

Estimate the density of benzene, C_6H_6, at 0°C using Grain's method. For benzene, M = 78.12 g/mol. T_b = 80.1°C (Lide, 1994).

1. Estimate the molar volume at the normal boiling point using Schroeder's method, Equation 4.4. The group contributions are given in Table 4.3. The features listed in Table 4.3 exhibited by the benzene molecule are six carbon atoms, six hydrogen atoms, one ring, and three double bonds.

$$V_b = \Sigma_i \ n_i \cdot v_{b,i}$$

$$V_b = (6)(7) + (6)(7) + (1)(-7) + (3)(7) \ \text{cm}^3/\text{mol}$$

$$= 98 \ \text{cm}^3/\text{mol}$$

2. Estimate liquid density at 0°C using Equation 4.3. Benzene is an aromatic hydrocarbon, and n = 0.29. T = 273.2 K. T_b = 353.2 K.

$$\rho_1 = (M/V_b) \left[3 - 2(T/T_b) \right]^n$$

$$= \left[(78.12 \ \text{g/mol})/(98 \ \text{cm}^3/\text{mol}) \right] \left[3 - 2(273.2/353.2) \right]^{0.29}$$

$$= 0.888 \ \text{g/cm}^3$$

The measured value of the density of benzene at 0°C is 0.899 g/cm³ (Lide, 1994). The error of the estimate is 1.2%.

4.2.4   Constantinou et al.'s Method

This group-contribution method produces fairly accurate estimates of the molar volumes of organic liquids at 25°C. If greater accuracy is needed, second-order estimates can also be made. The molar volume is given by (Constantinou et al., 1995):

$$V_m = 0.0121 \frac{m^3}{kmol} + \sum_i n_i \cdot v_{ml,i} + W \sum_j m_j v_{m2,j} \tag{4.5}$$

n = 312
W = 0, s = 0.00236 m³/kmol, AAE = 0.00139 m³/kmol, AAPE = 1.16%
W = 1, s = 0.00192 m³/kmol, AAE = 0.00105 m³/kmol, AAPE = 0.89%

where V_m = liquid molar volume at 25°C
       n_i = number of features i exhibited by molecule

$v_{m1,i}$ = first-order contribution of feature i to liquid molar volume
W = 0 for first-order estimate, 1 for second-order estimate
m_j = number of features j exhibited by molecule
$v_{m2,j}$ = second-order contribution of feature to liquid molar volume
n = number of chemicals in the training set
s = standard deviation, m^3/mol
AAE = average absolute error, m^3/mol
AAPE = average absolute percent error, %

Values of $v_{m1,i}$ and $v_{m2,j}$ are listed in Tables 4.4 and 4.5, respectively. The group contribution values were determined with a training set consisting of 134 hydrocarbons; 108 alcohols, aldehydes, ketones, acids, and esters; 27 amides and amines; 8 thioles and thialkanes; and 35 halogenated compounds.

Example

Estimate the molar volume and density of liquid 2-propyl cyclohexanone, 2-$C_3H_7(C_6H_8O)$, at 25°C using the method of Constantinou et al. The molar mass of 2-propyl cyclohexanone is 140.23 g/mol.

1. Identify the first-order group contributions using Table 4.4.

Group	n_i	$v_{n1,i}$
CH_3	2	0.02614 m³/kmol
CH_2	4	0.01641
CH	1	0.00711
C	1	−0.00380
$CH_2C=O$	1	0.02816

$$\Sigma_i n_i \cdot v_{m1,i} = (2)(0.02614) + (4)(0.01641) + 0.00711 - 0.00380 + 0.02816$$

$$= 0.1494$$

2. Identify the second-order group contributions using Table 4.5.

Group	m_j	$v_{m2,j}$
$(CH_3)_2CH$	1	0.00133 m³/kmol
6-Membered ring	1	0.00063
Alicyclic side chain	1	−0.00107
$C_{cyclic}=O$	1	−0.00111

$$\Sigma_j m_j \cdot v_{m2,j} = 0.00133 + 0.00063 - 0.00107 - 0.00111$$

$$= -0.0002$$

Table 4.4   First-Order Group Contributions to V_m^{25}, m^3/kmol

Group	$V_{m1,i}$	Group	$V_{M1,i}$
CH_3	0.02614	CH_2Cl	0.03371
CH_2	0.01641	$CHCl$	0.02663
CH	0.00711	CCl	0.02020
C	−0.00380	$CHCl_2$	0.04682
$CH_2{=}CH$	0.03727		
$CH{=}CH$	0.02692	CCl_3	0.06202
$CH_2{=}C$	0.02697	$aCCl$	0.02414
$CH{=}C$	0.01610	CH_2NO_2	0.03375
$C{=}C$	0.00296	$CHNO_2$	0.02620
$CH_2{-}C{=}CH$	0.04340	$aCNO_2$[a]	0.02505
aCH[a]	0.01317	CH_2SH	0.03446
aC[a]	0.00440	I	0.02791
$aCCH_3$[a]	0.02888	Br	0.02143
$aCCH_2$[a]	0.01916		
$aCCH$[a]	0.00993	$C{-}C$	0.01451
OH	0.00551	$Cl(C{-}C)$	0.01533
$aCOH$[a]	0.01133	aCF[a]	0.01727
CH_3CO	0.03655		
CH_2CO	0.02816		
CHO	0.02002		
CH_3COO	0.04500		
CH_2COO	0.03567	COO	0.01917
$HCOO$	0.02667	CCL_2F	0.05384
CH_3O	0.03274		
CH_2O	0.02311	$CCIF_2$	0.05383
$CH{-}O$	0.01799		
FCH_2O	0.02059		
CH_2NH_2	0.02646		
$CHNH_2$	0.01952		
CH_3NH	0.02674	$CON(CH_3)_2$	0.05477
CH_2NH	0.02318		
$CHNH$	0.01813		
CH_3N	0.01913	$C_2H_5O_2$	0.04104
CH_2N	0.01683		
$aCNH_2$[a]	0.01365	CH_3S	0.03484
C_5H_4N	0.06082	CH_2S	0.02732
C_5H_3N	0.05238		
CH_2CN	0.03313		
$COOH$	0.02232		

[a] The symbol "a" denotes an aromatic group.

From Constantinou et al., 1995.

3. Calculate V_m^{25} using Equation 4.5.

$$V_m^{25} = \Sigma_i n_i \cdot v_{m1,i} + W \cdot \Sigma_j m_j \cdot v_{m2,j} \quad m^3/kmol$$

$$= 0.01494 + (1)(-0.0002)$$

$$= 0.1492 \ m^3/kmol$$

Table 4.5 Second-Order Group Contributions to V_m^{25}, $m^3/kmol$

Group	$V_{m2,j}$	Group	$V_{m2,j}$
$(CH_3)_2CH$	0.00133	$CH_3C(=O)OCH$, $CH_3C(=O)OC$	0.00083
$(CH_3)_3C$	0.00179	$COCH_2C(=O)O$, $C(=O)CHC(=O)O$, $C(=O)CC(=O)O$	0.00036
$CH(CH_3)CH(CH_3)$	−0.00203		
$CH(CH_3)CH(CH_3)_2$	−0.00243		
$C(CH_3)_2C(CH_3)_2$	−0.00744	$CO-O-CO$	0.00198
5-Membered ring	0.00213	$aCC(=O)O$	0.00001
6-Membered ring	0.00063	$CHOH$	−0.00092
7-Membered ring	−0.00519	COH	0.00175
$CH_n=CH_m-C_p=C_k$ k,n,m,p ε(0,2)	−0.00188	$CH_m(OH)C_n(OH)$ m,n ε(0,2)	0.00235
$CH_3-CH_m=CH_n$ m,n ε(0,2)	0.00009	$CH_{m\ cyclic}-OH$ m ε(0,1)	−0.00250
$CH_2-CH_m=CH_n$ m,n ε(0,2)	0.00012	$CH_m(OH)CH_n(NH_p)$ m,n,p ε(0,2)	0.00046
$CH-CH_m=CH_n$, $C-CH_m=CH_n$ m,n ε(0,2)	0.00142	$CH_{m\ cyclic}-NH_p-CH_{n\ cyclic}$ m,n,p ε(0,2)	−0.00179
Alycyclic side chain $C_{cyclic}Cm$, m > 1	−0.00107	$CH_m-O-CH_m-CH_p$ m,n,p ε(0,2)	0.00206
$CHCH=O$, $CCH=O$	−0.00009	$aC-O-CH_m$ [a] m(0,3)	0.01203
$CH_3C(=O)CH_2$	−0.00030	$CH_{m\ cyclic}-S-CH_{n\ cyclic}$ m,n ε(0,2)	−0.00023
$CH_3C(=O)CH$, $CH_3C(=O)C$	−0.00108	CH_m-CH_n-Br m,n ε(0,2)	−0.0058
$C_{cyclic}=O$	−0.00111	$aCBr$ [a]	0.00178
$aCCH=O$ [a]	−0.00036	aCl [a]	0.00171
$CHC(=O)OH$, $CC(=O)OH$	−0.00050		
$aCC(-O)OH$ [a]	0.00777		

[a] The symbol "a" denotes an aromatic group.
From Constantinou et al., 1995.

4. The density is obtained from V_m^{25} using Equation 4.1.

$$d^{25} = \left(1/V_m^{25}\right)\left(1 \times 10^{-3}\right)(140.23) \ g/cm^3$$

$$= 0.9399 \ g/cm^3$$

The measured value of the density of 2-propyl cyclohexanone at 25°C is 0.923 g/ml (Dean, 1992). The predictive error is 1.8%.

4.2.5 Immirzi and Perini's Method

A method of estimating the density of a solid organic chemical is reported by Immirzi and Perini (1977) and discussed by Nelken (1990). The molecular volume of the crystalline solid is estimated with a group contribution method like the method of Girolami described above. The molecular volume is

Table 4.6   Group Contributions to V_S, Å³/Molecule

Structural Feature	v_i, Å³/molecule	No. in Training Set
–H	6.9	5228
=C–, –C≡	15.3	74
>C=	13.7	453
>C<	11.0	1165
=O	14.0	649
–O–	9.2	468
N≡	16.0	30
–N=	12.8	68
>N–	7.2	354
S	23.8	92
–F	12.8	14
–Cl	26.7	134
–Br	33.0	120
–I	45.0	26
Benzene ring[a]	75.2	443
Naphthalene ring[a]	123.7	0
Aliphatic ring	–3.0	0

[a] Value for the carbon skeleton only.

From Immirzi, A. and Perini, B., 1977. *Acta Crystallogr. Sect. A,* **33**, 216–218.

$$V_S = \sum_i m_i v_i \tag{4.6}$$

where V_S = molecular volume in solid, Å³/molecule
$\quad$ m_i = number of groups i exhibited by molecule
$\quad$ v_i = volume of group i, Å³/group

Group contribution values, v_i, are given in Table 4.6.

The molecular volume estimated with Equation 4.6 is used to calculate the density of the solid:

$$\rho = \frac{1.660M}{V_S} \tag{4.7}$$

where ρ = crystal density of chemical, g/cm³
$\quad$ M = molecular weight of chemical
$\quad$ V_S = molecular volume in solid, Å³/molecule

Immirzi and Perini's group contribution values were calculated with a training set of 500 chemicals ranging in molecular weight from 50 to 1000 g/mol. The method produced an average absolute error of 2% and a maximum error of +6.9% when applied to a test set of 53 solid chemicals of various types (Nelken, 1990).

The method accounts for more structural features than Girolami's method, but it does not apply to as many chemicals, and it is probably no more accurate. Immirzi and Perini's method does not apply to chemicals that are liquids at "room temperature". Derivatives of benzene and naphthalene are the only cyclic compounds included in the training set. Crystals with structural disorder or that contain solvent molecules other than water are excluded.

Example

Estimate the density of solid 4-bromophenol, HO–C$_6$H$_4$–Br, at 15°C using the method of Immirzi and Perini.

1. Estimate the molecular volume using Equation 4.6 and Table 4.6. The structure has one benzene carbon frame, five –H, one –O–, and one –Br.

$$V_S = \Sigma_i \ m_i \cdot v_i$$

$$= (1) \ (75.2) + (5) \ (6.9) + (1) \ (9.2) + (1) \ (33.0) \ \text{Å}^3/\text{molecule}$$

$$= 151.9 \ \text{Å}^3/\text{molecule}$$

2. Estimate the crystal density with Equation 4.7. M = 173.02 amu.

$$\rho = 1.660 \ M/V_S$$

$$= (1.660) \ (173.02)/(151.9) \ \text{g}/\text{cm}^3$$

$$= 1.891 \ \text{g}/\text{cm}^3$$

The measured density of crystalline 4-bromophenol is 1.840[15] (Lide, 1994). The error of the estimate is 2.8%.

4.3 SENSITIVITY TO ENVIRONMENTAL PARAMETERS AND METHOD ERROR

The density of an organic solid is not sensitive to changes in temperature or pressure. The density of an organic liquid does not vary significantly with change in pressure, but it is sensitive to temperature change. Girolami's method does not accurately specify the temperature to which the estimate applies and does not account for the temperature dependence of the density of liquid organic chemicals. Grain's method is more useful in this regard, since it does account for temperature.

Neither Girolami's method nor Grain's method distinguishes between chemical isomers. Neither method accounts for features such as chain branching that determine the molecular shape of aliphatic hydrocarbons and probably also determine, in part, their liquid density.

The group contribution method of Constantinou et al. employs a complex, non-linear model. The second-order estimation method demands proficiency at identifying large and complex fragments that may be present in the molecular structure. The method is probably best used as a computer application.

Estimation methods that apply to specific types of chemicals usually offer better accuracy than the general methods described in this chapter. Simple but accurate molecular connectivity models to estimate the $d^{20/4}$ values of alkanes, aliphatic alcohols, ethers, and acids are available (Kier and Hall, 1976; Needham et al., 1988). A simple, linear correlation between V_m^{20} and molecular surface area, estimated with contact atomic radii, is reported by Grigoras (1990) for a variety of saturated, unsaturated, and aromatic hydrocarbons: alcohols, acids, esters, amines, and nitriles. Grigoras' method is easy to apply if molecular modeling software is available.

REFERENCES

Dean, J.A., 1992. *Lange's Handbook of Chemistry,* 14th Edition. McGraw-Hill, New York.

Girolami, G.S., 1994. A simple "back of the envelope" method for estimating the densities and molecular volumes of liquids and solids. *J. Chem. Ed.,* **11**, 962–964.

Gordon, A.J. and Ford, R.A., 1972. *The Chemist's Companion.* John Wiley & Sons, New York.

Grigoras, S., 1990. A structural approach to calculate physical properties of pure organic substances: the critical temperature, critical volume, and related properties. *J. Comp. Chem.*, **11**, 493–510.

Immirzi, A. and Perini, B., 1977. Prediction of density in organic crystals. *Acta Crystallogr. Sect. A,* **33**, 216–218.

Kier, L.B. and Hall, L.H., 1976. *Molecular Connectivity in Chemistry and Drug Research.* Academic Press, New York.

Lide, D.R., 1994. *CRC Handbook of Chemistry and Physics.* 74th Edition. CRC Press, Boca Raton, FL.

Needham, D.E., Wei, I.-C., and Seybold, P.G., 1988. Molecular modeling of the physical properties of the alkanes. *J. Am. Chem. Soc.*, **110**, 4186–4194.

Nelken, L.H., 1990. Densities of Vapors, Liquids and Solids. Chapter 19. In *Handbook of Chemical Property Estimation Methods. Environmental Behavior of Organic Compounds.* Lyman, W.J., Reehl, W.F., and Rosenblatt, D.H., Eds. American Chemical Society, Washington, D.C.

Pauling, L., 1960. *The Nature of the Chemical Bond.* Cornell University Press, Ithaca, NY.

Reid, R.C., Prausnitz, J.M., and Poling, B.E., 1987. *The Properties of Gases and Liquids.* Fourth Edition. McGraw-Hill, New York.

CHAPTER 5

Surface Tension and Parachor

5.1 INTRODUCTION

Cohesive intermolecular forces produce a net attraction that pulls matter at the surface of a fluid into the bulk of the liquid. In order to add matter to the surface, increasing the surface area, work must be done to overcome the net attractive force exerted by the bulk liquid. The **surface tension**, γ (J/m^2, N/m) is *the ratio of the work done to expand the surface divided by the increase in surface area*. Surface tension values are used to estimate the vapor pressure of liquids in aerosols and in soil capillaries, among other things.

Measured values of surface tension of organic liquids that are not hydrogen bonded lie in the range of 0.020 to 0.040 N/m at 20°C. Values of surface tension as high as 0.065 N/m are observed for hydrogen-bonding organic liquids. The surface tension of water is 0.0728 N/m at 20°C.

The **parachor**, P ($m^3 \cdot kg^{1/4}/s^{1/2} \cdot mol$), is defined as:

$$P = \frac{M\gamma^{\frac{1}{4}}}{\rho_l - \rho_g} \tag{5.1}$$

where M (kg/mol) is the molar mass, ρ_l (kg/m^3) is the liquid density, and ρ_g (kg/m^3) is the vapor density of a chemical. The parachor changes by no more than a few percent over a temperature range of 100°C (Sugden, 1924; Exner, 1967), and we may consider its value to be fixed in the normal ambient range of −40 to +40°C.

This chapter describes two methods of estimating the surface tension of liquids. Surface tension is a control of the vapor pressure of liquids in aerosols and in soil capillaries. It is also related to the amount of material absorbed at interfaces. The relationship between surface tension and vapor pressure is discussed in Chapter 6.

The parachor is of little intrinsic interest, but the parameter provides a convenient way of estimating surface tension. The value of the parachor can be estimated with acceptable accuracy using the simple group contribution schemes described below.

The value of surface tension is estimated using the values of parachor and liquid density at the ambient temperature. Various methods of estimating liquid density are described in Chapter 4.

5.2 ESTIMATION METHODS

5.2.1   Sugden's Method

Sugden (1924) introduced the parachor parameter and first suggested a group contribution method of estimating its value. Based on Sugden's suggestions, Quale (1953) and Reid et al. (1987) developed the empirical method described here. While limited in scope, the method is both accurate and widely used. It requires a good understanding of chemical structure to apply correctly, however. The parachor is estimated as the sum of contributions of structural features:

$$P = \sum_i n_i \cdot p_i \tag{5.2}$$

where P = parachor, $cm^3 \cdot g^{1/4}/s^{1/2} \cdot mol$
       n_i = number of features of type i in molecule
       p_i = contribution of feature i to parachor, $cm^3 \cdot g^{1/4}/s^{1/2} \cdot mol$

Values of p_i reported by Reid et al. (1987) and by Grain (1990) are listed in Table 5.1.

As Equation 5.1 shows, a liquid's surface tension can be estimated from values of the parachor, molar mass, liquid density, and gas density of a chemical. Since ambient temperatures differ greatly from the critical temperatures of organic fluids, we may assume that ρ_l is much larger than ρ_g. Then $P \approx M\gamma^{1/4}/\rho_l$, and

$$\gamma = \left(\frac{P \cdot \rho_l}{M} \right)^4 \tag{5.3}$$

where γ = liquid surface tension, dyn/cm
       P = parachor, $cm^3 \cdot g^{1/4}/s^{1/2} \cdot mol$
       ρ_l = liquid density, g/cm^3
       M = molar mass, g/mol

Example

Estimate the parachor and surface tension of chlorobenzene, C_6H_5Cl, at 20°C using Sugden's method.

Table 5.1 **Group Contributions to the Parachor, P_i**
$cm^3 \cdot g^{1/4}/s^{1/2} \cdot mol$

Structural Feature	P_i	Structural Feature	P_i
Hydrocarbon Increments		**Special Groups**	
C	9.0	–COO–	63.8
H	15.5	–COOH	73.8
CH_3–	55.5	–OH	29.8
CH_2 in $-(CH_2)_n-$, n < 12	40.0	$-NH_2$	42.5
CH_2 in $-(CH_2)_n-$, n > 12	40.3	–O–	20.0
1-Methylethyl-	133.3	$-NO_2$	74
1-Methylpropyl-	171.9	$-NO_3$	93
1-Methylbutyl-	211.7	$-CO(NH_2)$	91.7
2-Methylpropyl-	173.3	–CHO	66
1-Ethylpropyl-	209.5	R–CO–R′, R + R′=2	22.3
1,1-Dimethylethyl-	170.4	R–CO–R′, R + R′=3	20.0
1,1-Dimethylpropyl-	207.5	R–CO–R′, R + R′=4	18.5
1,2-Dimethylpropyl-	207.9	R–CO–R′, R + R′=5	17.3
1,1,2-Trimethylpropyl-	243.5	R–CO–R′, R + R′=6	17.3
>C=C<, terminal position	19.1	R–CO–R′, R + R′=7	15.1
>C=C<, 2,3-position	17.7	R–CO–R′, R + R′=8	14.1
>C=C<, 3,4-position	16.3	R–CO–R′, R + R′=9	13.0
–C≡C–	40.6	R–CO–R′, R + R′=10	12.6
Three-membered ring	12	O (except as above)	20
Four-membered ring	6.0	N (except as above)	17.5
Five-membered ring	3.0	S	49.1
Six-membered ring	0.8	P	40.5
Seven-membered ring	4.0	F	26.1
C_6H_5–	189.6	Cl	55.2
Disubstituted benzene[a]		Br	68.0
		I	90.3
		Se	63
		As	54

[a] Di-substituted benzene isomers have slightly different values of parachor, $P_{ortho} < P_{meta} < P_{para}$. The range of differences between values is $1.8 < P_{meta} - P_{ortho} < 3.4$, $0.2 < P_{para} - P_{meta} < 0.5$, and $2.0 < P_{para} - P_{ortho} < 3.8$.

From Reid, R.C., Prausnitz, J.M., and Poling, B.E., 1987. *The Properties of Gases and Liquids.* Fourth Edition. McGraw-Hill, New York; Quale, O.R., 1953. *Chem. Rev., 53,* 439–589.

1. Estimate the parachor using Equation 5.2. The values of P_i for the phenyl group, C_6H_5–, and the chlorine atom, Cl, are given in Table 5.1.

$$P = \Sigma_i n_i \cdot p_i$$

$$= 189.6 + 55.2$$

$$= 244.8 \ cm^3 \cdot g^{1/4}/s^{1/2} \cdot mol$$

2. Estimate the surface tension using Equation 5.3. For chlorobenzene, M = 112.56 and $\rho_1 = 1.1063$ g/cm^3 at 20°C (Dean, 1996).

$$\gamma = \left(P \cdot \rho_1/M\right)^4 \text{ dyn/cm}$$

$$= \left[(244.8)(1.1063)/(112.56)\right]^4$$

$$= 33.52 \text{ dyn/cm}$$

The measured surface tension of chlorobenzene is 0.03359 N/m (33.59 dyn/cm) at 20°C (Reid et al., 1987). The relative absolute error of our estimate is 0.2%.

5.2.2 McGowan's Method

McGowan (1952) suggests a simple group-contribution method to estimate the parachor value that is more widely applicable than Sugden's. It is both accurate and easily applied. McGowan's method is discussed by Rechtensteiner (1990). It is a component of at least one computer software package for chemical property estimation (Kohlenbrander et al., 1995). The molecular parachor is estimated as the sum of atomic parachors:

$$P = \sum_i n_i A_i - 19 N_{bonds} \qquad (5.4)$$

where P   = parachor, cm$^3 \cdot$g$^{1/4}$/s$^{1/2} \cdot$mol
       n_i = number of atoms of type i in molecule
       A_i = contribution of atom i, cm$^3 \cdot$g$^{1/4}$/s$^{1/2} \cdot$mol
       N_{bonds} = number of chemical bonds in molecule

Atomic parachors, A_i, are given in Table 5.2. In counting the number of bonds, N_{bonds}, single, double, or triple bonds are treated equally. With the exception of strained ring structures, 19 dyn/cm is deducted from the value of parachor for each bond in a molecule, regardless of type.

Example

Estimate the parachor and surface tension of chlorobenzene at 20°C using McGowan's method.

1. Use Table 5.2 to calculate the sum of group contributions to P. C$_6$H$_5$Cl has 12 bonds of various types connecting atoms.

$$\Sigma_i n_i \cdot A_i = n_c A_c + n_H A_H + n_{Cl} A_{Cl}$$

$$= (6)(47.6) + (5)(24.7) + (1)(62.0)$$

$$= 471.1 \text{ cm}^3 \cdot \text{g}^{1/4}/\text{s}^{1/2} \cdot \text{mol}$$

Table 5.2 Atomic Contributions to the Parachor, A_i

Atom	Value, $cm^3 \cdot g^{1/4}/s^{1/2} \cdot mol$
As	87.6
B	53.4
Br	76.1
C	47.6
Cl	62.0
F	30.5
H	24.7
I	98.9
N	41.9
O	36.2
P	73.5
S	67.7
Se	81.9

From McGowan, J.C., 1952. *Chem. Ind. (London)*, 495–496.

2. Estimate the parachor using Equation 5.4.

$$P = \Sigma_i n_i \cdot A_i - 19 \cdot N_{bonds}$$

$$= 471.1 - (19)(12)$$

$$= 243.1 \;\; cm^3 \cdot g^{1/4}/s^{1/2} \cdot mol$$

3. Estimate surface tension using Equation 5.3. For chlorobenzene, M = 112.56 and ρ_1 = 1.1063 g/cm³ at 20°C (Dean, 1996).

$$\gamma = \left(P \cdot \rho_1 / M\right)^4 \;\; dyn/cm$$

$$= \left[(243.1)(1.1063)/(112.56)\right]^4$$

$$= 32.59 \;\; dyn/cm$$

The measured surface tension of chlorobenzene is 0.03359 N/m (33.59 dyn/cm) at 20°C (Reid et al., 1987). The relative absolute error of our estimate is 2.9%.

5.3 SENSITIVITY TO ENVIRONMENTAL PARAMETERS AND METHOD ERROR

The error of estimating surface tension with the methods described above is probably no greater than the error of measuring the parameter. Reid et al. (1987) report that Sugden's method produces a 5% error on average in estimating the parachor of a diverse set of chemicals. The error of an individual estimate may be as much as 30%. Estimating the surface tension from the value of parachor produces an average error of 4.5% and a maximum error of −20% with a test set of 28 polar, nonpolar, and hydrogen-bonding liquids (Reid et al., 1987). McGowan (1952) reports estimates of the parachor of 38 simple organic and inorganic chemicals with similar accuracy.

Since surface tension is proportional to ρ_l^{-4}, the method is extremely sensitive to the value of liquid density chosen. A 1% error in the value of ρ_l can lead to a 30% error in the surface tension estimate. The surface tension estimate is equally sensitive to error in the parachor value.

Surface tension decreases with increasing temperature in going from the triple point to the critical point of a liquid. The decrease is roughly linear and can be estimated with the Othmer equation (Yaws et al., 1991). Over the normal ambient temperature range of −40 to 40°C, the parachor varies by about 1%, and the variation is usually ignored.

McGowan's method takes little account of the structural features of molecules that influence intermolecular forces. It does not distinguish between single and multiple bonds and does not account for differences in bonding between different types of atoms. Sugden's method takes account of some structural features, but it is also incomplete and overall appears not to be significantly more accurate than McGowan's method.

The effect of molecular geometry may be accounted for with a group exchange calculation using the parachor of a reference chemical that is structurally similar to the chemical of interest. Then, as we saw in Chapter 2,

$$q_{R-X} = q_{R-Y} + \Pi \tag{5.5}$$

where q_{R-X} is the parachor of the chemical of interest, q_{R-Y} is the parachor of the reference chemical, and Π is the group exchange factor, the difference between fragment contributions and correction factors attributable to groups x and y. Care must be taken that the appropriate group contribution values are used. For instance, the contribution of the H hydrocarbon increment in Table 5.1, 15.5 $cm^3 \cdot g^{1/4}/s^{1/2} \cdot mol$, differs from the contribution of acidic hydrogen atoms to the parachor of a molecule.

Example

Estimate the parachor of p-nitrotoluene, $C_7H_7NO_2$, using the accepted value of parachor of o-nitrotoluene. The parachor of o-nitrotoluene is 297.7 at ambient temperatures (McGowan, 1952).

1. The value of Π is the difference between the group contributions of ortho- and para-disubstituted benzene isomers. The footnote to Table 5.1 shows that $2.0 < \Pi = P_{para} - P_{ortho} < 3.8$. Use the average value of 2.9.
2. Estimate the parachor of p-nitrotoluene using Equation 5.4.

$$\text{Parachor}_{\text{p-}C_7H_7NO_2} = \text{Parachor}_{\text{o-}C_7H_7NO_2} + \Pi$$

$$= 297.7 + 2.9$$

$$= 300.6 \ \text{cm}^3 \cdot g^{1/4} / s^{1/2} \cdot \text{mol}$$

The accepted value of the parachor of p-nitrotoluene is 302.8 $\text{cm}^3 \cdot g^{1/4}/s^{1/2} \cdot \text{mol}$ (McGowan, 1952). The relative absolute error is 0.7%.

The group exchange estimate of the parachor is not always as accurate as the simple group-contribution estimate. You can demonstrate this for yourself by estimating the parachor of m-nitrobenzene both ways. The accepted value is 297.0 (McGowan, 1952).

REFERENCES

Dean, J.A., 1996. *Lange's Handbook of Chemistry*, 14th Edition. McGraw-Hill, New York.

Exner, O., 1967. Additive properties. III. Re-examination of the additive character of parachor. *Collect. Czechoslov. Chem. Commun.*, **32**, 24–54.

Grain, C.F., 1990. Surface Tension. Chapter 20. *Handbook of Chemical Property Estimation Methods. Environmental Behavior of Organic Compounds.* Lyman, W.J., Reehl, W.F., and Rosenblatt, D.H., Eds. American Chemical Society, Washington, D.C.

Kohlenbrander, J.P., Drefahl, A., and Reinhard, M., 1995. *DESOC (Data Evaluation System for Organic Compounds) User's Guide.* Stanford Bookstore, Stanford, CA 94305-3079.

Lide, D.R., 1994. *CRC Handbook of Chemistry and Physics.* 74th Edition. CRC Press, Boca Raton, FL.

McGowan, J.C., 1952. Molecular volumes and the periodic table. *Chem. Ind. (London)*, 495–496.

Quale, O.R., 1953. The parachors of organic compounds. *Chem. Rev.*, **53**, 439–589.

Rechtensteiner, C.E., 1990. Boiling Point. Chapter 12. *Handbook of Chemical Property Estimation Methods. Environmental Behavior of Organic Compounds.* Lyman, W.J., Reehl, W.F., and Rosenblatt, D.H., Eds. American Chemical Society, Washington, D.C.

Reid, R.C., Prausnitz, J.M., and Poling, B.E., 1987. *The Properties of Gases and Liquids.* Fourth Edition. McGraw-Hill, New York.

Sugden, S., 1924. The influence of the orientation of surface molecules on the surface tension of pure liquids. *J. Chem. Soc.*, **125**, 1177–1189

Yaws, C.L., Yang, H.-C., and Pan, X., 1991. 633 organic chemicals: surface tension data. *Chem. Eng.*, March, 140–150.

CHAPTER 6

Vapor Pressure

6.1 INTRODUCTION

This chapter describes methods of estimating the vapor pressure of a pure chemical and of chemicals in mixtures. The **(saturation) vapor pressure**, P^S (Pa), is *the pressure of a pure chemical vapor that is in equilibrium with the pure liquid or solid*. It is an important control of a chemical's partitioning between air, water, and soil and of its volatilization rate. For vapor–solid equilibrium, P^S is sometimes called the **sublimation pressure**. The *vapor pressure of a chemical in a mixture of volatile chemicals* is its **partial vapor pressure**.

Many different units of pressure are widely used in addition to the pascal, the SI unit. They include the standard atmosphere (1 atm = 101.325 kPa), the bar (1 bar = 100 kPa), the torr (1 atm = 760 torr), and the millimeter of mercury (1 mmHg ≈ 1 torr). The mmHg and the torr are almost exactly equal and can be used interchangeably.

The reported saturation vapor pressures of chemicals at ordinary temperatures range from 760 to below 1×10^{-9} torr. Many hazardous chemicals exhibit vapor pressures of less than 1 torr in the normal ambient temperature range of −40 to +40°C. Vapor pressures below 1 torr are difficult to measure, and reliable values are hard to find in the literature. Computational methods that offer reasonably accurate estimates of such low vapor pressures are particularly useful, therefore, and of particular interest to environmental specialists. The subject was reviewed by Burkhard et al. (1985) and Reid et al. (1987). The literature prior to 1981 was reviewed by Mackay et al. (1982) and by Grain (1990a).

6.2 A VAPOR PRESSURE MODEL

Saturation vapor pressure is a sensitive function of molecular structure and ambient temperature. Molecular structure determines the type and strength of the attractive intermolecular forces that a chemical exhibits. The attractive forces are, in order of increasing strength: the London dispersion forces exhibited by all molecules, polar

55

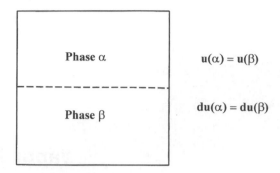

Figure 6.1   So long as phase α and phase β coexist, the chemical potential of a chemical partitioned between them is the same in both phases regardless of any change in temperature and pressure; $\mu_\alpha = \mu_\beta$ and $d\mu_\alpha = d\mu_\beta$.

interactions exhibited by all asymmetric molecules, and hydrogen bonding exhibited by molecules containing O–H, N–H, and F–H bonds. The energy required to escape the liquid phase and the vapor pressure depends on the sum of all forces exhibited by a chemical.

There is a large difference in strength between the attractive intermolecular interactions. The strength of the van der Waals (London and dipole) forces varies from 0.1 to 10 kJ/mol. Hydrogen-bond strengths span the range from 10 to 40 kJ/mol. The quantitative differences are so large that separate vapor pressure models are used for molecules that are nonpolar, for those that are polar, and for those that exhibit hydrogen bonding.

The relationship between vapor pressure and temperature is obtained in the following way. Imagine a chemical partitioned between two phases, α and β, as shown in Figure 6.1. So long as the two phases are in equilibrium, $\mu_\alpha = \mu_\beta$. If the pressure, P (Pa), and temperature, T (K), change while phase equilibrium is maintained, $d\mu_\alpha = d\mu_\beta$. In Section 2.2.1 we saw that $d\mu_\phi = -S_{m,\phi}dT + V_{m,\phi}dP$, where $S_{m,\phi}$ (J/mol·K) and $V_{m,\phi}$ (m³/mol) are molar entropy and molar volume, respectively, of chemical in phase ϕ. If we combine the last two relationships, we obtain $-S_{m,\alpha}dT + V_{m,\alpha}dP = -S_{m,\beta}dT + V_{m,\beta}dP$. Rearranging terms,

$$\frac{dP}{dT} = \frac{\Delta S_{m,tr}(T)}{\Delta V_{m,tr}} \qquad (6.1)$$

where $\Delta S_{m,tr} = S_{m,\beta} - S_{m,\alpha}$ and $\Delta V_{m,tr} = V_{m,\beta} - V_{m,\alpha}$. Equation 6.1 is the **Clapeyron equation**. It describes the relationship between the vapor pressure and temperature of a chemical that is distributed between two phases at equilibrium. Since $\Delta S_{m,tr} = \Delta H_{tr}/T$, where $\Delta H_{tr}(T)$ (J/mol) is the (latent) heat of transition between phases at temperature T, we can also write

$$\frac{dP}{dT} = \frac{\Delta H_{tr}(T)}{T\Delta V_{m,tr}} \qquad (6.2)$$

The **Clausius–Clapeyron equation** is an approximate form of the Clapeyron equation that applies when one of the two phases is a gas. It is derived as follows. For a liquid in equilibrium with its vapor, $\Delta H_{tr} = \Delta H_{VAP}(T)$ (J/mol), the (latent) heat of vaporization of the liquid at temperature T, and $\Delta V_{m,tr} \approx V_{m,g} \approx nRT/P^S$, where $V_{m,g}$ (m^3/mol) is the molar volume of gas, R (8.3145 $m^3 \cdot Pa$/mol·K) is the ideal gas constant, and P^S (Pa) is the vapor pressure of the liquid. Making these substitutions in Equation 6.2, we find that the temperature dependence of the vapor pressure is

$$\frac{d \ln P^S}{dT} = \frac{\Delta H_{vap}(T)}{RT^2} \tag{6.3}$$

The heat of vaporization of most chemicals decreases with increasing temperature. (Majer et al. [1985] reviewed this subject.) The **normal boiling point,** T_b (K), *the temperature at which the vapor pressure equals 1 atm*, is a convenient reference point for liquids. Over the range of temperature from ambient to T_b,

$$\Delta H_{VAP}(T) = \Delta H_{VAP}(T_b) + \Delta C_P(T_b)(T - T_b) \tag{6.4}$$

where $\Delta C_P(T_b)$ (J/K·mol) is the difference in heat capacity between the vapor and liquid at the boiling point. We are assuming that $\Delta C_P(T)$ does not vary significantly over the range of temperature between T and T_b.

By combining Equations 6.3 and 6.4 and integrating from T_b to T (from 1 atm to P^S), we find that

$$\ln P^S = -\frac{\Delta H_{VAP}(T_b)}{R}\left(\frac{1}{T} - \frac{1}{T_b}\right) - \frac{\Delta C_P(T_b)}{R}\left(\frac{T_b}{T} - 1\right) - \frac{\Delta C_P(T_b)}{R}\ln\left(\frac{T_b}{T}\right) \tag{6.5}$$

For a reversible phase change, $\Delta H_{tr} = T \cdot \Delta S_{tr}$, where ΔS_{tr} (J/K·mol) is the entropy change of the transition. Substituting $T_b \cdot \Delta S_{VAP}(T_b)$ for $\Delta H_{VAP}(T_b)$ in Equation 6.5 gives

$$\ln P^S = -\left(\frac{\Delta S_{VAP}(T_b) - \Delta C_P(T_b)}{R}\right)\left(\frac{T_b}{T} - 1\right) - \frac{\Delta C_P(T_b)}{R}\ln\left(\frac{T_b}{T}\right) \tag{6.6}$$

Equation 6.6 describes the relationship between the vapor pressure of a liquid and the ambient temperature. If the chemical is a solid at temperature T, Equation 6.6 gives the vapor pressure of the **supercooled liquid**, an *(unstable) liquid that has been cooled below its freezing point*. As illustrated in Figure 6.2, the vapor pressure of the supercooled liquid is higher than that of the solid. The vapor pressure of the solid is proportional to ΔS_{SUB}, while the vapor pressure of the supercooled liquid is proportional to $\Delta S_{VAP} = \Delta S_{SUB} + \Delta S_{FUS}$, where ΔS_{FUS} (J/mol·K) is the entropy of fusion.

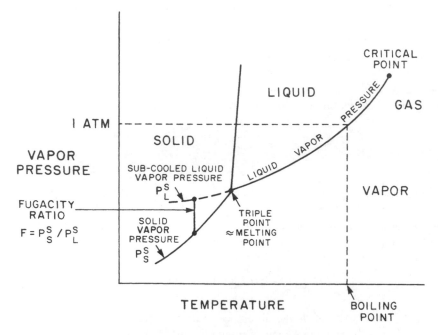

Figure 6.2   The phase diagram of a pure chemical illustrating the relationship between the vapor pressure of the supercooled (or subcooled) liquid and the vapor pressure of the solid. (From Mackay, D., 1991. *Multimedia Environmental Models. The Fugacity Approach.* Lewis Publishers, Boca Raton, FL. With permission.)

The vapor pressure of a solid chemical, P_S^S, is given by the Clausius–Clapeyron equation with $\Delta H_{SUB}(T)$ (J/mol), the (latent) heat of sublimation, substituted for $\Delta H_{VAP}(T)$. Alternatively, the vapor pressure of the solid is related to that of the supercooled liquid, P_L^S, given by Equation 6.6 with the following approximation (Prausnitz, 1969):

$$\ln P_S^S = \ln P_L^S - \frac{\Delta S_{FUS}(T_m)}{R}\left(\frac{T_m}{T} - 1\right) \tag{6.7}$$

where $\Delta S_{FUS}(T_m)$ (J/mol·K) is the entropy of fusion at the melting point T_m (K). Yalkowski (1979) suggests that, for many organic chemicals, $\Delta S_{FUS}(T_m) \approx 56$ J/mol·K.

6.3 METHODS OF ESTIMATING SATURATION VAPOR PRESSURE

In this section, the theoretical framework described above is used to develop semi-empirical methods of estimating the value of a chemical's vapor pressure. We require that all data needed to apply the method be readily available to environmental specialists and that the method's estimates be reasonably accurate and reliable. First, some simple relationships are derived from Equations 6.6 and 6.7.

Instead of independently estimating $\Delta C_P(T_b)$ and $\Delta S_{VAP}(T_b)$ in Equation 6.6, we incur relatively little error by relating the two terms and estimating only one of them. The value of the ratio $\Delta C_P(T_b)/\Delta S_{VAP}(T_b)$ ranges from -0.6 for small, nonpolar hydrocarbons to -1.0 for polar chemicals (Schwarzenbach et al., 1993). Substituting the average value $-0.8\Delta S_{VAP}(T_b)$ for $\Delta C_P(T_b)$ in Equation 6.6, we obtain

$$\ln P^S = -\frac{\Delta S_{VAP}(T_b)}{R}\left\{1.8\left(\frac{T_b}{T}-1\right)-0.8\left[\ln\left(\frac{T_b}{T}\right)\right]\right\}$$

This relationship applies only to liquids. We want a model that applies to both liquids and solids. Experiment shows that the value of $\Delta S_{FUS}(T_m)$ in Equation 6.7 is about 56.5 J/mol·K on average (Yalkowski, 1979; Mackay et al., 1982). Combining the foregoing relationship with Equation 6.7 and setting $\Delta S_{FUS}(T_m)/R = 56.5/8.3145 = 6.80$, we find that

$$\ln P^S = -\frac{\Delta S_{VAP}(T_b)}{R}\left[1.8\left(\frac{T_b}{T}-1\right)-0.8\ln\left(\frac{T_b}{T}\right)\right]-6.8\left(\frac{T_m}{T}-1\right) \qquad (6.8)$$

where P^S = vapor pressure at temperature T, atm
         $\Delta S_{VAP}(T_b)$ = entropy of vaporization at T_b, J/mol·K
         T_b = normal boiling point, K
         T_m = melting point, K
         T = temperature, K

The subscripts "S" and "L" that appear in Equation 6.7 are dropped here with the understanding that Equation 6.8 applies to both liquids and solids. The term containing T_m is omitted if $T < T_m$.

In the absence of measured values, we need a convenient way of evaluating $\Delta S_{VAP}(T_b)$. **Trouton's rule** is $\Delta S_{VAP}(T_b) \approx 10.5R = 87$ J/mol·K. Trouton's rule is fairly accurate for nonpolar liquids. Another empirical estimate, variously known as the **Trouton–Hildebrand–Everett rule** or the **Kistiakowsky equation**, is $\Delta S_{VAP}(T_b) \approx 4.4R + R\ln T_b$. The Kistiakowsky equation is significantly more accurate than Trouton's rule. The $R\ln T_b$ term accounts for the variation of molar volume with changing T_b. The average predictive error of the Kistiakowsky equation is reported to be 1.4% and the maximum error to be 3.4% when applied to a test set of 31 nonpolar chemicals (Fishtine, 1963). Trouton's rule and the Kistiakowsky equation are identical for chemicals with $T_b \approx 400$ K. Both fail for highly polar chemicals and hydrogen-bonded liquids.

Fishtine (1963) modified the Kistiakowsky equation with empirical correction factors, making it applicable to polar and hydrogen-bonded chemicals. His modification of the Kistiakowsky equation is given in Table 6.1, Equation 6.9. Values of the empirical correction factors, K_F, are given in Table 6.2 for various classes of chemicals. The predictive error of the modified Kistiakowsky equation, Equation 6.9, was evaluated by estimating the latent heats of vaporization of a test set of 32 polar

Table 6.1   Relationships for Estimating $\Delta S_{VAP}(T_b)$ (J/mol·K)

Hydrocarbons

$$\Delta S_{VAP}\left(T_b\right) = K_F\left(4.4R + R\ln T_b\right) \tag{6.9}$$

Alcohols, Acids, Methyamines

$$\Delta S_{VAP}\left(T_b\right) = 81.119 + 13.083\ \log\ T_b \tag{6.10}$$

$$- 25.769\frac{T_b}{M} + 0.146528\frac{T_b^2}{M} - 2.1362 \times 10^{-4}\frac{T_b^3}{M}$$

Other Polar Hydrocarbons[a]

$$\Delta S_{VAP}\left(T_b\right) = 44.367 + 15.33\ \log\ T_b \tag{6.11}$$

$$+ 0.39137\frac{T_b}{M} + 4.330 \times 10^{-3}\frac{T_b^2}{M} - 5.627 \times 10^{-6}\frac{T_b^3}{M}$$

where	$\Delta S_{VAP}(T_b)$	= entropy of vaporization at T_b, J/mol·K
	K_F	= Fishtine correction factor given in Table 6.2
	T_b	= normal boiling point, K
	M	= molar mass, g/mol

[a] For esters, multiply the calculated value of $\Delta S_{VAP}(T_b)$ by 1.03.

and hydrogen-bonded liquids. The average deviation from the (often unreliable) probable value was reported to be 1.6%.

Fishtine's correction factors for rigid aromatic and heterocyclic compounds incorporate values of the molecular dipole moment of these chemicals. The dipole moment of a molecule can be estimated by vector addition of the dipole moments of chemical bonds and groups. Chemical bond moments (in Debye) are roughly equal to the electronegativity difference of the bonded atoms. The calculation requires good understanding of chemical structure. It is discussed by Fishtine (1963), by Minkin et al. (1970), and by Nelken and Birkett (1990). Dean (1996) and McClellan (1963) published extensive tabulations of molecular dipole moments. Also, molecular dipole moment is a standard output of all computational chemistry computer programs.

Vetere (1986) proposed some convenient modifications of the Kistiakowsky equation for polar liquids. Vetere's equations provide improved estimates of the vapor pressure of polar and hydrogen-bonded liquids. The estimates usually lie within ±3% of measured values and almost always within ±5% (Reid et al., 1987). Vetere's equations, as reported by Reid et al. (1987), are given in Table 6.1, Equations 6.10 and 6.11.

In order to estimate the value of $\Delta S_{VAP}(T_b)$, the chemical's molar mass and normal boiling point are needed. Several good tabulations of boiling and melting point values are available (see, for instance, Dean, 1996; Lide, 1994; and U.S. Department of Commerce, 1995).

Table 6.2   Fishtine Correction Factors for $\Delta S_{VAP}(T_b)$

Chemical Class	Number of Carbon Atoms in Molecule					
	2	4	6	8	10	12–20
n- and cyclic alkanes,perfluoroalkanes	1.00	1.00	1.00	1.00	1.00	1.00
Branched alkanes	0.99	0.99	0.99	0.99	0.99	0.99
Olefins	1.01	1.01	1.01	1.01	1.01	1.01
Substituted benzenes[a]	$1 + 2\mu/100$					
Substituted naphthalenes					$1 + \mu/100$	
Alkyl chlorides	1.04	1.03	1.03	1.03	1.02	1.01
Alkyl bromides	1.03	1.02	1.02	1.02	1.01	1.01
Alkyl iodides	1.02	1.02	1.02	1.01	1.01	1.01
Haloalkanes, incompletely substituted	1.05	1.04	1.04	1.03	1.02	1.01
Haloalkanes, completely substituted	1.01	1.01	1.01	1.01	1.01	1.01
Esters, aldehydes		1.08	1.06	1.04	1.03	1.01
Ketones		1.07	1.06	1.04	1.03	1.01
Ethers	1.03	1.02	1.02	1.01	1.01	1.01
Oxides	1.08	1.06	1.05	1.03	1.01	1.01
Nitriles, nitroalkanes	1.07	1.06	1.05	1.04	1.03	1.01
Tertiary amines		1.01	1.01	1.01	1.01	1.01
Sulfides, mercaptans	1.03	1.01	1.01	1.01	1.01	1.01
Primary amines	1.13	1.11	1.10	1.09	1.07	1.05[b]
Secondary amines	1.09	1.08	1.07	1.05	1.04	1.03[b]
Anilines			1.09[c]			
Naphthylamines					1.06[d]	
Alcohols[e]	1.31	1.31	1.30	1.28	1.26	1.24[b]
Cyclic alcohols			1.20	1.21	1.26	
Phenols			1.15[f]			
Naphthols					1.09[g]	

[a] μ is the molecular dipole moment (in Debye).
[b] Value for n = 12 only.
[c] The value given is for a single –NH$_2$ group. K_F = 1.01 for N-substituted anilines. K_F = 1.14 for anilines with multiple –NH$_2$ groups.
[d] The value given is for a single –NH$_2$ group. K_F = 1.03 for N-substituted naphthylamines.
[e] The value given is for a single –OH group. K_F = 1.33 for diols (n < 9). K_F = 1.38 for triols (2 < n < 6).
[f] The value given is for a single –OH group. K_F = 1.23 for multiple –OH groups.
[g] The value given is for a single –OH group.
Notes: (1) Count an attached phenyl group as a single carbon atom; (2) count a metal atom in organometallics as a carbon atom; (3) use the same K_F value for all isomers; (4) K_F = 1.06 for chemicals not included in this table.

After Fishtine, S.H., 1963. *Ind. Eng. Chem.*, **55**, 47–56; Schwartzenbach et al., 1993.

6.3.1   Nonpolar Hydrocarbons

Mackay et al. (1982) describe a semi-empirical estimation method that is useful for nonpolar hydrocarbons and halocarbons that have low vapor pressures. The value of $\Delta S_{VAP}(T_b)$ in Equation 6.8 is estimated using the Kistiakowsky model, Equation 6.9, with K_F = 1.00. The model's regression coefficients are determined using nonlinear regression analysis and a training set of 72 halogenated and non-halogenated hydrocabons that boil above 100°C. Both aromatic and aliphatic hydro-carbons are included in the training set. The vapor pressure values of the training set span the range from 6.10×10^{-2} to 8.74×10^{-7} atm. The result is

$$\ln P^S = -\left(4.4 + \ln T_b\right) \times \left[1.803\left(\frac{T_b}{T} - 1\right) - 0.803\ln\left(\frac{T_b}{T}\right)\right] - 6.8\left(\frac{T_m}{T} - 1\right) \quad (6.12)$$

where P^S = vapor pressure at temperature T, atm
  T_b = normal boiling point, K
  T_m = melting point, K
  T = temperature, K

T_b must be greater than 373.2 K. The term containing T_m is omitted if $T < T_m$.

Mackay's model tends to overestimate vapor pressure. It produces a mean predictive error of a factor of 1.25 when applied to the training set of 72 hydrocarbons and halocarbons (Mackay et al., 1982). An average absolute predictive error of a factor of 7.38 is observed when the model is applied to a set of 15 polychlorinated biphenyls with vapor pressure values in the range from 1.01 to 5.30×10^{-8} Pa (Burkhard et al., 1985). Very few sublimation pressure estimates are reported, but it appears that significant additional error is not incurred when estimating the vapor pressure of solids. The method is not recommended for polar chemicals, although model estimates lie within an order of magnitude of reported vapor pressure values of a small set of oxygen, nitrogen, and sulfur chemicals (Mackay et al., 1982).

Example

Estimate the vapor pressure of 2-chlorobiphenyl at 25.00°C (298.2 K) using the Mackay–Kistiakowsky method. For 2-chlorobiphenyl, $C_{12}H_9Cl$, M = 188.66 g/mol, T_b = 274°C (547 K), and T_m = 34°C (307 K) (Lide, 1994).

1. The vapor pressure is estimated using Equation 6.12.

$$\ln P^S = -\left(4.4 + \ln T_b\right)\left\{1.803\left[(T_b/T) - 1\right] - 0.803\ln\left(T_b/T\right)\right\} - 6.8\left[(T_m/T) - 1\right]$$

$$= -\left[4.4 + \ln(547)\right]\left\{1.803\left[(547/298) - 1\right] - 0.803\ln(547/298)\right\}$$

$$\quad - 6.81\left[(307/298) - 1\right]$$

$$= -10.70(1.02) - 0.21$$

$$= -11.12$$

$$P^S = 1.48 \times 10^{-5} \text{ atm}$$

$$= 1.48 \times 10^{-5} \text{ atm}\left(101325 \text{ Pa/atm}\right) = 1.49 \text{ Pa}$$

The measured saturation vapor pressure of 2-chlorobiphenyl is 1.53 Pa at 25°C (Burkhard et al., 1985).

6.3.2 Polar and Hydrogen-Bonded Hydrocarbons

(1) Vetere's Method

Vapor pressures of polar chemicals and of chemicals that exhibit hydrogen bonding can be estimated using Vetere's estimates of $\Delta S_{VAP}(T_b)$, Equations 6.10 and 6.11, along with Equation 6.8.

As mentioned above, estimates of $\Delta S_{VAP}(T_b)$ for polar chemicals made with Vetere's equations appear to be reliable. The average predictive error of Vetere's equations is less than ±3% and the error is usually within ±5% (Reid et al., 1987).

Example

Estimate the vapor pressure of phenol at 25.00°C (298.2 K) using Vetere's modification. For phenol C_6H_6O, M = 94.11 g/mol, $T_b = 182°C$ (455 K), and $T_m = 43°C$ (316 K) (Lide, 1994). $P^S = 0.341$ torr (45.46 Pa) at 25°C (GCA Corp., 1985).

1. Estimate $\Delta S_{VAP}(T_b)$ of phenol using Equation 6.10.

$$\Delta S_{VAP}(T_b) = 81.119 + 13.0831\log T_b - 25.769(T_b/M)$$

$$+ 0.146528(T_b^2/M) - 2.1362 \times 10^{-4}(T_b^3/M)$$

$$= 81.119 + 13.0831\log\ 455 - 25.769(455/94.11)$$

$$+ 0.146528(455^2/94.11) - 2.1362 \times 10^{-4}(455^3/94.11)$$

$$= 99.83\ \text{J/mol} \cdot \text{K}$$

$$\Delta S_{VAP}(T_b)/R = (99.83\ \text{J/mol} \cdot \text{K})/(8.3145\ \text{J/mol} \cdot \text{K})$$

$$= 12.01$$

2. The vapor pressure of phenol is estimated using Equation 6.8.

$$\ln\ P^S = -[\Delta S_{VAP}(T_b)/R]\{1.8[(T_b/T)-1] - 0.8\ln(T_b/T)\} - 6.8[(T_m/T)-1]$$

$$= -(12.01)\{(1.8)[(455/298)-1] - 0.8\ln(455/298)\} - 6.8[(314/298)-1]$$

$$= -(12.01)(0.61) - 0.37$$

$$= -7.69$$

$$P^S = e^{-7.69}\ \text{atm}$$

$$P^S = 4.56 \times 10^{-4}\ \text{atm}\ (46.1\ \text{Pa})$$

The measured saturation vapor pressure of phenol is 0.341 torr (45.46 Pa) at 25°C (GCA Corp., 1985). An estimate made using Mackay's model, Equation 6.12, gave a value of 72 Pa.

(2) Fishtine's Method

An alternative method of estimating the vapor pressure of polar chemicals employs the modified Kistiakowsky equation, Equation 6.9 with K_F values listed in Table 6.2. The average deviation of the estimate from the probable value was reported to be 1.6% when the method was applied to a test set of 32 polar and hydrogen-bonded liquids (Fishtine, 1963). The method is widely used (Grain, 1990a), but good knowledge of chemical structure is required to apply it correctly. The value of K_F can be difficult to estimate and, at times, informed but subjective judgment is called for.

Example

Estimate the vapor pressure of phenol at 25.00°C (298.2 K) using Fishtine's correction to the Kistiakowsky equation. For phenol C_6H_6O, M = 94.11 g/mol, T_b = 182°C (455 K), and T_m = 43°C (316 K) (Lide, 1994). K_F = 1.15 for phenol (Fishtine, 1963).

1. Estimate $\Delta S_{VAP}(T_b)$ using Equation 6.9.

$$\Delta S_{VAP}(T_b)/R = K_F\left(4.4 + \ln\ T_b\right)$$

$$= (1.15)(4.4 + \ln\ 455)$$

$$= 12.10$$

2. Estimate P^S using Equation 6.8.

$$\ln\ P^S = -K_F\left[\Delta S_{VAP}(T_b)/R\right]\left\{1.8\left[(T_b/T)-1\right]-0.8\ln(T_b/T)\right\}-6.8\left[(T_m/T)-1\right]$$

$$\ln\ P^S = -(12.10)\left\{1.8\left[(455/298)-1\right]-0.8\ln(455/298)\right\}-6.8\left[(314/298)-1\right]$$

$$= -(11.45)(0.61)-0.37$$

$$= -7.75$$

$$P^S = e^{-7.75}\ \text{atm}$$

$$= 4.30\times10^{-4}\ \text{atm}\ (433.6\ \text{Pa})$$

The measured saturation vapor pressure of phenol is 0.341 torr (45.46 Pa) at 25°C (GCA Corp., 1985). The estimate using Vetere's equation, Equation 6.9, is 46.1 Pa, and the estimate using Mackay's equation, Equation 6.12, is 72 Pa.

6.3.3 Other Useful Methods

A UNIFAC group-contribution method of estimating vapor pressure values is reported for hydrocarbons, chlorocarbons, PCBs, alcohols, ketones, and acids in the molar mass range below 500 g/mol and with vapor pressures from 10 to 760 torr (Jensen et al., 1981; Burkhard et al., 1985). Another method of estimating vapor pressure worth noting is the solvatochromic structure–vapor pressure relationship reported for hydrocarbons, halocarbons, alcohols, dialkyl ethers, and some nitrogen and sulfur compounds with vapor pressures in the range of 1×10^{-8} to 760 torr (Banerjee et al., 1990). Computational software is available for these methods.

6.4 VAPOR PRESSURE AND AIR/LIQUID PARTITIONING OF CHEMICALS IN MIXTURES

The partial vapor pressure of a component in a mixture of miscible liquids is proportional to its mole fraction and to its saturation vapor pressure. In mathematical terms, **Raoult's law** is

$$p_j = x_j P_j^S \tag{6.13}$$

where p_j is the partial pressure of component j in the vapor that is in equilibrium with the liquid solution, and x_j is the mole fraction of component j in the solution. Note that the value of P_j approaches P_j^S as x_j approaches 1.

Raoult's law applies exactly to ideal solutions and is reasonably accurate for nearly ideal solutions, solutions in which the components are structurally similar. Raoult's law can be used to estimate the vapor pressure of the solvent of a dilute solution of nonionic chemicals. If an ionic solute is present in solution or if nonionic solute levels are not low, the chemical activity, a_j, of the solvent should be substituted for x_j in Equation 6.13.

Example

Calculate the partial pressure of benzene (C_6H_6) in equilibrium at 25°C with a liquid that is 26.6% by weight benzene and 73.4% by weight toluene (C_7H_8). At 25°C, the vapor pressure of pure benzene is 95.2 torr (GCA Corp., 1995). $M_{C_6H_6} = 78.1$ g/mol, and $M_{C_7H_8} = 92.4$ g/mol.

1. In 100 g of mixture, there is 26.6 g/78.1 g $mol^{-1} = 0.341$ mol of benzene and 73.4 g/92.4 g $mol^{-1} = 0.794$ mol toluene. The mole fraction of benzene is

$$X_{C_6H_6} = 0.341/(0.341 + 0.794)$$

$$= 0.300$$

2. The partial pressure of benzene is estimated using Equation 6.13.

$$P_{C_6H_6} = X_{C_6H_6} P_{C_6H_6}^S$$

$$= (0.300)(95.2 \text{ torr})(101325 \text{ Pa}/760 \text{ torr})$$

$$= 3.81 \times 10^3 \text{ Pa}$$

Raoult's law is frequently used to estimate the air–liquid partition coefficient of a volatile organic component of an oil mixture, K_{EQ}. If the molar concentration of chemical j in air is C_{gj} (mol/m^3) and the molar concentration of chemical j in liquid is C_{lj} (mol/m^3), $K_{EQ} = C_{gj}/C_{lj}$. Now, $C_{gj} = y_j \cdot \rho_a/M_a$, where y_j is the mole fraction of chemical j in air, ρ_a (g/m^3) is the density of air, and M_a (g/mol) is the molar mass of air. **Dalton's law** for an ideal gas mixture is $y_j = P_j/P_0$, where P_0 (Pa) is the total atmospheric pressure. Combining Dalton's law and Raoult's law, $y_j = x_j \cdot P^S/P_0$, and $C_{gj} = x_j \cdot P^S \cdot \rho_a/P_0 \cdot M_a$. $C_{lj} = x_j \cdot \rho_l/M_l$, where ρ_l (g/m^3) is the liquid density and M_l (g/mol) is the molar mass of liquid. Combining the expressions for C_{gj} and C_{lj}, we find that

$$K_{EQ} = \frac{P_j^S \rho_a M_l}{P_0 \rho_l M_a} \tag{6.14}$$

where K_{EQ} = gas–liquid partition coefficient
            P_j^S = saturation vapor pressure of component j, Pa
            ρ_a = air density, g/m^3
            M_l = average molar mass of oil, g/mol
            P_0 = total pressure, Pa
            ρ_l = oil density, g/m^3
            M_a = molar mass of air, 28.8 g/mol

Equation 6.14 applies only to miscible components of a liquid mixture. The partition coefficients of organic chemicals in water solutions are estimated with Henry's law. This is discussed in Chapter 8.

Example

Estimate K_{EQ} of chloroform in an oil film at 25°C (298 K) and 1 atm. Laboratory tests show that M_l = 280 g/mol and ρ_l = 0.898 g/cm^3. At 25°C, $P_{CHCl_3}^S$ = 208 mmHg (2.773 × 10^4 Pa) for chloroform (GCA Corp., 1985), and ρ_a = 1.1845 g/dm^3 (Lide, 1994).

1. The K_{EQ} of an oil film is estimated using Equation 6.14.

$$K_{EQ} = (2.773 \times 10^4 \text{ Pa})(1.1845 \times 10^3 \text{ g/m}^3)(280 \text{ g/mol})/(101325 \text{ Pa})$$

$$(0.898 \times 10^6 \text{ g/m}^3)(28.8 \text{ g/mol})$$

$$= 3.51 \times 10^{-3}$$

Equation 6.14 is used to estimate K_{EQ} of a chemical component of an oil film present at the surface of a soil or water column. A slightly different relationship is used to estimate K_{EQ} of a component of oily waste buried in a soil column. In this case, the ideal gas law states that $C_{gj} = P_j \cdot a/RT$, where a (m^3_{air}/m^3_{soil}) is the volumetric air content of the soil–waste mixture. Given Raoult's law, $C_{gj} = x_j \cdot P_j^S \cdot a/RT$. $C_{lj} = x_j \cdot L/M_l$, where L (g oil/m^3 soil) is the loading of oil in the soil column. Combining terms, we find that for a miscible component of buried organic waste

$$K_{EQ} = \frac{P_j^S a M_l}{RTL} \tag{6.15}$$

where K_{EQ} = gas–liquid partition coefficient
      P_j^S = saturation vapor pressure of component j, Pa
      a   = volumetric air content of soil/waste, (m^3_{air}/m^3_{soil})
      M_l = average molar mass of oil, g/mol
      R   = gas constant, 8.3145 Pa·m^3/mol·K
      T   = absolute temperature, K
      L   = oil loading in soil, g oil/m^3 soil

Example

Estimate K_{EQ} of chloroform in oily waste found in a landfill at 25°C (298 K). Laboratory tests show that M_l = 150 g/mol, L = 0.46 g oil/cm^3 soil, and a = 0.30 m^3/m^3. At 25°C, $P^S_{CHCl_3}$ = 208 mmHg (2.773 × 10^4 Pa) for chloroform (GCA Corp., 1985).

1. The K_{EQ} of a component of oily waste in a soil column is estimated using Equation 6.15.

$$K_{EQ} = P^S_{CHCl_3} \cdot M_l \cdot a/R \cdot T \cdot L$$

$$= \left(2.773 \times 10^4 \ Pa\right)\left(150 \ g/mol\right)\left(0.30 \ m^3/m^3\right)/\left(8.3145 \ Pa \cdot m^3/mol \cdot K\right)$$

$$\left(298 \ K\right)\left(0.46 \times 10^6 \ g/m^3\right)$$

$$= 1.1 \times 10^{-3}$$

6.5 SENSITIVITY TO ENVIRONMENTAL PARAMETERS AND METHOD ERROR

In general, the methods of estimating vapor pressure described in this chapter yield good results in the vapor-pressure range of 10 to 760 torr (Grain, 1990a; Lynch et al., 1991; Schwarzenbach et al., 1993). Below 10 torr, the methods appear to offer no better than order-of-magnitude estimates. The method evaluations are often done with estimated rather than measured boiling and melting points, however. Available procedures for estimating T_b and T_m are not highly accurate, and this contributes to the reported predictive error.

In deriving Equation 6.7, the value of the ratio $\Delta C_P(T_b)/\Delta S_{VAP}(T_b)$ was set equal to -0.8. While the error in estimating the value of $\Delta C_P(T_b)$ is avoided, the error incurred in making this approximation ranges from about $\pm 5\%$ for chemicals that boil at or below $100°C$ to more than $\pm 100\%$ for chemicals that boil at or above $400°C$ (Schwarzenbach et al., 1993). This may not be satisfactory for chemicals that boil at very high temperatures. Methods of estimating $C_P(T_b)$ of gases and liquids are reviewed by Reid et al. (1987) and by Birkett (1990).

The value of $\Delta S_{FUS}(T_m)$ may be substantially different from the 56.5 J/mol·K used in the vapor-pressure models for solids, and it is particularly difficult to estimate. Its value depends upon the relative disorder of the liquid chemical compared to that of the solid and is larger for asymmetric, flexible molecules than for symmetric, rigid molecules of the same size. To account for molecular flexibility, Yalkowski (1979) suggests setting $\Delta S_{FUS}(T_m) = [56.6 + 10.5(n - 5)]$ J/mol·K where n is the total number of atoms exclusive of protons on flexible chains. Better estimates are obtained by accounting for both rotational symmetry and molecular flexibility (Mishra and Yalkowsky, 1991).

The value of $\Delta S_{FUS}(T_m)$ of straight-chain aliphatic hydrocarbons increases irregularly with chain length. Straight-chain aliphatic hydrocarbons with an odd number of carbon atoms in the chain obey different vapor pressure–temperature relationships than straight-chain aliphatic hydrocarbons with an even number of carbon atoms in the chain (Burrows, 1992). This **odd–even effect** is produced by the different stereochemical limitations that are imposed on the packing arrangements of odd and even straight-chain hydrocarbons in the solid state.

None of the methods described in this chapter distinguishes between chemical isomers. This can result in significant discrepancies between estimated and measured vapor pressure values. Constantinou and Gani (1994) describe a group contribution method of estimating the standard enthalpy of vaporization at 298 K that does distinguish among isomers. The method produces an absolute error of 1.11 KJ/mol, an absolute percent error of 2.57%, when applied to a test set of 225 diverse chemicals.

Don't be unduly pessimistic about the reliability of the vapor–pressure estimation methods. For chemicals with low vapor pressures, different methods of measuring vapor pressures agree to within a factor of 2 or 3, and the results of interlaboratory comparisons vary by as much as an order of magnitude (Spencer and Cliath, 1983). The greatest difficulty and uncertainty arises when determining the vapor pressure of chemicals of extremely low volatility, i.e., those with vapor pressures below 0.01 torr. The reported predictive error of the estimation methods described in this chapter are comparable to measurement errors.

For most chemicals of interest, the heat of vaporization does not vary greatly within the temperature range from -40 to $+40°C$. Equation 6.3 can be integrated to give a relationship that is valid when temperature remains within a narrow range and ΔH_{VAP} and ΔZ are roughly constant. The relationship is

$$\ln P^S = -\frac{B}{T} + A \qquad (6.16)$$

where B is $\Delta H_{VAP}/\Delta Z$ and A is the integration constant.

Equation 6.16 suggests that $\ln P^S$ is a linear function of 1/T. This is adequate if temperature is confined to a narrow range. If temperature varies over a large range, variation of $\Delta H_{VAP}/\Delta Z$ may be significant, and the **Antoine equation** is more accurate than Equation 6.16. The Antoine equation is

$$\ln P^S = A - \frac{B}{T+C} \qquad (6.17)$$

where P^S = vapor pressure at temperature T, torr
       T   = temperature, K
      A,B,C = empirical constants

The coefficients of Equations 6.16 and 6.17 are tabulated in various publications and data bases. Each set of coefficients is the result of a regression analysis, and coefficients from different sets cannot be mixed. The units vary according to the tabulation, torr, bar, or atm and °C or K. A temperature range is always given for which the model is valid. As with any empirical model, do not extrapolate beyond the temperature range specified. Also, it pays to check if the data apply to the solid or liquid phase of the chemical. Models developed with chromatographic retention times may apply to the vapor pressure of the supercooled liquid.

Example

Estimate the vapor pressure of tetrachloroethylene at 25.00°C using the Antoine equation. In the normal ambient temperature range, A = 6.976, B = 1386.92, and C = 217.53 (GCA Corp., 1985).

1. Use Equation 6.17, where T is temperature in °C.

$$\log_{10} P^S = A - B/(T+C)$$

$$= 6.976 - 1386.92/(25.00 + 217.53)$$

$$= 1.26$$

$$P^S = 10^{1.26} = 18.1 \ \text{torr}$$

The measured value is 19 torr at 25°C (GCA Corp., 1985).

The thermodynamic relationships used in this chapter apply to liquid and solid chemicals under a total applied pressure equal to their vapor pressure. Vapor pressure depends slightly on applied external pressure. The relationship is

$$P = P^S e^{\frac{V_{m,l}\Delta P_{app}}{RT}}$$ (6.18)

where P   = partial vapor pressure, Pa
       P^S = the partial vapor pressure when $\Delta P = 0$, Pa
       ΔP_{app} = change in applied pressure, Pa
       $V_{m,l}$ = molar volume of the liquid, m³/mol
       R   = 8.3145 J/mol·K
       T   = temperature, K

Chemical molar volumes are in the 0.1-L/mol range, and an increase of pressure of as much as 1 atm will result in a negligible increase in vapor pressure. The vapor pressure of a chemical is roughly constant under normal fluctuations of atmospheric pressure.

Example

Calculate the vapor pressure of water in air at 25°C and 1.000 atm pressure. The vapor pressure of water is 23.76 torr at 25°C (298.15 K) in the absence of an external pressure. The molar volume of water is 18.05×10^{-6} m³/mol at 25°C.

1. $V_m^{(1)} \Delta P/RT = \left(18.05 \times 10^{-6}\ \text{m}^3/\text{mol}\right)\left(760\ \text{torr} - 24\ \text{torr}\right)$

    $\left(101325\ \text{Pa}/760\ \text{torr}\right)/\left(8.3145\ \text{J/mol}\cdot\text{K}\right)\left(298.15\ \text{K}\right)$

    $= 7.14 \times 10^{-4}$

2. The vapor pressure of water is calculated with Equation 6.18:

    $$P = \left(23.76\ \text{torr}\right)\exp\left(7.14 \times 10^{-4}\right)$$

    $$= 23.78\ \text{torr}$$

The vapor pressure increases by 0.07%.

You may need to correct for the effect of applied pressure when dealing with volatile chemicals sorbed on submicron aerosols. The correction is made with Equation 6.18. P is the vapor pressure of chemical in aerosol with radius of curvature r (m), and P^S is the saturation vapor pressure of a flat pool of liquid chemical (r = ∞). If the aerosol droplet has a surface tension of γ (N/m), the additional applied pressure is ΔP_{app} = 2γ/r. For example, the additional pressure inside a droplet of chlorobenzene of radius 0.01 μm at 20°C and surface tension 0.03359 N/m is (2) (0.03359 N/m)/1.00 × 10⁻⁸ m = 6.72×10^6 Pa (66.3 atm). Using Equation 6.18, the vapor pressure of chlorobenzene is estimated to increase by 13%, enhancing evaporation and redistribution to larger aerosol droplets. Since thermodynamic relationships do not apply to molecular-scale systems, this calculation is not valid when r is much smaller than 1×10^{-8} m.

In a soil capillary, the radius of curvature of a liquid meniscus is usually negative, the effective applied pressure on the liquid is reduced, and the liquid's vapor pressure is reduced. Assume, for example, that liquid chlorobenzene found in a fine soil pore has a meniscus with a radius of curvature of 0.01 μm; then the liquid's vapor pressure is reduced by about 13% and accumulation of chlorobenzene in the soil is enhanced.

Methods of estimating the surface tension of organic liquids are described in Chapter 5. Some older methods are reviewed by Reid et al. (1987) and by Grain (1990b). Yaws et al. (1991) has published reference data with which to estimate the parachor of 633 organic chemicals.

REFERENCES

Banerjee, S., Howard, P.H., and Lande, S.S., 1990. General structure–vapor pressure relationships for organics. *Chemosphere,* **21**, 1173–1180.

Birkett, J.D., 1990. Heat Capacity. Chapter 23. In *Handbook of Chemical Property Estimation Methods. Environmental Behavior of Organic Compounds.* Lyman, W.J., Reehl, W.F., and Rosenblatt, D.H., Eds. American Chemical Society, Washington, D.C.

Burkhard, L.P., Andren, A.W., and Armstrong, D.E., 1985. Estimation of vapor pressures for polychlorinated biphenyls: a comparison of eleven predictive methods. *Environ. Sci. Technol.,* **19**, 500-507.

Burrows, H.D., 1992. Studying odd-even effects and solubility behavior using α,ω-dicarboxylic acids. *J. Chem. Ed.,* **69**, 69–73.

Constantinou, L. and Gani, R., 1994. New group contribution method for estimating properties of pure compounds. *AIChE J.,* **40**, 1697–1710.

Dean, J.A., 1996. *Lange's Handbook of Chemistry,* 14th Edition. McGraw-Hill, New York.

Fishtine, S.H., 1963. Reliable latent heats of vaporization. *Ind. Eng. Chem.,* **55**, 47-56.

Grain, C.F., 1990a. Vapor Pressure. Chapter 14. In *Handbook of Chemical Property Estimation Methods. Environmental Behavior of Organic Compounds.* Lyman, W.J., Reehl, W.F., and Rosenblatt, D.H., Eds. American Chemical Society, Washington, D.C.

Grain, C.F., 1990b. Surface Tension. Chapter 20. In *Handbook of Chemical Property Estimation Methods. Environmental Behavior of Organic Compounds.* Lyman, W.J., Reehl, W.F., and Rosenblatt, D.H., Eds. American Chemical Society, Washington, D.C.

Jensen, T., Fredenslund, A., and Rasmussen, P., 1981. Pre-component vapor pressures using UNIFAC group contribution. *Ind. Eng. Chem. Fundam.,* **20**, 239-246.

Lide, D.R., 1994. *CRC Handbook of Chemistry and Physics.* 74th Edition. CRC Press, Boca Raton, FL.

Lynch, D.G., Tirado, N.F., Boethling, R.S., Huse, G.R, and Thom, G.C., 1991. Performance of on-line chemical property estimation methods with TSCA premanufacture notice chemicals. *Sci. Total Environ.,* **109/110**, 643–648.

Mackay, D., 1991. *Multimedia Environmental Models. The Fugacity Approach.* Lewis Publishers, Boca Raton, FL.

Mackay, D., Bobra, A.M., Chan, D., and Shiu, W.Y., 1982. Vapor pressure correlations for low volatility environmental chemicals. *Environ. Sci. Technol.,* **16**, 645-649.

Majer, V., Svoboda, V., and Kehiaian, H.V., 1985. *Enthalpies of Vaporization of Organic Compounds. A Critical Review and Data Compilation.* Chemical Data Series No. 32. Blackwell Scientific, Oxford.

McClellan, A.L., 1963. *Tables of Experimental Dipole Moments.* W.H. Freeman, San Francisco.

Minkin, V.I., Osipov, O.A., and Zhdanov, Y.A., 1970. *Dipole Moments in Organic Chemistry.* Hazard, B.J., translator. Plenum Press, New York.

Mishra, D.S. and Yalkowsky, S.H., 1991. Estimation of vapor pressure of some organic compounds. *Ind. Eng. Chem. Res.*, **30**, 1609–1612.

Nelken, L.H. and Birkett, J.D., 1990. Dipole Moment. Chapter 25. In *Handbook of Chemical Property Estimation Methods. Environmental Behavior of Organic Compounds.* Lyman, W.J., Reehl, W.F., and Rosenblatt, D.H., Eds. American Chemical Society, Washington, D.C.

Prausnitz, J.M., 1969. *Molecular Thermodynamics of Fluid-Phase Equilibria.* Prentice-Hall, Englewood Cliffs, NJ.

Reid, R.C., Prausnitz, J.M., and Poling, B.E., 1987. *The Properties of Gases and Liquids.* Fourth Edition. McGraw-Hill, New York.

Schwarzenbach, R.P., Gschwend, P.M., and Imboden, D.M., 1993. *Environmental Organic Chemistry.* John Wiley & Sons, New York.

Spencer, W.F. and Cliath, M.M., 1983. Measurement of pesticide vapor pressures. *Residue Rev.*, **85**, 57–71.

U.S. Department of Commerce, 1995. DIPPR Data Compilation of Pure Compound Properties, 1995. Version 10.0. Project sponsored by the Design Institute for Physical Property Data. American Institute of Chemical Engineers, NIST Standard Reference Database 11, U.S. Department of Commerce, Gaithersburg, MD 20899.

Vetere, A., 1986. An empirical correlation for the calculation of vapor pressures of pure compounds. *Chem. Eng. J.*, **32**, 77–86.

Yalkowski, S.H., 1979. Estimation of entropies of fusion of organic compounds. *Ind. Eng. Chem. Fundam.*, **18**, 108–111.

Yaws, C.L., Yang, H.-C., and Pan, X., 1991. 633 Organic chemicals: surface tension data. *Chem. Eng.*, **March**, 140–150.

CHAPTER 7

Aqueous Solubility and Activity Coefficient

7.1 INTRODUCTION

This chapter describes methods of estimating the solubility and activity coefficient of liquid and solid hydrophobic chemicals in water, primarily. The solubility of gases is discussed in more detail in Chapter 8.

A **saturated solution** is *a solution of a chemical solute that is in equilibrium with the pure undissolved chemical.* The **solubility** is the *concentration of solute in a saturated solution.* Aqueous solubility, C_W^S, is commonly expressed either in terms of solute **molarity** (mol solute/L solution), **mass concentration** (g solute/L solution), or **mass ratio** (g solute/kg solvent). Occasionally, it is expressed in terms of solute **molality** (mol solute/kg solvent), **mole percent** (mol solute/100 mol solution), or **mole fraction** (mol solute/mol solution). In addition, solubility is often expressed in semiquantitative terms such as **miscible or infinitely soluble** ($C_W^S = \infty$), **soluble** ($C_W^S > 0.1$ mol/L), **moderately soluble** (0.1 mol/L $> C_W^S > 0.01$ mol/L), and **insoluble** ($C_W^S < 0.01$ mol/L). Many organic chemicals are said to be **sparingly soluble** in water ($C_W^S \leq 1$ g/L).

Plenty of data are available on the aqueous solubility of organic chemicals, but, unfortunately, much of it is semiquantitative. Some useful collections of quantitative values are published by IUPAC (1987), Howard (1989), Mackay et al. (1995), Penning et al. (1990), Verschueren (1983), and Yalkowsky and Dannenfelser (1990). The solubility of a chemical is usually measured in air-saturated distilled water and reported for a defined temperature. The range of solubility of organic chemicals in air-saturated water at 20°C extends from infinitely soluble (completely miscible) to below 1×10^{-12} mol/L.

Aqueous solubility determines, in large measure, the speed with which chemicals travel and biodegrade in the environment. It is an important control of the partitioning of chemicals between air, water, soil, and biota (see Chapters 8, 10, and 11). Since the aqueous solubility is the highest equilibrium concentration a chemical can reach in water, it is an important control of environmental chemical and biological kinetic processes.

This chapter describes estimation methods for aqueous solubility and activity coefficient that are applicable to hydrophobic chemicals that are of concern to environmental specialists. It begins with some background information illustrating the strengths and weaknesses of the estimation methods. Then each method, its range of applicability, and its accuracy are described. Finally, the sensitivity of aqueous solubility and activity coefficient to environmental factors is discussed.

7.2 BACKGROUND

How is aqueous solubility related to the other properties of a chemical? We address this question by first examining the correlation between solubility and activity coefficient. Imagine a two-phase system in which a liquid organic chemical is in equilibrium with its saturated aqueous solution. The fugacity of the chemical in the organic phase is f_O (Pa) $= \gamma_O x_O f_O^0$, where γ_O is the activity coefficient of the chemical in the organic phase, x_O is its mole fraction, and f_O^0 (Pa) is the reference fugacity of the pure liquid organic chemical, the standard state of the chemical at the temperature of the system. The fugacity of the chemical in the aqueous phase is f_W (Pa) $= \gamma_W^S x_W^S f_O^0$, where γ_W^S is the activity coefficient of the organic solute in saturated aqueous solution, and x_W^S is the mole fraction solubility of organic chemical in water.

Assume that the chemical is hydrophobic and that water is completely excluded from the organic phase. Since the organic liquid is pure, $\gamma_O = 1$, $x_O = 1$, and the fugacity of the liquid is the reference fugacity, $f_O = f_O^0$. The criterion for equilibrium is that the fugacity of the chemical is the same in both phases; $f_O = f_W$. Substituting the terms above for f_O and f_W, we see that $f_O^0 = \gamma_W^S x_W^S f_O^0$ or

$$x_W^S = \frac{1}{\gamma_W^S} \qquad (7.1)$$

for organic liquids.

Contrary to our assumption, water is soluble to some degree in all liquids, and $\gamma_O x_O$ will differ from 1 even for hydrophobic chemicals. Equation 7.1 is an approximate relationship that is most accurate for highly hydrophobic chemicals.

For hydrophobic chemicals, the solute level in saturated solution is low enough that $x_W^S \approx n_C/n_W = C_W^S V_{ml}$, where n_C/n_W is the mole ratio of chemical to water in the aqueous phase, C_W^S (mol/L) is the molar solubility of solute, and V_{ml} is the partial molar volume of water. At low concentrations of nonionic organic solutes, we can assume that $V_{ml} \approx 0.018$ L/mol, the molar volume of water, and the relationship between molar solubility and activity coefficient is $C_W^S \approx 55.6/\gamma_W^S$.

The process of making an aqueous solution of a solid chemical is similar in some respects to melting the solid. Imagine that the first step is to convert the solid to the supercooled liquid state. In this case, Equation 7.1 must be corrected to account for the change in state of the chemical (Mackay and Shiu, 1977). The corrected relationship for organic solids is

$$x_W^S = \frac{1}{\gamma_W^S} \frac{P^S(S)}{P^0(L)}$$

(7.2)

where $P^S(S)$ (Pa) is the vapor pressure of the pure solid chemical and $P^0(L)$ (Pa) is the vapor pressure of the pure supercooled liquid.

Prausnitz (1969) suggests that $\ln P^S(S)/P^0(L) \approx -(\Delta S_{FUS}/R)[(T_m - T) - 1]$, where ΔS_{FUS} (J/mol·K) is the entropy of fusion of the solid chemical, R (8.31451 J/mol·K) is the gas constant, T_m (K) is the melting point of the solid, and T (K) is the ambient temperature. Yalkowsky (1979) suggests that, for many organic chemicals, $\Delta S_{FUS}(T_m) \approx 56$ J/mol·K. Making this substitution, setting T = 298 K, and re ing the relationship, we find that $\ln P^S(S)/P^0(L) \approx -0.023(T_m - 298)$.

Similarly, we imagine that the first step in preparing an aqueous solution gaseous chemical is to convert the gas into a superheated liquid. The corr ed relationship for a gaseous organic chemical is

$$x_W^S = \frac{1}{\gamma_W^S} \frac{P_g}{P^0(L)}$$

(7.3)

where P_g (Pa) is the gas pressure and $P^0(L)$ is the vapor pressure of the pure superheated liquid. The solubilities of gases are usually reported for a gas pressure, P_g, of 1 atm.

Next, we examine the correlation between a chemical's aqueous solubility and its octanol–water distribution function. Imagine that the chemical is partitioned between n-octanol solution and water solution. At equilibrium, the fugacity of the chemical is the same in both phases: $f_O = f_W$. As before, f_O is the chemical's fugacity in the organic phase, now an octanol solution. Substituting the appropriate expressions for the chemical's fugacity, we find that $\gamma_O x_O = \gamma_W x_W$ at equilibrium, where γ_O is the activity coefficient of solute in octanol and x_O is the mole fraction of solute in octanol.

In dilute solution, $x_O \approx n_c/n_O - C_O \cdot V_{m,O}$, where n_c/n_O is the mole ratio of chemical to octanol in the octanol phase, C_O is the molar concentration of chemical in octanol, and $V_{m,O}$ is the partial molar volume of octanol. Similarly, for the water phase, $x_W \approx C_W \cdot V_{m,W}$, where C_W is the molar concentration of chemical in water and $V_{m,W}$ is the partial molar volume of the water. Combining these relationships, we see that $\gamma_O \cdot C_O \cdot V_{m,O} = \gamma_W \cdot C_W \cdot V_{m,W}$.

The octanol–water partition coefficient, K_{OW}, is defined as the ratio of solute concentration in the octanol phase to solute concentration in the water phase. We show, by rearranging the relationship derived in the preceding paragraph, that K_{OW} is

$$K_{OW} = \frac{C_O}{C_W} = \frac{\gamma_W}{\gamma_O} \frac{V_{m,W}}{V_{m,O}}$$

(7.4)

Values of K_{OW} are always reported for dilute solutions at 25°C, and the partial molar volumes of the octanol and water phases are roughly the same for every measurement. A 2.3 M solution of H_2O in octanol has $V_{m,O} = 0.12$ L/mol, and a 4.5 × 10^{-3} M solution of octanol in water has a $V_{m,W} = 0.018$ L/mol. Assuming that the presence of low levels of organic solute does not significantly alter the molar volumes, $V_{m,W}/V_{m,O} \approx 0.15$ for all K_{OW} measurements and $K_{OW} \approx 0.15 \, \gamma_W/\gamma_O$.

Most organic chemicals form nearly ideal solutions with octanol. Measured values of γ_O span a range from 1 to 10 (Schwarzenbach et al., 1993). On the other hand, measured values of γ_W for organic chemicals span a range from 10 to 1×10^{11}. The variation in K_{OW} from chemical to chemical is due almost entirely to variation in γ_W: $K_{OW} \approx \text{const} \cdot \gamma_W$.

For nonionic solutes at low concentration, we can assume that γ_W is roughly independent of chemical concentration in the water phase. If γ_W is not affected by the presence of octanol in the water phase, then $\gamma_W \approx \gamma_W^S$, where γ_W^S is the activity coefficient in the saturated water solution. Referring to Equation 7.1 for organic liquids, $\gamma_W^S = 1/x_W^S = 1/C_W^S \cdot V_{ml}$ and $K_{OW} = A/C_W^S \cdot V_{ml}$, where A is a constant. Converting to log functions and rearranging,

$$\ln C_W^S = A - \log K_{OW} \qquad (7.5)$$

Using a line of reasoning similar to the foregoing, Yalkowsky and Valvani (1980) suggest that $A \approx 0.8$.

How is chemical solubility related to chemical structure? The general rule of solubility is "like dissolves like." Substances that are similar to water are more soluble than substances that are not. Water is a polar molecule, and polar interactions between water and the polar portion of solute molecules serve to enhance solubility. The greater the size of the polar to the nonpolar portion of the solute molecule, the more soluble the chemical, in general. This is understood in terms of the maximization of entropy during mixing of solute and solvent. Mixing is a spontaneous process, and the total entropy must increase to a maximum during the process. The increase is moderated, however, by a decrease in entropy of the solvent in forming a hydration shell of water molecules around the solute molecules in solution (Israelchvili, 1992). Since water molecules in the hydration shell are more randomly oriented around polar solute molecules than around nonpolar solute molecules, the decrease in entropy on forming the hydration shell is smaller for polar solutes than for nonpolar solutes.

The ability to form hydrogen bonds with water enhances the solubility of the solute, also. The hydrogen bond is a moderately attractive force occurring between hydrogen atoms on one small, highly electronegative atom such as oxygen, nitrogen, or fluorine, and the lone pair electrons on another small, highly electronegative atom such as oxygen, nitrogen, or fluorine. Either the solute or water can donate the hydrogen atom.

Aqueous solubility tends to decrease with increasing molecular size (McAuliffe, 1966; Amidon and Anik, 1976). Molecular features such as unsaturation, chain branching, and molecular flexibility that serve to minimize molecular size help increase solubility. This can also be understood in terms of the maximization of entropy during mixing. The larger the solute molecule, the greater the decrease in entropy of the solvent in forming the hydration shell, and the less favorable the solution process, in general.

These qualitative solubility rules are formalized in the quantitative linear solvation energy relationships introduced by Kamlet and co-workers (Kamlet et al., 1988a; Yalkowsky et al., 1988). Aqueous solubility and related properties are correlated with molecular volume, solute polarity/polarizability, hydrogen-bond donor acidity, and hydrogen-bond acceptor basicity. The rules governing group contributons to the molecular properties of the solute are given in two papers by Kamlet et al. (1988a, 1988b).

Group contribution methods of estimating aqueous solubility values arc based on the assumption that aqueous solubility is an additive property. (See Section 2.3.2.) Individual molecular fragments contribute quantifiable amounts to the size of a molecule and to the polar interactions between solute and water molecules, and they make quantifiable contributions to the value of log C_W^S. To some degree, solubility is also a constitutive property. Intramolecular interactions of polar groups in the molcule minimize molecular size and reduce overall molecular polarity. In this case, the presence of topological features that enhance intramolecular interactions requires a correction to the group-contribution estimates of log C_W^S values.

7.3 METHODS OF ESTIMATING AQUEOUS SOLUBILITY

Yalkowsky and Banerjee (1992) critically reviewed the subject of estimation methods for aqueous solubility. Estimation methods described in the literature prior to 1981 are reviewed by Lyman (1990). Three types of methods are found to be reasonably accurate and broadly applicable: group contribution methods, the Universal Quasi-Chemical Functional Group Activity Coefficient (UNIFAC) method, and quantitative correlations with log K_{OW}. The correlations of aqueous solubility with log K_{OW} are the most accurate when applied to complex molecules and are the most widely used. A hybrid method involving a log C_W^S–log K_{OW} correlation and group contributions has also been proposed. All of these methods are described here.

7.3.1 Group-Contribution Methods

Irmann (1965) proposed the first group-contribution method of estimating the solubility of organic solutes in water. Several others have been proposed since. Methods that are both broadly applicable and reasonably accurate have been reported by Wakita et al. (1986), Klopman et al. (1992), and Myrdal et al. (1992, 1993). The method of Wakita et al. (1986) has been implemented in the computer program CHEMCALC2 (Suzuki, 1991). The method of Klopman et al. (1992) has been implemented in the computer program DESOC (Kohlenbrander et al., 1995). Kuhne et al. (1995) critically evaluated the group contribution methods of Klopman et al., Suzuki, and Wakita et al. along with the molecular connectivity–aqueous solubility relationship of Nirmalakhandan and Speece (1989). These were used as the basis of a refined model. It is described below. Klopman's method is extremely simple to apply and reliable. It is also discussed here.

(1) Klopman et al.'s Method

Klopman et al. (1992) propose two group-contribution methods to estimate the aqueous solubility of nonionic organic chemicals: one that is more accurate and one that is more broadly applicable. Both methods employ the following model:

$$\log S = C_0 + \sum_i n_i G_i \qquad (7.6)$$

Model I: n = 469, s = 0.458, r^2 = 0.961, F(33 449) = 229.1
Model II: n = 483, s = 0.526, r^2 = 0.948, F(33 449) = 245.3

where   S   = aqueous solubility, weight percent g/g%
        C_0 = constant
        n_i = number of groups i in molecule
        G_i = contribution of group i to solubility
        n   = number of chemicals in training set
        s   = standard deviation, log units
        r   = correlation coefficient
        F   = F ratio

The group contributions, G_i, are given in Table 7.1 for model I and in Table 7.2 for model II. The training set consists of diverse chemicals with log C_W^S values in the range from 2 to -8. Cross-validation studies of model I were carried out using a randomly selected 5% of the data base (about 25 chemicals) as a test set and 95% of the data base as a training set. An average standard deviation of 0.503 log units and an average r^2 of 0.946 were obtained with ten different test sets. A similar cross-validation study of model II produced an average standard deviation of 0.546 log units and an average r^2 of 0.953 with ten different test sets.

Klopman's models were compared with 20 other important models using a data set of 21 complex and multifunctional chemicals, primarily pharmaceutical chemicals and pesticides. The same test set was used by Yalkowsky and Banerjee (1992) in evaluating solubility estimation methods. With this test set, model I proved to be the most accurate. Also, it is more broadly applicable than all but two quantitative solubility–log K_{OW} relationships. Model II proved to be the most broadly applicable of all solubility estimation methods evaluated. Its accuracy compares favorably with the accuracy of the other models evaluated by Yalkowsky and Banerjee.

Solubility is determined, in part, by the interaction of polar functional groups in the solute. Polar group interactions are not explicitly accounted for in Klopman's model, however. The problem may not be reflected in the test statistics, since most chemicals in Klopman's data base are monofunctional.

Table 7.1 Group Contributions to Log C_W^S for Model I.[a] C_W^S is Measured in Weight Percent

Group	n	Freq	G_i[b]	Comments
–CH$_3$	249	449	–0.3361	
–CH$_2$–	200	565	–0.5729	
–CH–(–)	81	95	–0.6057	
–C–(–)(–)	30	35	–0.7853	
=CH$_2$	15	17	–0.6870	
=CH–	30	41	–0.3230	
=C–(–)	83	104	–0.3345	Not =C=
–C≡CH	6	7	–0.6013	
–C$_r$H$_2$–[c]	18	41	–0.4568	
–C$_r$H–(–)[c]	10	40	–0.4072	
–C$_r$–(–)(–)[c]	11	34	–0.3122	
=C$_r$H–[c]	227	1180	–0.3690	
=C$_r$–(–)[c]	234	758	–0.4944	Not =C$_r$=[c]
–F	13	41	–0.4472	Linked to sp^3 C
–F	6	9	–0.1773	Linked to other atom
–Cl	59	156	–0.4293	Linked to sp^3 C
–Cl	92	266	–0.6318	Linked to other atom
–Br	25	38	–0.6321	Linked to sp^3 C
–Br	17	33	–0.9643	Linked to other atom
–I	6	7	–1.2391	Linked to sp^3 C
–I	4	4	–1.2597	Linked to other atom
–OH	27	27	1.4642	Primary alcohol
–OH	19	19	1.5629	Secondary alcohol
–OH	16	16	1.0885	Tertiary alcohol
–OH	23	26	1.1919	Linked to non–sp^3 C not in COOH
–OH	0	0	ND	Linked to N, O, P, or S
–O$_r$–[c]	7	8	–0.2991	
–O–	23	35	0.8515	
–CHO	9	9	0.4476	Aldehyde
–COOH	25	26	0.2653	Conjugated acid
–COOH	15	24	1.1695	Nonconjugated acid
–COO–	32	35	0.8724	Ester
–CONH–	3	3	0.1931	Not –CONH$_2$
–CO–	7	7	1.3049	
–C$_r$O–[c]	3	3	1.5413	
–SO–	3	6	0.5826	
–NH$_2$	15	15	0.6935	
–NH–	7	8	0.9549	Not –N$_r$H–[c]
–C≡N	5	5	0.6262	
HN≡	0	0	ND	
–N$_r$=[c]	4	7	–0.3722	Not –N=
–NO$_2$	17	19	–0.2647	
–SH	3	3	–0.5118	
S=P–	5	5	–2.4096	
S=	4	4	–1.3197	Not –S– or –S$_r$–, not in S=P
Alkane	6	6	–1.5397	
Hydrocarbon	76	76	–0.2598	Not alkane
Constant[d]				
C_0			3.5650	

Table 7.1 (continued)   Group Contributions to Log C_W^S for Model I.[a] C_W^S is Measured in Weight Percent

[a] n is the number of chemicals in the training set. Freq is the number of times the group occurs in training set. Open valences in the group structures are not filled by hydrogens atoms.
[b] ND denotes an undetermined value.
[c] Atoms with subscript r are in ring structures.
[d] C_0 is applied only once.

After Klopman, G., Wang, S., and Balthasar, D.M., 1992. *J. Chem. Inf. Comput. Sci.*, **32**, 474–482.

Example

Estimate the aqueous solubility of pentachlorobenzene using Klopman's method.

1. Identify fragments in the molecular structure. The fragment contributions for model I are given in Table 7.1.

Fragment	Contribution
C_0	3.5650
=C˙–(–)	−0.4944
=C˙H(–)	−0.3690
–Cl (not connected to sp³ carbon)	−0.6318

2. Estimate log S using Equation 7.6.

$$\log\ S = C_0 + \Sigma_i n_i G_i$$

$$= 3.5650 + (5)(-0.4944) + (1)(-0.3690) + (5)(-0.6318)$$

$$= -2.4350$$

The estimated aqueous solubility, S_W^S, is 3.67×10^{-3} g/g% or about 1.46×10^{-5} mol/L. Two measured values of the aqueous solubility of pentachlorobenzene are 2.24×10^{-6} (shake-flask method) and 3.32×10^{-6} mol/L (generator-column method) (Doucette and Andren, 1988).

(2)   Kuhne et al.'s Method

Kuhne et al. (1995) describe a group contribution method which is a refinement of the methods of Klopman et al. and Wakita et al. (1986). In Kuhne et al.'s method, Equation 7.6 is used with the revised set of group contribution values and correction terms, G_i, given in Table 7.3. The correction terms 50 to 57 are applied in addition to the group contributions 1 to 49. The melting point correction for solids is multiplied by $T_m - 25$, where T_m is the melting point in °C. It is applied only when $T_m > 25$°C. Aqueous solubility is calculated in terms of mol/L instead of weight percent.

Table 7.2   Group Contributions to Log C_w^s for Model II. C_w^s is Measured in Weight Percent

Group	n^a	$Freq^a$	G_i^b	Comments
CH$_3$	258	458	−0.4169	sp^3 C
CH$_2$	223	628	−0.5199	sp^3 C
CH	93	137	−0.3057	sp^3 C
C	58	88	−0.1616	sp^3 C
=CH$_2$	15	17	−0.7788	sp^2 C
CH=	258	1262	−0.3843	sp^2 C
C=	286	890	4.5085	sp^2 C
C or CH	13	22	−0.4711	sp C
NH$_2$	15	15	0.6184	sp^3 N
NH	11	12	0.7796	sp^3 N
N$_r$H	1	1	0.7974	sp^3 N
N	3	3	1.0734	sp^3 N
N$_r$	2	2	0.3906	sp^3 N
N=	2	2	−0.8015	sp^2 N
N$_r$=	5	8	−0.3677	sp^2 N
N	5	5	1.0026	sp N
NO$_2$	18	21	−2.2003	
OH	122	138	1.0910	sp^3 O
O	67	93	0.4452	sp^3 O
O=	112	158	0.9545	sp^3 O
S	8	8	−0.6161	sp^3 S
S	8	8	−0.3648	sp^2 S
S	3	3	−1.9783	Other S
P	9	9	−0.9139	Any P atom
F	20	53	−0.5862	
C1	143	430	−0.6292	
Br	42	71	−0.9190	
I	10	11	−1.4676	
COO	36	40	−0.4537	Ester
COOH	40	50	−1.2440	Acid
CONH$_n$ (n = 0,1,2)	6	6	−0.5531	
Alkane	6	6	−1.8549	
Hydrocarbon	77	77	−0.2168	Not alkane

Constant^d

C_0	3.7253

^{a} n is the number of chemicals in the training set. Freq is the number of times the group occurs in the training set.
^{b} ND denotes an undetermined value.
^{c} Atoms with subscript r are in ring structures.
^{d} C_0 is applied once.

From Klopman, G., Wang, S., and Balthasar, D.M., 1992. *J. Chem. Inf. Comput. Sci.*, **32**, 474–482. Reprinted with permission of the American Chemical Society.

When applied to a training set of 694 diverse organic nonelectrolytes (351 liquids and 343 solids), the method produced a correlation coefficient (r^2) of 0.95 and an average absolute error of 0.38 log units. The training set is given in Table 7.4.

Table 7.3   Group Contributions to Log C_W^S at 25°C.[a] C_W^S is Measured in mol/L

Group	Freq[a]	G_i	Description[b]
1	679	0.0727	H attached to any C or to nonaromatic N bonded to aromatic C
2	5	0.0	C triple bonded to another C
3	62	−0.5610	C double bonded to another C
4	477	−0.6113	C aliphatic single bonded
5	371	−0.4257	C nonfused aromatic
6	48	−0.3803	C fund aromatic
7	6	−0.2327	Fluorine attached to nonaromatic C
8	65	−0.5201	Chlorine attached to nonaromatic C
9	20	−0.6409	Bromine attached to nonaromatic C
10	6	−1.2959	Iodine attached to nonaromatic C
11	4	0.0	Fluorine attached to aromatic C
12	132	−0.5694	Chlorine attached to aromatic C
13	15	−0.9387	Bromine attached to aromatic C
14	3	−1.4597	Iodine attached to aromatic C
15	34	1.0917	Primary OH group attached to nonaromatic C
16	29	1.2120	Secondary OH group attached to nonaromatic C
17	17	1.0736	Tertiary OH group attached to nonaromatic C
18	44	1.3169	OH group attached to aromatic C
19	4	0.5479	OH group attached to S
20	58	0.9212	−O− in aliphatic ethers
21	33	0.5668	−O− between aromatic C and any other atom
22	4	−0.7242	−O− in rings between aromatic rings (e.g., in dioxines)
23	49	1.1042	O double bonded to any other atom
24	10	−0.4591	CH in aldehyde group attached to nonaromatic C
25	3	−0.8240	CH in aldehyde group attached to aromatic C
26	31	0.7538	COOH group attached to nonaromatic C
27	18	0.4747	COOH group attached to aromatic C
28	35	0.4694	COO group, C attached to nonaromatic C
29	14	0.3610	COO group, C attached to aromatic C
30	9	0.0	C≡N
31	23	0.5814	NH_2 (primary amino group) attached to nonaromatic C
32	11	0.8909	NH (secondary amino group) attached to nonaromatic C
33	19	1.0124	N (tertiary amino group) attached to nonaromatic C
34	49	0.8308	N nonaromatic attached to aromatic C
35	6	−1.7814	N nonaromatic, double bonded to any other atom
36	12	2.1701	N in aromatic rings (as in pyridine)
37	2	−0.9504	Aromatic ring NCNCCC (pyrimidine type)
38	7	−2.7665	Aromatic ring NCNCNC (triazine type)
39	6	1.2685	NH_2CO_2 (carbamates)
40	7	0.0	$O=CNH_2$
41	25	0.0	$O=CNH$
42	10	0.4489	$O=CN$
43	6	0.0	NO_2 attached to nonaromatic C
44	37	−0.2657	NO_2 attached to aromatic C
45	20	0.0	S single bonded
46	22	−1.0613	S with one double bond
47	7	−1.7472	Any other S
48	4	0.5766	NH_2SO_2
49	22	−1.9164	Any P

Table 7.3 (continued)   Group Contributions to Log C_W^S at 25°C.[a] C_W^S is Measured in mol/L

Group	Freq[a]	G_i	Description[b]
			Correction Terms[c]
50	236	0.2288	Branch from nonaromatic C to any non-H except COO
51	65	0.2990	Nonaromatic ring
52	31	−0.1839	CH_2 group in nonaromatic hydrocarbons
53	26	−0.4299	CH_3 group in nonaromatic hydrocarbons
54	4	−1.1063	2 OH groups bonded to adjacent C
55	2	−0.5774	4 halogens at one C
			Melting Point Correction for Solid Compounds[d]
56	57	−0.00305	Nonaromatics
57	288	−0.00589	Aromatics and 5-rings with 6 π electrons
			Constant[e]
58	694	0.4273	Constant, C_0

[a] Freq is the number of times the group occurs in the training set.
[b] Aromatic groups are restricted to 6 atom rings containing 6 π electrons.
[c] Corrections are applied in addition to the group contributions.
[d] The melting point correction is multiplied by $(T_m - 25)$, where T_m is the melting point in °C. Apply only to chemicals with $T_m > 25$°C.
[e] C_0 is applied once.

Reprinted from *Chemosphere*, **30**, Kuhne, R., Ebert, R.-U., Kleint, F., Schmidt, G., and Schuurmann, G., Group contribution methods to estimate water solubility of organic chemicals, pp. 2061–2077. Copyright 1995, with kind permission from Elsevier Science Ltd., The Boulevard, Langford Lane, Kidlington OX5 1GB, U.K.

7.3.2   Quantitative Solubility–K_{OW} Relationships

Quantitative C_W^S–K_{OW} relationships are widely used to estimate the aqueous solubility of organic chemicals. The models are components of the important computer programs CHEMEST (Lyman and Potts, 1987), DTEST in E4CHEM (Bruggemann and Munzer, 1988), AUTOCHEM in the GEMS Graphical Exposure Modeling System, PCCHEM in the PCGEMS system (GEMS, 1986), and WSKOWWIN (Meylan et al., 1996). Critical reviews of quantitative C_W^S–K_{OW} relationships were published by Isnard and Lambert (1989), Muller and Klein (1992), and Yalkowsky and Banerjee (1992).

As we saw in Section 7.2, the aqueous solubility of nonionic organic solutes is inversely correlated with the octanol–water partition coefficient. The general relationship is

$$\log C_W^S = A + B \log K_{OW} \tag{7.7}$$

where C_W^S = aqueous solubility, mol/L
   K_{OW} = octanol–water partition coefficient
   A,B = constants

Table 7.4 Training Set for the Group Contribution Scheme Given in Table 7.3

No.	Substance	T_M [°C][a]	log C_w^s [mol/L][b]	
			exp.	calc.
1	pentane		-3.18	-3.17
2	cyclopentane		-2.64	-2.53
3	hexane		-3.84	-3.83
4	2-methylpentane		-3.74	-3.65
5	3-methylpentane		-3.68	-3.65
6	2,2-dimethylbutane		-3.55	-3.67
7	cyclohexane		-3.10	-3.18
8	methylcyclopentane		-3.30	-3.01
9	heptane		-4.53	-4.48
10	2,4-dimethylpentane		-4.26	-4.14
11	methylcyclohexane		-3.85	-3.66
12	octane		-5.24	-5.13
13	2,2,4-trimethylpentane		-4.74	-4.80
14	1,2-dimethylcyclohexane		-4.30	-4.15
15	1-pentene		-2.68	-2.60
16	trans-2-pentene		-2.54	-2.85
17	cyclopentene		-2.10	-2.21
18	1-hexene		-3.23	-3.25
19	4-methyl-1-pentene		-3.24	-3.08
20	cyclohexene		-2.59	-2.86
21	trans-2-heptene		-3.82	-4.15
22	methylcyclohexene		-3.27	-3.52
23	1-octene		-4.44	-4.56
24	1-nonene		-5.05	-5.21
25	1,4-pentadiene		-2.09	-2.03
26	1,5-hexadiene		-2.68	-2.68
27	2-methyl-1,3-butadiene		-2.03	-2.05
28	1-pentyne		-1.64	-1.63
29	1-hexyne		-2.36	-2.28
30	1-heptyne		-3.01	-2.93
31	1-octyne		-3.66	-3.58
32	1-nonyne		-4.24	-4.23
33	benzene		-1.64	-1.69
34	toluene		-2.21	-2.16
35	o-xylene		-2.80	-2.62
36	ethylbenzene		-2.77	-2.62
37	p-xylene		-2.77	-2.62
38	m-xylene		-2.82	-2.62
39	propylbenzene		-3.37	-3.09
40	1,2,4-trimethylbenzene		-3.31	-3.09
41	1,3,5-trimethylbenzene		-3.40	-3.09
42	1,2,3-trimethylbenzene		-3.20	-3.09
43	1-ethyl-2-methylbenzene		-3.21	-3.09
44	1-ethyl-4-methylbenzene		-3.11	-3.09
45	isopropylbenzene		-3.27	-2.86
46	indane		-3.04	-2.93
47	1,2,4,5-tetramethylbenzene	80.0	-4.59	-3.88
48	butylbenzene		-4.06	-3.55
49	1,4-diethylbenzene		-3.75	-3.55
50	p-isopropyltoluene		-3.77	-3.33
51	t-butylbenzene		-3.66	-3.10
52	isobutylbenzene		-4.12	-3.33
53	2-butylbenzene		-3.89	-3.33
54	pentamethylbenzene	50.8	-4.00	-4.17
55	pentylbenzene		-4.64	-4.02
56	t-amylbenzene		-4.15	-3.56

No.	Compound			
57	hexylbenzene		-5.21	-4.49
58	styrene		-2.82	-2.67
59	biphenyl	70.5	-4.31	-4.22
60	diphenylmethane	25.9	-4.08	-4.43
61	fluorene	116.0	-5.00	-4.80
62	bibenzyl	52.0	-4.62	-5.04
63	1-methylfluorene	87.0	-5.22	-5.10
64	t-stilbene	124.0	-5.80	-5.51
65	naphthalene	80.3	-3.60	-3.48
66	2-methylnaphthalene	34.6	-3.77	-3.68
67	1-methylnaphthalene		-3.70	-3.62
68	1-ethylnaphthalene		-4.17	-4.09
69	2-ethylnaphthalene		-4.29	-4.09
70	1,5-dimethylnaphthalene	81.0	-4.74	-4.42
71	2,3-dimethylnaphthalene	103.0	-4.72	-4.55
72	acenaphthylene	90.0	-3.96	-4.30
73	1,3-dimethylnaphthalene		-4.29	-4.09
74	1,4-dimethylnaphthalene		-4.14	-4.09
75	2,6-dimethylnaphthalene	109.0	-4.89	-4.58
76	acenaphthene	95.0	-4.63	-4.35
77	1,4,5-trimethylnaphthalene	58.0	-4.92	-4.75
78	anthracene	216.3	-6.35	-5.75
79	phenanthrene	100.0	-5.26	-5.07
80	9-methylanthracene	79.0	-5.89	-5.41
81	2-methylanthracene	204.0	-6.96	-6.14
82	pyrene	152.0	-6.19	-6.13
83	9,10-dimethylanthracene	182.0	-6.57	-6.48
84	fluoranthene	111.0	-6.00	-5.89
85	1,2,3,6,7,8-hexahydropyrene	33.0	-5.96	-6.58
86	benzo(a)fluorene	187.0	-6.68	-6.69
87	benzo(b)fluorene	209.0	-8.04	-6.82
88	naphthalene	341.0	-8.60	-7.95
89	triphenylene	199.0	-6.74	-7.11
90	benzo(a)pyrene	179.0	-8.19	-7.76
91	7,12-dimethylbenz(a)anthracene	122.0	-7.02	-7.59
92	benzo(e)pyrene	178.5	-7.80	-7.75
93	benzo(b)fluoranthene	167.0	-8.23	-7.69
94	benzo(j)fluoranthene	165.5	-8.00	-7.68
95	benzo(k)fluoranthene	217.0	-8.49	-7.98
96	3-methylcholanthrene	179.0	-7.92	-8.24
97	1,2 5,6-dibenzanthracene	269.0	-8.66	-8.99
98	coronene	439.0	-9.33	-10.81
99	iodomethane		-1.0	-1.26
100	dichloromethane		-0.63	-1.08
101	bromochloromethane		-0.80	-1.20
102	dibromomethane		-1.17	-1.32
103	trichloromethane		-1.17	-1.44
104	bromodichloromethane		-1.54	-1.56
105	chlorodibromomethane		-1.90	-1.68
106	tribromomethane		-1.91	-1.81
107	tetrachloromethane		-2.31	-2.38
108	tetrabromomethane	90.1	-3.14	-3.07
109	bromoethane		-1.09	-1.07
110	iodoethane		-1.60	-1.73
111	1,1-dichloroethane		-1.29	-1.32
112	1,2-dichloroethane		-1.06	-1.54
113	1-chloro-2-bromoethane		-1.32	-1.67
114	1,2-dibromoethane		-1.68	-1.79
115	1,1,1-trichloroethane		-2.00	-1.68
116	1,1,2-trichloroethane		-1.48	-1.91
117	1,1,2,2-tetrachloroethane		-1.74	-2.27
118	1,1,1,2-tetrachloroethane		-2.18	-2.27
119	1,1,2,2-tetrabromoethane		-2.72	-2.76
120	pentachloroethane		-2.60	-2.64

Table 7.4 (continued)   Training Set for the Group Contribution Scheme Given in Table 7.3

No.	Substance	T_M [°C][a]	log C_w^S [mol/L][b] exp.	calc.	No.	Substance	T_M [°C][a]	log C_w^S [mol/L][b] exp.	calc.
121	1,1,2-trichlorotrifluoroethane		-3.04	-2.14	149	2,3-dichloro-2-methylbutane		-2.69	-2.26
122	hexachloroethane	186.8	-3.67	-3.49	150	1-chlorohexane		-3.12	-2.81
123	1-chloropropane		-1.47	-1.42	151	γ-hexachlorocyclohexane	112.5	-4.59	4.52
124	2-chloropropane		-1.41	-1.19	152	δ-hexachlorocyclohexane	138.5	-4.51	-4.60
125	1-bromopropane		-1.73	-1.54	153	mirex	485.0	-6.80	-8.45
126	2-bromopropane		-1.59	-1.31	154	1,1-dichloroethylene		-1.64	-1.36
127	1-iodopropane		-2.29	-2.19	155	1,2-dichloroethylene		-1.30	-1.59
128	2-iodopropane		-2.09	-1.96	156	1,2-dibromoethylene		-1.32	-1.83
129	1,3-dichloropropane		-1.62	-2.01	157	1,2-diiodoethylene	73.0	-3.22	-3.29
130	1,2-dichloropropane		-1.60	-1.78	158	trichloroethylene		-1.96	-1.95
131	1-bromo-3-chloropropane		-1.85	-2.13	159	tetrachloroethylene		-2.54	-2.32
132	1,3-dibromopropane		-2.08	-2.25	160	3-chloropropylene		-1.36	-1.46
133	1,2-dibromopropane		-2.15	-2.02	161	chlordane	105.0	-6.86	-6.51
134	1,2-dibromo-3-chloropropane		-2.38	-2.62	162	heptachlor	95.5	-6.32	-6.16
135	1,2,3-trichloropropane		-1.92	-2.37	163	aldrin	104.3	-7.33	-6.44
136	1-chlorobutane		-2.03	-1.88	164	hexachloro-1,3-butadiene		-4.92	-4.02
137	2-chlorobutane		-1.96	-1.65	165	hexachlorocyclopentadiene		-5.18	-3.88
138	1-bromobutane		-2.37	-2.00	166	1,1,3,4,4-pentachloro-1,2-butadiene		-4.23	-3.71
139	1-iodobutane		-2.96	-2.66	167	fluorobenzene		-1.80	-1.76
140	1-chloro-2-methylpropane		-2.00	-1.65	168	chlorobenzene		-2.38	-2.33
141	1-bromo-2-methylpropane		-2.43	-1.78	169	bromobenzene		-2.55	-2.70
142	1,1-dichlorobutane		-2.40	-2.25	170	1-fluoro-4-iodobenzene		-3.13	-3.30
143	2,3-dichlorobutane		-2.70	-2.02	171	1,4-dichlorobenzene	53.1	-3.27	-3.14
144	1-chloropentane		-2.73	-2.35	172	1,2-dichlorobenzene		-3.05	-2.97
145	2-chloropentane		-2.63	-2.12	173	1,3-dichlorobenzene		-3.04	-2.97
146	3-chloropentane		-2.63	-2.12	174	1-bromo-2-chlorobenzene		-3.19	-3.34
147	2-chloro-2-methylbutane		-2.51	-1.89	175	1-bromo-3-chlorobenzene		-3.21	-3.34
148	1-bromo-3-methylbutane		-2.89	-2.24	176	1-bromo-4-chlorobenzene	65.0	-3.63	-3.58

No.	Compound	mp (°C)		
177	1,4-dibromobenzene	87.3	-4.07	-4.08
178	1,2-dibromobenzene		-3.50	-3.71
179	1,3-dibromobenzene		-3.54	-3.71
180	1,4-diiodobenzene	131.0	-5.37	-5.38
181	1,2,4-trichlorobenzene		-3.59	-3.62
182	1,2,3-trichlorobenzene	52.6	-4.00	-3.78
183	1,3,5-trichlorobenzene	63.5	-4.48	-3.84
184	1,2,3-tribromobenzene	87.8	-5.04	-5.09
185	1,2,4-tribromobenzene	44.5	-4.50	-4.84
186	1,3,5-tribromobenzene	119.6	-5.60	-5.28
187	1,2,3,5-tetrafluorobenzene		-2.31	-1.98
188	1,2,4,5-tetrafluorobenzene		-2.38	-1.98
189	1,2,4,5-tetrachlorobenzene	139.5	-5.56	-4.93
190	1,2,3,5-tetrachlorobenzene	50.7	-4.63	-4.41
191	1,2,3,4-tetrachlorobenzene	46.8	-4.57	-4.39
192	1,2,4,5-tetrabromobenzene	173.5	-6.98	-6.61
193	pentachlorobenzene	84.5	-5.65	-5.25
194	hexachlorobenzene	228.7	-7.68	-6.74
195	2-chlorotoluene		-3.52	-2.80
196	alpha-chlorotoluene		-2.39	-2.75
197	p-chlorotoluene		-3.08	-2.80
198	p-chlorobiphenyl	75.4	-5.20	-4.89
199	o-chlorobiphenyl	32.1	-4.54	-4.64
200	m-chlorobiphenyl		-4.88	-4.60
201	4,4'-PCB	149.0	-6.56	-5.97
202	2,2'-PCB	61.0	-5.27	-5.45
203	2,4'-PCB	43.0	-5.28	-5.34
204	3,3'-PCB	29.0	-5.80	-5.26
205	3,4-PCB	49.5	-7.44	-5.38
206	2,4-PCB		-5.25	-5.24
207	2,6-PCB	34.7	-5.21	-5.30
208	2,2',5-PCB	44.0	-6.02	-5.99
209	2,4,5-PCB	76.3	-6.27	-6.18
210	2,4',5-PCB	67.0	-6.25	-6.13
211	2,4,6-PCB	62.5	-6.14	-6.10
212	2,3',5-PCB	40.0	-6.01	-5.97
213	2',3,4-PCB	60.0	-6.29	-6.09
214	2,3,4'-PCB	69.0	-6.26	-6.14
215	2,4,4'-PCB	57.0	-6.21	-6.07
216	2,3,6-PCB	49.0	-6.29	-6.02
217	2,2',3,3'-PCB	121.0	-7.28	-7.09
218	2,3,4,5-PCB	90.7	-7.16	-6.91
219	2,3',4,4'-PCB	128.0	-7.80	-7.13
220	3,3',4,4'-PCB	179.9	-8.72	-7.43
221	2,2',3,5'-PCB	47.0	-6.47	-6.65
222	2,2',4,5'-PCB	63.9	-6.57	-6.75
223	2,2',5,5'-PCB	87.0	-7.00	-6.89
224	2,2',6,6'-PCB	103.0	-6.80	-6.98
225	2,3',4',5-PCB	198.0	-8.03	-7.54
226	2,2',4,4'-PCB	104.0	-7.25	-6.99
227	2,2',4,5,5'-PCB	83.0	-6.51	-6.86
228	2,2',3,3',4-PCB	76.5	-7.33	-7.47
229	2,2',3,4,5'-PCB	119.0	-7.05	-7.72
230	2,2',4,6,6'-PCB	112.0	-7.91	-7.68
231	2,3',4,4',5-PCB	85.0	-7.32	-7.52
232	2,2',3,4,5-PCB	109.0	-7.39	-7.66
233	2,2',3,4,6-PCB	100.0	-7.21	-7.61
234	2,3,4,5,6-PCB	100.0	-7.43	-7.61
235	2,2',4,4',5,5'-PCB	124.4	-7.92	-7.75
236	2,2',3,3',6,6'-PCB	112.9	-8.56	-8.32
237	2,2',3,3',5,6-PCB	114.0	-8.65	-8.33
238	2,2',3,4,4',5'-PCB	100.0	-8.60	-8.25
239	2,2',3,4,4',5,5'-PCB	80.5	-8.32	-8.13
240	2,2',3,4,5,5'-PCB	85.0	-7.68	-8.16

Table 7.4 (continued)   Training Set for the Group Contribution Scheme Given in Table 7.3

No.	Substance	T_M [°C]^a	log C_w^s [mol/L]^b		No.	Substance	T_M [°C]^a	log C_w^s [mol/L]^b	
			exp.	calc.				exp.	calc.
241	2,2',3,5,5',6-PCB	100.0	-7.42	-8.25	269	2-methyl-l-butanol		-0.47	-0.51
242	2,2',4,4',6,6'-PCB	112.5	-8.71	-8.32	270	isopentanol		-0.52	-0.51
243	2,3,3',4,4',5-PCB	127.0	-7.82	-8.41	271	2,2-dimethyl-1-propanol	53.0	-0.40	-0.37
244	2,3,3',4,4',6-PCB	107.0	-7.66	-8.29	272	3-methyl-2-butanol		-0.20	-0.16
245	2,2',3,3',4,4'-PCB	151.7	-9.01	-8.55	273	1-hexanol		-1.24	-1.20
246	2,2',3,3',4,5-PCB	85.0	-8.78	-8.16	274	3-hexanol		-0.80	-0.85
247	2,2',3,4,4',5,6-PCB	83.0	-7.92	-8.79	275	2-hexanol		-0.89	-0.85
248	2,2',3,3',4,4',6-PCB	122.2	-8.30	-9.02	276	2,3-dimethylbutanol		-0.39	-0.75
249	2,2',3,4,5,5',6-PCB	147.0	-8.94	-9.17	277	2-methyl-3-pentanol		-0.70	-0.63
250	2,2',3,3',5,5',6,6'-PCB	160.6	-9.15	-9.89	278	4-methyl-2-pentanol		-0.80	-0.63
251	2,2',3,3',4,4',5,5'-PCB	156.0	-9.16	-9.86	279	2-methyl-2-pentanol		-0.49	-0.76
252	2,2',3,3',4,4',5,5',6-PCB	204.5	-10.26	-10.79	280	2-methyl-1-pentanol		-1.11	-0.97
253	2,2',3,3',4,5,5',6,6'-PCB	182.6	-10.41	-10.66	281	4-methyl-l-pentanol		-1.14	-0.97
254	2,2',3,3',4,4'5,5',6,6'-PCB	304.5	-11.62	-12.02	282	2,2-dimethyl-l-butanol		-1.04	-0.75
255	DDD	109.5	-7.20	-7.40	283	3,3-dimethyl-l-butanol		-0.50	-0.75
256	DDT	108.5	-7.15	-7.75	284	2-ethyl-l-butanol		-1.17	-0.97
257	DDE	89.0	-6.90	-7.32	285	3-methyl-2-pentanol		-0.72	-0.63
258	2-chloronaphthalene	59.5	-4.14	-4.00	286	3,3-dimethyl-2-butanol		-0.62	-0.40
259	1-chloronaphthalene		-3.93	-3.80	287	3-methyl-3-pentanol		-0.38	-0.76
260	1-bromonaphthalene		-4.35	-4.17	288	2,3-dimethyl-2-butanol		-0.41	-0.53
261	2-bromonaphthalene	57.0	-4.40	-4.36	289	cyclohexanol		-0.44	-0.70
262	1-butanol		0.00	-0.27	290	1-heptanol		-1.81	-1.67
263	2-butanol		0.43	0.08	291	2-heptanol		-1.55	-1.32
264	2-methylpropanol		0.04	-0.04	292	3-heptanol		-1.47	-1.32
265	1-pentanol		-0.60	-0.74	293	4-heptanol		-1.40	-1.32
266	2-pentanol		-0.29	-0.39	294	2,4-dimethyl-2-pentanol		-0.92	-1.00
267	3-pentanol		-0.24	-0.39	295	2,2-dimethyl-1-pentanol		-1.52	-1.21
268	2-ethyl-2-propanol		0.08	-0.30	296	2,4-dimethyl-1-pentanol		-1.60	-1.21

No.	Compound			
297	4,4-dimethyl-1-pentanol		-1.55	-1.21
298	5-methyl-2-hexanol		-1.38	-1.09
299	2-methyl-2-hexanol		-1.08	-1.23
300	2,2-dimethyl-3-pentanol		-1.15	-0.86
301	2,4-dimethyl-3-pentanol		-1.22	-0.86
302	3-methyl-3-hexanol		-1.00	-1.23
303	2,3-dimethyl-2-pentanol		-0.89	-1.00
304	2,3-dimethyl-3-pentanol		-0.85	-1.00
305	3-ethyl-3-pentanol		-0.85	-1.23
306	2,3,3-trimethyl-2-butanol	83.0	-0.72	-0.95
307	1-octanol		-2.39	-2.13
308	2-octanol		-2.09	-1.79
309	2-ethyl-l-hexanol		-2.11	-1.91
310	3-methyl-2-heptanol		-1.72	-1.56
311	3-methyl-3-heptanol		-1.60	-1.70
312	2,2,3-trimethyl-3-pentanol		-1.27	-1.24
313	2-methyl-2-heptanol		-1.72	-1.70
314	phenol	40.9	0.00	-0.54
315	1,3-benzenediol	110.0	0.81	0.30
316	1,2-benzenediol	105.0	0.62	0.33
317	1,4-benzenediol	172.0	-0.17	-0.07
318	2-methylphenol	30.9	-0.62	-0.95
319	phenylmethanol		-0.40	-1.14
320	4-methylphenol	34.7	-0.70	-0.97
321	3-methylphenol		-0.68	-0.91
322	salicyl alcohol	86.0	-0.29	-0.25
323	1-phenylethanol		-0.92	-1.37
324	2,4-dimethylphenol	64.0	-1.19	-1.38
325	3,5-dimethylphenol		-1.40	-1.61
326	thymol	51.5	-2.22	-2.24
327	p-t-butylphenol	101.0	-2.41	-2.30
328	p-phenylphenol	164.5	-3.48	-3.53
329	4,4'-isopropylidenediphenol	152.5	-2.82	-3.16
330	1-naphthol	96.0	-2.22	-2.33
331	2-naphthol	121.0	-2.28	-2.48
332	naphthalene-1,5-diol	260.0	-2.92	-2.05
333	propionaldehyde		0.58	0.21
334	butyraldehyde		-0.01	-0.25
335	valeraldehyde		-0.85	-0.72
336	capronaldehyde		-1.30	-1.18
337	enanthaldehyde		-1.70	-1.65
338	caprylaldehyde		-2.36	-2.12
339	pelargonaldehyde		-3.17	-2.58
340	acrylaldehyde		0.57	0.17
341	t-crotonaldehyde		0.32	-0.30
342	benzaldehyde		-1.19	-1.48
343	methyl ethyl ketone		0.52	-0.10
344	methyl propyl ketone		-0.19	-0.57
345	methyl butyl ketone		-0.80	-1.03
346	methyl isobutyl ketone		-0.74	-0.81
347	cyclohexanone		-0.60	-0.88
348	methyl pentyl kentone		-1.42	-1.50
349	diisopropyl ketone		-1.30	-1.04
350	methyl hexyl ketone		-2.05	-1.97
351	dibutyl ketone	-2.59	-2.43	
352	methyl heptyl ketone	-2.58	-2.43	
353	propyl vinyl ketone	-0.83	-1.08	
354	progesterone:	129.5	-4.43	-5.42
355	acetophenone		-1.28	-1.43
356	benzophenone	48.5	-3.12	-3.37
357	anthraquinone	284.0	-5.19	-3.88
358	acetic acid	2.00	0.79	
359	oxalic acid	189.5	0.38	1.43
360	butyric acid		-0.19	-0.14

CHEMICAL PROPERTY ESTIMATION: THEORY AND PRACTICE

Table 7.4 (continued) Training Set for the Group Contribution Scheme Given in Table 7.3

No.	Substance	T_M [°C][a]	log C_W^S [mol/L][b] exp.	calc.	No.	Substance	T_M [°C][a]	log C_W^S [mol/L][b] exp.	calc.
361	succinic acid	185.0	-0.20	0.52	389	isopropyl acetate		-0.55	-0.59
362	glutaric acid	96.5	1.00	0.32	390	butyl acetate		-1.24	-1.29
363	capronic acid		-1.06	-1.08	391	ethyl butyrate		-1.28	-1.29
364	adipic acid	152.0	-0.82	-0.32	392	methyl valerate		-1.36	-1.29
365	caprylic acid		-2.30	-2.01	393	isobutyl acetate		-1.21	-1.06
366	caprinic acid	31.4	-3.44	-2.96	394	malonic acid diethylester		-0.82	-0.82
367	vulvic acid	45.0	-4.62	-3.93	395	amyl acetate		-1.89	-1.75
368	methacrylic acid		0.00	0.04	396	propyl butyrate		-1.92	-1.75
369	sorbic acid	134.5	-1.77	-1.50	397	methyl capronate		-2.00	-1.75
370	benzoic acid	122.4	-1.55	-1.86	398	ethyl valerate		-1.75	-1.75
371	m-toluic acid	112.0	-2.14	-2.27	399	isoamyl acetate		-1.92	-1.52
372	p-toluic acid	180.0	-2.60	-2.67	400	hexyl acetate		-2.46	-2.22
373	o-toluic acid	107.0	-2.06	-2.24	401	ethyl capronate		-2.31	-2.22
374	phenylacetic acid	76.5	-0.89	-1.78	402	methyl caprylate		-3.39	-2.68
375	o-phthalic acid	191.0	-2.11	-1.86	403	ethyl heptylate		-2.71	-2.68
376	m-phthalic acid	348.5	-3.22	-2.79	404	glyceryl triacetate		-0.60	-0.59
377	p-phthalic acid	425.0	4.05	-3.24	405	ethyl caprylate		-3.39	-3.15
378	cinnamic acid	133.0	-2.48	-2.62	406	ethyl pelargonate		-3.80	-3.62
379	methyl formate		0.58	0.50	407	ethyl caprinate		-4.10	-4.08
380	methyl acetate		0.52	0.11	408	methyl acrylate		-0.22	-0.40
381	ethyl formate		0.15	0.04	409	ethyl acrylate		-0.74	-0.87
382	ethyl acetate		-0.04	-0.36	410	methyl methacrylate		-0.80	-0.64
383	propyl formate		-0.49	-0.43	411	methyl benzoate		-1.85	-1.80
384	methyl propionate		-0.14	-0.36	412	ethyl benzoate		-2.32	-2.26
385	propyl acetate		-0.72	-0.82	413	dimethyl terephthalate	141.0	-4.01	-2.58
386	ethyl propionate		-0.66	-0.82	414	dimethyl phthalate		-1.66	-1.90
387	methyl butyrate		-0.82	-0.82	415	propyl benzoate		-2.67	-2.73
388	isobutyl formate		-1.01	-0.67	416	diethyl phthalate		-2.35	-2.83

No.	Compound			
417	o-dibutyl phthalate		-4.40	-4.70
418	diisobutyl phthalate		-4.66	-4.24
419	di(2-ethylhexyl))-phthalate		-6.96	-7.96
420	benzyl butyl phthalate		-5.64	-5.56
421	dimethoxymethane		0.48	0.82
422	diethyl ether		-0.09	-0.47
423	methyl propyl ether		-0.39	-0.47
424	methyl isopropyl ether		-0.06	-0.24
425	methyl butyl ether		-0.99	-0.94
426	ethyl propyl ether		-0.66	-0.94
427	methyl t-butyl ether		-0.24	-0.48
428	ethyl isopropyl ether		-0.55	-0.71
429	tetrahydropyran		-0.03	-0.78
430	dipropyl ether		-1.62	-1.40
431	diisopropyl ether		-1.10	-0.94
432	propylisopropylether		-1.34	-1.17
433	1,5-dimethoxydiethylether		0.88	0.24
434	1,2-diethoxyethane		-0.77	-0.58
435	1,1-diethoxyethane		-0.43	-0.35
436	dibutyl ether		-1.85	-2.33
437	ethyl vinyl ether		-0.85	-0.51
438	methoxybenzene		-1.85	-1.59
439	diphenyl ether	28.0	-3.96	-3.40
440	dibenzo-p-dioxine	122.5	-5.31	-5.82
441	ditolyl ether		-4.85	-4.32
442	propylene oxide		-0.59	0.38
443	styrene oxide		-1.60	-1.42
444	tetrahydrofurane		1.15	-0.32
445	2-methyltetrahydrofurane		0.11	-0.55
446	furane	83.0	-0.82	-0.41
447	dibenzofurane		-4.60	-4.03
448	2-butoxyethanol		-0.42	-0.38
449	diethyleneglycolemonoethylether		0.85	0.44
450	citric acid	153.0	0.51	1.18
451	glucose	83.0	0.74	0.75
452	fructose	103.0	0.64	0.69
453	diethyleneglycolemonobutylether		0.79	-0.49
454	testosterone	155.0	4.02	4.62
455	cortisone	222.0	-3.11	-2.27
456	hydrocortisone	218.5	-3.09	-2.08
457	furfural		-0.10	0.40
458	salicylic acid	158.3	-1.82	-0.83
459	o-methoxyphenol	28.0	-1.96	-0.36
460	p-hydroxybenzoic acid	214.5	-1.41	-1.16
461	p-hydroxybenzaldehyde	116.0	-0.96	-0.77
462	2-phenoxyethanol		-0.70	-1.04
463	p-methoxybenzaldehyde		-1.49	-1.38
464	ethyl-p-hydroxybenzoate	116.0	-2.35	-1.55
465	2-acetylsalicylic acid	135.0	-1.72	-1.93
466	salicin	199.0	-0.85	-1.15
467	phenyl salicylate	42.0	-3.15	-2.45
468	phenolphthalein	260.0	-2.89	-4.61
469	2,2,2-trichloro-1,1-ethanediol	57.0	0.72	0.49
470	1,3-dichloro-2-propanol		-0.11	-0.64
471	1,1,1-trifluoro-2-propanol		0.30	0.08
472	2-chlorophenol		-1.06	-1.09
473	4-chlorophenol	43.2	-0.70	-1.20
474	3-chlorophenol	32.8	-0.70	-1.13
475	4-bromophenol	63.5	-1.09	-1.68
476	2,4-dichlorophenol	42.0	-1.55	-1.83
477	2,4,6-trichlorophenol	69.0	-2.34	-2.63
478	2,4,5-trichlorophenol	68.0	-2.21	-2.63
479	pentachlorophenol	189.0	-4.28	-4.62
480	4-chloro-3-methylphenol	66.0	-1.57	-1.80

Table 7.4 (continued)　Training Set for the Group Contribution Scheme Given in Table 7.3

No.	Substance	T_M [°C][a]	log C_w^s [mol/L][b]	
			exp.	calc.
481	chloroacetic acid	56.0	1.81	0.10
482	trichloroacetic acid	57.5	0.60	-0.63
483	p-chlorobenzoic acid	243.0	-3.31	-3.21
484	o-chlorobenzoic acid	142.0	-1.89	-2.62
485	m-chlorobenzoic acid	158.0	-2.59	-2.71
486	2,2'-dichloroethylether		-1.12	-1.65
487	2-chloro-dibenzodioxine	88.5	-5.82	-6.26
488	2,7-dichlorodibenzodioxine	209.0	-7.92	-7.62
489	TCDD	305.0	-10.22	-9.47
490	methoxychlor	89.0	-6.89	-6.15
491	dieldrin	175.0	-6.29	-5.48
492	2,4-dichlorophenoxyaceticacid	140.5	-2.40	-2.87
493	dicamba	115.0	-1.70	-3.00
494	diclofopmethyl	40.0	-3.82	-4.89
495	ethylenediamine		-2.77	0.66
496	triethylamine		-0.14	-1.14
497	dipropylamine		-0.46	-1.33
498	aniline		-0.41	-0.79
499	o-phenylenediamine	103.5	-0.42	-0.35
500	p-phenylenediamine	142.0	-0.38	-0.57
501	o-methylaniline		-0.85	-1.25
502	N-methylaniline		-1.28	-1.25
503	p-methylaniline	44.5	-1.21	-1.37
504	m-methylaniline		-0.85	-1.25
505	benzylamine		-1.54	-1.65
506	N-ethylaniline		-1.70	-1.72
507	N,N-dimethylaniline		-1.92	-1.72
508	N,N-diethylaniline		-3.03	-2.65
509	p,p'-biphenyidiamine	128.0	-2.70	-2.75
510	diphenylamine	53.5	-3.51	-3.22
511	di-(p-aminophenyl)methane	89.0	-2.30	-2.99
512	o-tolidine	130.0	-2.21	-3.70
513	1-naphthylamine	49.2	-1.92	-2.40
514	carbazole	245.0	-5.27	-4.91
515	pyridine		0.76	0.83
516	2-ethyl pyridine		0.51	-0.10
517	4-ethyl pyridine		0.83	-0.10
518	2,3-dimethylpyridine		0.38	-0.10
519	2,4-dimethylpyridine		0.38	-0.10
520	2,5-dimethylpyridine		0.40	-0.10
521	2,6-dimethylpyridine		0.45	-0.10
522	3,4-dimethylpyridine		0.36	-0.10
523	3,5-dimethylpyridine		0.38	-0.10
524	quinoline		-1.33	-0.63
525	acetonitrile		0.26	0.03
526	propionitrile		0.28	-0.43
527	acrylonitrile		0.15	-0.48
528	benzonitrile		-1.00	-1.76
529	phthalonitrile	140.0	-2.38	-2.51
530	hydrazobenzene	131.0	-2.92	-2.77
531	azobenzene	68.5	-2.75	-2.55
532	cyanoguanidine	209.0	-0.31	-0.75
533	diisopropanolamine	45.0	0.81	1.34
534	o-aminophenol	172.0	-0.72	-0.41
535	p-aminophenol	190.0	-0.80	-0.51
536	glycine	233.0	0.52	0.66

No.	Compound				No.	Compound			
537	(L)-alanine	315.0	0.26	0.18	569	P-hydroxyacetanilide	167.0	-1.03	-1.52
538	(L)-valine	315.0	-0.30	-0.53	570	(L)-tyrosine	316.0	-2.59	-1.67
539	(L)-leucine	294.0	-0.80	-0.93	571	propoxur	91.5	-2.05	-2.58
540	(L)-isoleucine	284.0	-0.59	-0.90	572	morphine	197.0	-3.28	-2.75
541	o-aminobenzoic acid	145.0	-1.52	-1.09	573	codeine	157.0	-1.52	-3.65
542	p-aminobenzoic acid	187.0	-0.40	-1.34	574	nitromethane		0.26	0.03
543	(L)-phenyl alanine	283.0	-0.92	-2.72	575	nitroethane		-0.22	-0.43
544	(L)-tryptophan	293.0	-1.28	-2.84	576	1-nitropropane		-0.80	-0.90
545	O-methyl carbamate	52.0	0.97	1.22	577	2-nitropropane		-0.62	-0.67
546	O-ethyl carbamate	49.0	0.85	0.76	578	nitrobenzene		-1.80	-2.03
547	O-butyl carbamate	51.0	-0.66	-0.17	579	1,3-dinitrobenzene	89.9	-2.29	-2.75
548	O-isobutyl carbamate	67.0	-0.30	0.01	580	1,2-dinitrobenzene	118.5	-3.10	-2.92
549	O-t-butyl carbamate	105.0	0.10	0.12	581	1,4-dinitrobenzene	174.0	-3.39	-3.24
550	O-benzyl carbamate	87.0	-0.35	-1.33	582	1,3,5-trinitrobenzene	122.5	-2.89	-3.28
551	ethyl-p-aminobenzoate	92.0	-2.10	-1.75	583	o-nitrotoluene		-2.33	-2.49
552	carbaryl	142.0	-3.28	-3.74	584	m-nitrotoluene		-2.44	-2.49
553	urea	132.7	0.96	0.91	585	p-nitrotoluene	51.8	-2.49	-2.65
554	acetamide	81.0	1.58	0.09	586	2,4-dinitrotoluene	69.6	-2.82	-3.10
555	N,N-dimethylacetamide		1.11	-0.07	587	2,6-dinitrotoluene	66.0	-3.00	-3.07
556	acrylamide	84.5	0.95	-0.43	588	2,4,6-trinitrotoluene	80.1	-3.22	-3.50
557	benzamide	128.0	-0.96	-2.14	589	1-nitronaphthalene	59.5	-3.54	-3.70
558	phthalamide	228.0	-2.92	-2.57	590	p-nitrophenol	113.0	-0.74	-1.30
559	acetanilide	114.0	-1.33	-2.45	591	o-nitrophenol	44.0	-1.74	-0.90
560	phthalimide	238.0	-2.62	-2.14	592	m-nitrophenol	97.0	-1.01	-1.21
561	fenuron	131.0	-1.60	-1.93	593	dinoseb	40.0	-3.38	-2.85
562	phenobarbital	176.0	-2.32	-2.49	594	p-nitrobenzoic acid	242.4	-2.80	-2.91
563	phenytoin	296.5	-3.89	-5.25	595	m-nitrobenzoic acid	142.0	-1.68	-2.32
564	uracil	335.0	-1.49	-1.92	596	o-nitroanisole	335.0	-1.96	-1.93
565	antipyrine	114.0	0.72	-1.90	597	p-nitroanisole	54.0	-2.41	-2.10
566	(L)-serine	220.0	-0.02	-1.48	598	m-nitroaniline	112.5	-2.19	-1.64
567	(L)-glutamine	185.0	-0.55	0.26	599	o-nitroaniline	71.5	-1.96	-1.40
568	o-hydroxybenzamide	142.0	-1.82	-0.98	600	p-nitroaniline	147.8	-2.37	-1.85

Table 7.4 (continued) Training Set for the Group Contribution Scheme Given in Table 7.3

No.	Substance	T_M [°C][a]	log C_w^s [mol/L][b] exp.	calc.	No.	Substance	T_M [°C][a]	log C_w^s [mol/L][b] exp.	calc.
601	guanidinoacetic acid	280.0	-1.51	-0.68	629	terbacil	176.0	-2.48	-2.17
602	(L)-arginine	238.0	0.00	-1.21	630	alachlor	39.5	-3.26	-4.06
603	cytosine	320.0	-1.14	-0.55	631	chloroxuron	151.0	-4.89	-4.39
604	theophyline	272.0	-1.39	0.54	632	chloropicrin		-2.00	-1.29
605	nitrofurantoin	268.0	-3.47	-2.62	633	p-chloronitrobenzene	83.3	-2.92	-3.01
606	p-chloroaniline	72.5	-1.66	-1.71	634	o-chloronitrobenzene	32.0	-2.55	-2.71
607	o-chloroaniline		-1.52	-1.43	635	m-chloronitrobenzene	46.0	-2.77	-2.79
608	m-chloroaniline		-1.37	-1.43	636	3,4-dichloronitrobenzene	41.2	-3.20	-3.41
609	3,4-dichloroaniline	71.0	-3.24	-2.34	637	2,3-dichloronitrobenzene	61.5	-3.48	-3.53
610	3-trifluoromethylaniline		-1.47	-1.71	638	2,5-dichloronitrobenzene	55.5	-3.32	-3.49
611	3,3'-dichlorobenzidine	132.4	-4.92	-4.06	639	pentachloronitrobenzene	146.0	-5.82	-5.95
612	2,6-dichlorobenzonitrile	145.0	-4.24	-3.75	640	nitrofen	70.5	-5.46	-5.28
613	chlorothalonil	250.5	-5.64	-5.73	641	trifluralin	48.5	-5.68	-5.32
614	simazine	227.0	-4.55	-4.01	642	diazepam	125.0	-3.74	-4.51
615	atrazine	176.0	-3.85	-3.95	643	triadimefon	82.3	-3.61	-3.63
616	trietazine	100.5	-4.06	-4.20	644	ethanethiol		-0.60	-0.43
617	cyanazine	166.5	-3.15	-3.73	645	butanethiol		-2.18	-1.36
618	propazine	211.0	-4.43	-4.39	646	thiophenol		-2.12	-1.76
619	fluridone	152.0	-4.44	-5.63	647	2,2-dichlorodiethylsulfide		-2.37	-2.48
620	chloropham	41.4	-3.38	-2.78	648	phenothiazin	185.9	-5.10	-3.84
621	monolinuron	81.5	-2.57	-1.46	649	ametryn	84.0	-3.04	-1.23
622	monuron	170.5	-2.89	-2.81	650	terbutryn	104.0	-4.00	-3.58
623	diuron	158.0	-3.80	-3.38	651	thiourea	176.0	0.32	-0.54
624	linuron	93.0	-3.52	-2.17	652	ethylenethiourea	203.0	-0.71	-0.64
625	propanil	91.5	-3.00	-4.07	653	N,N'-diethylthiourea	69.0	-1.46	-1.32
626	fluometuron	163.0	-3.43	-3.05	654	phenylthiourea	149.0	-1.77	-2.68
627	propachlor	77.0	-2.48	-3.47	655	1-naphthylthiourea	198.0	-2.53	-4.44
628	neburon	101.5	-4.77	-4.44	656	methylisothiocyanate	35.9	-1.00	-0.66

No.	Compound	mp			No.	Compound	mp		
657	2-mercaptobenzothiazol	181.0	-3.15	-2.64	676	tricresyl phosphate		-6.70	-6.65
658	2-thiouracil	340.0	-2.26	-0.92	677	phcrate		-4.11	-3.95
659	methionine	281.0	-0.42	-0.65	678	disulfoton		-4.23	-4.42
660	pebulate		-3.35	-3.10	679	fenthion		-4.57	-4.25
661	carboxine	94.0	-3.14	-3.04	680	prometryn	72.5	-4.10	-3.00
662	taurine	328.0	-0.09	0.16	681	methylchlorpyrifos	46.0	-4.82	-2.84
663	p-sulfoaniline	122.0	-1.12	-0.42	682	chlorpyrifos	41.5	-5.49	-3.75
664	2-amino-l-sulfonaphthalene	272.0	-1.170	-2.77	683	diazinon	120.0	-3.64	-4.98
665	1-amino-6-sulfonaphthalene	180.0	-2.35	-2.23	684	methylparathion	35.0	-3.68	-3.72
666	asulam	144.0	-1.66	-0.67	685	dicapthon	52.0	-4.31	-4.46
667	sulfanilamide	165.0	-1.34	-1.11	686	fenitrothion		-4.04	-4.12
668	saccharin	228.8	-1.64	-2.63	687	parathion		-4.66	-4.59
669	p-toluenesulfonamide	137.1	-1.74	-2.31	688	acephate	85.5	0.54	-0.70
670	o-toluenesulfonamide	156.3	-2.02	-2.43	689	malathion		-3.37	-3.48
671	sulfaguanidine	190.0	-1.99	-1.99	690	ethion		-5.54	-6.14
672	oryzalin	137.0	-5.16	-4.41	691	DEF		-5.14	-5.76
673	trichlorfon	83.5	-0.22	-0.87	692	bromophos	53.5	-6.09	-5.78
674	triethyl phosphate		0.43	-0.50	693	ronnel	41.0	-5.72	-5.34
675	triphenyl phosphate	49.0	-5.66	-5.40	694	iodophenphos	76.0	-6.62	-6.44

[a] Melting points are given for chemicals that are solids at room temperature; $T_m > 25°C$.
[b] Experimental C_W^S values are measured at 25°C. Calculated C_W^S values are obtained using Equation 7.6 with the group contributions listed in Table 7.3.

Reprinted from Chemosphere, **30**, Kuhne, R., Ebert, R.-U., Kleint, F., Schmidt, G., and Schuurmann, G., Group contribution methods to estimate water solubility of organic chemicals, pp. 2061–2077. Copyright 1995, with kind permission from Elsevier Science Ltd., The Boulevard, Langford Lane, Kidlington 0X5, 1GB, U.K.

Table 7.5 Aqueous Solubility–Octanol/Water Partition Coefficient Relationships of Organic Liquids

$(\log C_W^S = A + B \log K_{OW})$

Chemical Class	B	A	n	r	s	Ref.
Alcohols	−1.113	0.926	41	0.967	0.136	Hansch et al. (1968)
Ketones	−1.229	0.720	13	0.980	0.164	Hansch et al. (1968)
Esters	−1.013	0.520	18	0.990	0.201	Hansch et al. (1968)
Ethers	−1.182	0.935	12	0.938	0.160	Hansch et al. (1968)
Alkyl halides	−1.221	0.832	20	0.928	0.235	Hansch et al. (1968)
Alkynes	−1.294	1.403	7	0.953	0.319	Hansch et al. (1968)
Alkenes	−1.294	0.248	12	0.985	0.131	Hansch et al. (1968)
Benzene derivatives	−0.996	0.339	16	0.975	0.179	Hansch et al. (1968)
Alkanes	−1.237	−0.248	16	0.953	0.199	Hansch et al. (1968)
Aromatic hydrocarbons	−0.947	0.727	18	0.995		Tewari et al. (1982)
Unsaturated hydrocarbons	−1.101	0.275	6	0.993		Tewari et al. (1982)
Halogenated hydrocarbons	−1.103	0.356	13	0.993		Tewari et al. (1982)
Normal hydrocarbons	−1.029	0.481	4	0.999		Tewari et al. (1982)
Aldehydes, ketones	−0.927	0.431	8	0.989		Tewari et al. (1982)
Esters	−1.073	0.306	7	0.991		Tewari et al. (1982)
Alcohols	−0.971	0.338	6	0.997		Tewari et al. (1982)
Mixed	−1.016	0.515	111	0.931	0.421	Valvani et al. (1981)
Acids, bases, neutrals	−1.163	0.845	28	0.985		Hafkenscheid and Tomlinson (1983)
Halogenated alkanes	−0.962	1.50	9	0.937		Chiou and Freed (1977)
General (theoretical)	-1.00	0.8				Yalkowsky and Valvani (1980)

After Yalkowsky, S.H. and Banerjee, S., 1992. *Aqueous Solubility. Methods of Estimation for Organic Compounds.* Marcel Dekker, New York.

Hansch et al. (1968) were the first to report such a correlation for liquid organic chemicals. Many such correlations have been published since. The information, as summarized by Yalkowsky and Banerjee (1992), is shown in Table 7.5. Values of A and B are given along with the number of chemicals in the training set, n, the standard deviation, s, and the correlation coefficient, r. The correlations in Table 7.5 apply to aqueous solutions at "room temperature", 25 ± 5°C.

Equation 7.7 applies to chemicals that are liquid at "room temperature". For chemicals that are solids, Equation 7.7 applies to the supercooled liquid. A relationship that applies to both liquids and solids is obtained by modifying Equation 7.7 to account for a change of phase from supercooled liquid to solid. The correction is proportional to the melting point of the solid (Yalkowsky and Valvani, 1980; Yalkowsky, Valvani, and Mackay, 1983). A general relationship, applicable to medium molecular-weight crystalline compounds with rigid molecular structures, is

$$\log C_W^S - A + B \log K_{OW} + D(T_m - 25)$$ (7.8)

where C_W^S = aqueous solubility, mol/L
T_m = melting point of pure solute, °C

K_{OW} = octanol–water partition coefficient

A, B, D = constants

In using Equation 7.8, liquid chemicals are assigned a melting point of 25°C. The relationship has been tested with a variety of chemicals. The results are summarized in Table 7.6. Values of A, B, and D are given along with the number of chemicals in the training set, n, the standard deviation, s, and the correlation coefficient, r. Similar results are reported by Bowman and Sans (1983) for carbamate and organophosphorous insecticides.

Within experimental uncertainty, the experimental results validate the theoretical relationships. The training sets referenced in Tables 7.5 and 7.6 are small. The solubility values of the chemicals in the training sets usually range from 1×10^{-5} to 1 mol/L, although the solubility range is limited to 0.001 to 0.1 mol/L in some cases. Also, measured values of log K_{OW} greater than 5 used in the correlation analysis are unreliable. This suggests that the theoretical relationships, Equations 7.7 and 7.8, are as accurate as any of the class-specific models.

Bruggemann and Altschuh (1991) performed a validation study of many of the correlations listed in Tables 7.5 and 7.6. They found that the solubility of phenols, aromatic carboxylic acids, and long-chain aliphatic carboxylic acids are poorly estimated by most general C_S^W–K_{OW} correlations, and they should not be used to estimate the solubility of these classes of chemical.

Chiou et al. (1982) point out that when the absolute value of the constant B in some of the relationships in Tables 7.5 and 7.6 exceeds unity, the solubility of the solute is greater in octanol-saturated water than it is in pure water. As evidence they report that the aqueous solubilities of DDT and hexachlorobenzene are increased in the presence of octanol. On the other hand, Banerjee et al. (1980), Miller et al. (1985), and Yalkowsky, Valvani, and Roseman (1983) found that the effect of n-octanol on the aqueous solubility of a number of the compounds is negligible.

A general relationship for solids and liquids that is not restricted to rigid molecules is

$$\log C_W^S = -\frac{\Delta S_{FUS}}{2.303RT}\left(T_m - 25\right) - \log K_{OW} + 0.52 \tag{7.9}$$

where C_W^S = aqueous solubility, mol/L

ΔS_{FUS} = entropy of fusion, cal/K

R = gas constant, 1.9872 cal/mol·K

T = ambient temperature, 298 K

T_m = melting point of pure solute, °C

K_{OW} = octanol–water partition coefficient

Yalkowsky and Banerjee (1992) derived approximate values of ΔS_{FUS} for individual classes of chemicals. Some values are listed in Table 7.7. Only slight improvement in accuracy is obtained when experimental entropies of fusion are used in Equation 7.9 (Hafkenscheid and Tomlinson, 1983).

TABLE 7.6   Aqueous Solubility–Octanol/Water Partition Coefficient Relationships of Organic Liquids and Solids

$[\log C_W^S = A + B \log K_{OW} + D(T_m - 25)]$

Chemical Class	D	B	A	n	r	sd	Ref.
Polycyclic aromatics	−0.01	−0.88	−0.012	32	0.990	0.251	Yalkowsky and Valvani (1979)
Polyfunctional aromatics	−0.01	−1.00	0.01	38	0.910		Briggs (1981)
Halobenzenes	−0.0095	−0.987	0.718	35	0.995	0.137	Yalkowsky et al. (1979)
Methyl, haloaromatics	−0.010	−0.944	0.323	162	0.978	0.034	Yalkowsky, Valvani, and Mackay (1983)
Mixed drugs	−0.012	−1.13	1.32	36	0.955	0.402	Yalkowsky, Valvani, and Roseman (1983)
Polycyclic hydrocarbons, halobenzenes	−0.012	−1.05	0.87	155	0.989	0.308	Yalkowsky and Valvani (1980)
Polycyclic aromatics	−0.01	−0.88	−0.12	32	0.99	0.25	Yalkowsky and Valvani (1980)
Halobenzenes	−0.01	−0.99	0.72	35	0.995	0.14	Yalkowsky and Valvani (1980)
Steroids	−0.01	−0.88	0.08	19	0.85	0.31	Yalkowsky and Valvani (1980)
Alcohols	−0.01	−0.904	0.07	12	0.959	0.38	Yalkowsky and Valvani (1980)
Alkyl benzoates	−0.005	−1.14	0.633	140	0.991	0.27	Yalkowsky and Valvani (1980)
Priority pollutants	−0.017	−1.123	0.88	27	0.96		Banerjee et al. (1980)
Mixed	−0.011	−1.12	0.80	111	0.987	0.321	Yalkowski et al. (1988)
	−0.010	−1.000	0.254	45		0.216	Mackay et al. (1980)
Substituted benzenes	−0.007	−1.20	0.50	32	0.951		Hafkenscheid and Tomlinson (1981a)
Acids, bases, neutrals	−0.010	−0.99	0.72	86	0.912		Hafkenscheid and Tomlinson (1983a)
PCBs	−0.010	−1.00	0.05	20		0.28	Bruggeman et al. (1982)
Mixed	−0.0054	−1.26	1.00	300	0.965	0.582	Isnard and Lambert (1989)
	−0.0107	−1.045	1.475	1450	0.953	0.688	Meylan et al. (1996)
General (theoretical)	−0.01	−1.00	0.8				Yalkowsky and Valvani (1980)

After Yalkowsky, S.H. and Banerjee, S., 1992. *Aqueous Solubility. Methods of Estimation for Organic Compounds*. Marcel Dekker, New York.

Table 7.7   Some Estimated Values of the Entropy of Fusion

Molecular Type	ΔS_{FUS} (cal/K)
Small rigid	9
Small nonrigid	11
Large rigid	13.5
Large nonrigid[a]	13.5 + 2.5 (n − 5)

[a] For flexible molecules with n carbon atoms in a skeletal chain. n > 5.

From Yalkowsky, S.H. and Banerjee, S., 1992. *Aqueous Solubility. Methods of Estimation for Organic Compounds.* Marcel Dekker, New York.

Equation 7.9 overestimates C_W^S values by a factor of 10 or more for hydrophobic dyes (Baughman and Perenich, 1989; Yen, Perenich, and Baughman, 1989). Yalkowsky and Banerjee (1992) speculate that the dyes are not miscible with octanol, and the relationship should not be used to estimate solubility in this case.

Example

Estimate the aqueous solubility of pentachlorobenzene using Equation 7.9.

1. Pentachlorobenzene is a small, rigid molecule. Using Table 7.7, the estimated value of ΔS_{FUS} is 9 cal/K.
2. Estimate log C_W^S using Equation 7.9. For pentachlorobenzene, reported log K_{OW} values range from 4.88 to 5.69 (Sabljic, 1987). We will use the average, 5.29. Also, $T_m = 86°C$ (Lide, 1995).

$$\log \ C_W^S = \left(\Delta S_{FUS}/2.303RT\right)\left(T_m - 25\right) - \log \ K_{OW} + 0.52$$

$$= \left[9/(2.303)(1.9872)(298)\right]\left(86 - 25\right) - 5.29 + 0.52$$

$$= -4.37$$

The estimated aqueous solubility, C_W^S, is 4.29×10^{-5} mol/L. The measured value of aqueous solubility of pentachlorobenzene is reported as 2.24×10^{-6} mol/L (shake-flask method) and as 3.32×10^{-6} mol/L (generator-column method) (Doucette and Andren, 1988).

Estimation methods based on quantitative C_W^S–K_{OW} relationships are potentially the most accurate, provided they are used with reliable values of K_{OW} like those in the STARLIST data base (Leo, 1993) or in the LOGKOW data base (Sangster, 1993). Generally, applicable C_W^S–K_{OW} correlations may produce large estimation errors as the low correlation coefficients listed in Tables 7.5 and 7.6 demonstrate. (This is true, in particular, when the models are used with estimated values of log K_{OW} and T_m.) Correlations that are applicable to specific classes of chemicals are more reliable, but it is often not clear how to apply them to estimate the solubility of polyfunctional chemicals. If the chemical falls under more than one chemical class, there are no rules about deciding which regression formula to use.

All C_W^S–K_{OW} regression models are based on the assumption that only the C_W^S values are subject to significant measurement error. However, K_{OW} values are hard to measure accurately, especially when the value of log K_{OW} is greater than 5. Table 9.10 in Chapter 9 shows that reported values of log K_{OW} for individual chemicals vary by as much as 3.3 log units. Such large errors in the log K_{OW} values of the training sets used to derive the C_W^S–K_{OW} correlations will lead to significant uncertainty in the slope of the regressions models (Sabljic, 1991).

7.3.3   Quantitative Aqueous Solubility–Activity Coefficient Relationships

(1) Yalkowsky's Method

As Equation 7.1 shows, the aqueous solubility of a chemical is inversely proportional to its activity coefficient in water at saturation. For sparingly soluble nonionic solutes, $\gamma_W^S \approx \gamma_{W\infty}$ or $x_W^S \approx 1/\gamma_W^\infty$, where γ_W^∞ is the chemical's activity coefficient in water at infinite dilution (Schwarzenbach et al., 1993). Converting from mole fraction to molar solubility in this relationship, $C_W^S \approx 1/V_{ml}\gamma_W^\infty$, where V_{ml} (55.54 mol/L) is the molar volume of water. Reasoning along the lines described in Section 7.3.2 for C_W^S–K_{OW} relationships, a general correlation between aqueous solubility and activity coefficient is proposed by Yalkowsky (1993):

$$\log_{10} x_W^S = -\frac{\Delta S_f\left(T_m - T\right)}{2.303RT} - \log_{10} \gamma_W^\infty \qquad (7.10)$$

where x_W^S = mole fraction aqueous solubility
ΔS_f = entropy of fusion, cal/K
T_m = melting point of solid, K
T = ambient temperature, K
R = 1.9872 cal/mol·K
γ_W^∞ = infinite-dilution activity coefficient

The relationship applies to weak and nonelectrolytes.

(2) AQUAFAC Method

The AQUAFAC method (Myrdal et al., 1992, 1993, 1995; Lee et al., 1996) is based on a relationship that is similar to Equation 7.10, $C_W^S = C_{W,ideal}^S/\gamma_{W\infty}$, or

$$\log C_W^S = \log C_{W,ideal}^S - \log \gamma_W^\infty \qquad (7.11)$$

$n = 621$, $s = 0.47$, $r^2 = 0.98$, $F = 1523$

where C_W^S = aqueous solubility, mol/L
  γ_W^∞ = infinite-dilution activity coefficient
  n = number of chemicals in the training set
  s = standard deviation
  r = correlation coefficient
  F = F ratio

The ideal solubility is given by

$$\log C_{W,ideal}^S = -0.000733 \Delta S_{FUS}(T_m - 25) \qquad (7.12)$$

where $C_{W,ideal}^S$ = ideal solubility, mol/L
  ΔS_{FUS} = entropy of fusion, cal/K
  T_m = melting point, °C

The value 0.000733 is equal to $(2.303\,RT)^{-1}$ when $T = 298$ K $(25°C)$. The activity coefficient is estimated by summing group contributions.

$$\log \gamma_W^S = \sum_i n_i \cdot q_i \qquad (7.13)$$

where γ_W^∞ = infinite-dilution activity coefficient
  n_i = number of fragments i in molecular structure
  q_i = contribution of fragment i to $\log \gamma_W$

The fragment contributions, q_i, are given in Table 7.8. The method applies to hydrocarbons, halocarbons, and nonhydrogen bonding chemicals in the solubility range of 3.5×10^{-13} to 3.6 mol/L.

Example

Estimate the aqueous solubility of 1,2-difluorobenzene using the AQUAFAC method.

1. Identify groups in the molecular structure.

Group	Contribution
–CH= (aromatic)	0.321
>C= (aromatic)	0.525
Y–F	−0.141

2. Calculate $\log \gamma_W^\infty$ using Equation 7.13.

$$\log \gamma_W^\infty = \Sigma_i n_i q_i$$

$$= (4)(0.321) + (2)(0.525) + (2)(-0.141)$$

$$= 2.052$$

Table 7.8　AQUAFAC Group Contributions to Activity Coefficient

Group	Neighboring Groups[a]		
	X_n	X_nY	X_nY_2
$-CH_3$	0.706(735)[b]	0.204(238)	
$-CH_2-$	0.545(1207)	[0.030][c](278)	[0.149](21)
>CH–	0.305(276)	[0.085](60)	[−0.127](17)
>C<	0.019(89)	−0.308(73)	[−0.520](3)
$CH_2=(CH_2=X)$	0.579(42)		
–CH=(XCH=)	0.636(86)	0.321(3060)	
>C=(X_2CH=)	0.583(61)	0.525(1492)	0.319(885)
–C≡CH	0.438(14)		
–O–	−1.510(87)	−0.664(83)	[−0.017](42)
–OH	ND[d]	−1.1810(102)	
–OOCH	−1.283(14)	ND	
>C=O	−0.968(71)	−0.722(36)	−0.41(25)
–CHO	−1.111(11)	−0.772(16)	
–(COO) (ester)	−1.117(119)	−0.796(83)	
–COOH	ND	−1.419(49)	
>N–	−3.428(12)	0.379(13)	[0.320](4)
–N=(X–N=)	−0.668(3)	−0.969(95)	
>NH	−2.233(7)	[−0.110](8)	ND
$-NH_2$	−1.911(9)	−1.193(41)	
$-NO_2$	[−0.127](1)	[0.082](77)	
–C≡N	−0.619(10)	−0.427(9)	
–CON	−1.601(4)	ND	
–CONH	−1.509(19)	[−0.847](1)	
$-CONH_2$	−2.126(1)	[−0.508](1)	
–S–	ND	ND	[−0.310](1)
–N=C=S	1.203(5)	1.266(20)	
–F	0.251(26)	[−0.141](29)	
–Cl	0.389(179)	0.409(839)	
–Br	0.379(30)	0.645(62)	
–I	0.49(7)	0.887(20)	
$-NHCON(CH_3)_2$	−2.190(2)	−1.229(9)	
Epoxide	[−0.301] (5)		
>C=(aromatic)　0.525			
C (bridge head)　0.319			
–CH=(aromatic)　0.321			
–N=(aromatic)　−0.969			
C (ring　−0.063			
Orthobiphenyl halogen[e]　−0.123			

[a] X designates a neighboring sp[3]-hybridized atom (hydrogen, aliphatic carbon, amine nitrogen, ether oxygen, and halogen). Groups bonded exclusively to hydrogen are not included. Y designates a neighboring sp[2]- or sp-hybridized atom. Depending upon the number of bonds to the group, n = 0, 1, 2, 3, or 4.
[b] Number of occurrences in training set given in parentheses.
[c] Tentative values given in brackets.
[d] ND, group not yet defined.
[e] Halogen atom in 2, 2′, 6, or 6′ position of biphenyl ring.

From Myrdal, P., Ward, G.H., Dannenfelser, R.M., Mishra, D., and Yalkowsky, S.H., 1992. *Chemosphere,* **24**, 1047–1061; Lee, Y.-C., Myrdal, P.B., and Yalkowsky, S.H., 1996. *Chemosphere,* **33**, 2129–2144.

3. Estimate C_W^S using Equation 7.11. The chemical is a liquid at 25°C, and log $C_{W,ideal}^S = 0$.

$$\log C_W^S = \log C_{W,ideal}^S - \log \gamma_W^\infty$$

$$= 0 - 2.052$$

$$= -2.052$$

The reported value of log C_W^S is -2.00 (Myrdal et al., 1995). The estimate is in error by 2.6%.

(3) Kan and Tomson's Method

Kan and Tomson (1996) report a similar method which uses activity coefficients calculated with the UNIFAC procedure described below. Kan and Tomson's model is

$$\ln C_W^S = -0.023(T_m - T) - \log V_{ml} - \log \gamma_W^\infty \qquad (7.14)$$

where C_W^S = aqueous solubility, mol/L
$\quad V_{ml}$ = molar volume of water
$\quad \gamma_W^\infty$ = infinite-dilution activity coefficient
$\quad T$ = ambient temperature, °C

The term $-0.023(T_m - T)$ accounts for the phase change undergone by solids on forming a solution as discussed in Section 7.2. The value of T_m is set equal to 25°C for chemicals that are liquid at the ambient temperature.

The method applies to hydrocarbons, halocarbons, alcohols, ketones, alkylbenzenes, chlorobenzenes, polycyclic aromatic hydrocarbons, phthalates, phenols, anilines, polychlorobiphenyls, and organohalide insecticides for which log C_W^S values lie in the range of -12 to 0. The average absolute error of log C_W^S values estimated with the UNIFAC procedure is about 0.5 log unit.

(4) UNIFAC Calculation

For chemicals that exhibit weak intermolecular interactions (nonpolar and weakly polar chemicals), $\gamma_{Wi}^S \approx \gamma_{Wi}^\infty$, where γ_{Wi}^∞ is the activity coefficient of chemical i at infinite dilution. Making this substitution in Equation 7.1, we find:

$$x_W^S = \frac{1}{\gamma_W^\infty} \qquad (7.15)$$

where x_W^S = aqueous mole fraction solubility of chemical
$\quad \gamma_w^\infty$ = infinite-dilution activity coefficient

The value of the infinite-dilution activity coefficient can be estimated using the UNIFAC (Universal Quasi-Chemical Group Activity Coefficient) method.

The UNIFAC (UNIQUAC Functional Group Activity Coefficient) group contribution (Fredenslund et al. 1975, 1977) is widely used for predicting activity coefficients. It can be used to estimate the aqueous solubility of chemicals alone and in mixtures. While lengthy, the method is not too complex to apply by hand for simple molecules. Calculations involving complex molecules and cosolvent effects require the use of a computer. The method has been implemented in the program PC-UNIFAC 4.0 for personal computers (bri, 1996). Yalkowsky and Banerjee (1992) published the code and parameter values needed to implement the UNIFAC method on a personal computer.

The activity coefficient is treated as a typical additive-constitutive parameter. It is estimated as the product of a combinatorial part and a residual part, $\gamma_i = \gamma_i^C \cdot \gamma_i^R$ or

$$\ln \gamma_i = \ln \gamma_i^C + \ln \gamma_i^R \qquad (7.16)$$

where γ_i = activity coefficient of solute i
$\quad\quad\;\; \gamma_i^C$ = combinatorial part
$\quad\quad\;\; \gamma_i^R$ = residual part

The combinatorial part is related to the work required to form a cavity in water big enough to hold the solute molecule. The work required is proportional to the volume and surface area of the solute molecule. The residual part corrects for group interactions.

The combinatorial part is calculated as follows:

$$\ln \gamma_i^C = \ln \frac{\phi_i}{x_i} + 5q_i \ln \frac{\theta_i}{\phi_i} + l_i - \frac{\phi_i}{x_i} \sum_j x_j l_j \qquad (7.17)$$

$$l_i = 5(r_i - q_i) - (r_i - 1) \qquad (7.18)$$

$$\phi_i = \frac{r_i x_i}{\sum_j r_j x_j} \qquad (7.19)$$

$$\theta_i = \frac{q_i x_i}{\sum_j q_j x_j} \qquad (7.20)$$

The r_i and q_i in Equations 7.17 to 7.20 are given by

$$r_i = \sum_k v_k^i R_k \qquad (7.21)$$

$$q_i = \sum_k v_k^i Q_k \qquad (7.22)$$

where x_i = mole fraction of chemical i in the mixture
$\quad r_i$ = volume parameter
$\quad q_i$ = area parameter
$\quad \theta_i$ = area fraction of molecule
$\quad \Phi_i$ = volume fraction of molecule
$\quad v_k^i$ = the number of groups k in molecule i
$\quad R_k$ = volume parameter of group k
$\quad Q_k$ = area parameter of group k

The summations are taken over all components of the mixture.

The residual part is calculated as follows. Tabulated values of group area parameters, Q_k, and group interaction parameters, a_{mn}, are used to calculate the following parameters:

$$\ln \Gamma_k = Q_k \left[1 - \ln\left(\sum_m \theta_m \psi_{mk} \right) - \sum_m \frac{\theta_m \psi_{km}}{\sum_n \theta_n \psi_{nm}} \right] \qquad (7.23a)$$

$$\ln \gamma_i^R = \sum_k v_k^i \left(\ln \Gamma_k - \Gamma_k^i \right) \qquad (7.23b)$$

$$\psi_{mn} = e^{\left(-a_{mn}/T \right)} \qquad (7.24)$$

$$X_m = \sum_i \frac{x_i v_m^i}{\sum_j x_j \left(\sum_m v_m^j \right)} \qquad (7.25)$$

$$\Theta_m = \frac{Q_m X_m}{\left(\sum_p Q_p X_p \right)} \qquad (7.26)$$

where ψ_{mn} = interaction energy parameter
   a_{mn} = group interaction parameter, K
   T = temperature, K
   x_m = mole fraction of group m in mixture
   Θ_m = surface area fraction of group m in mixture

 In Equation 7.23b, the term $\ln \Gamma_k^i$ is the contribution to the activity coefficient of group k in pure liquid chemical i. The term is included so that the residual portion of the activity coefficient of a pure liquid chemical equals zero.

 A compilation of UNIFAC parameters was published by Gmehling et al. (1982). This was revised and extended by Macedo et al. (1983) and by Tiegs et al. (1987). Comparisons show that the compilation of Macedo et al. is more accurate than those published earlier (Campbell and Luthy, 1985; Arbuckle, 1986). Chen et al. (1993) recently published a revised set of interaction parameters.

 Sample UNIFAC calculations of activity coefficients and aqueous solubilities are given by Reid et al. (1987) and Valsaraj (1995). For example, $R_K = 3.390$ and $Q_K = 2.910$ for CCl_4 (Macedo et al., 1983). For H_2O, $R_K = 0.920$ and $Q_K = 1.400$. The interaction parameters for CCl_4 and H_2O are $a_{12} = 1201.0$ and $a_{21} = 497.5$ K (Chen et al., 1993). For an aqueous solution of CCl_4 at a mole fraction of 3.66×10^{-6}, the UNIFAC procedure gives $\ln \gamma_{CCl_4} = 9.3$ and $\ln \gamma_{H_2O} = 3.74$ (Valsaraj, 1995).

 Estimated values of γ^∞ are obtained by repeating the UNIFAC calculations for decreasing solute concentrations until a limiting solute activity coefficient is found. However, the UNIFAC group contribution values are derived with data taken at high solute concentrations, and the extrapolation to low solute concentrations may not be reliable. Banerjee (1985) noted that the UNIFAC method underpredicts the solubility of chemicals with values of $\log C_W^S$ less than -2. The following correlation was suggested to correct the problem.

$$\log C_W^S = 1.2 + 0.78 \log \left(\frac{\gamma_{org}}{\gamma_{aq}} \right) - 0.01 \left(T_m - 25 \right) \qquad (7.27)$$

n = 50, r = 0.98

where C_W^S = aqueous solubility, mol/L
   γ_{org} = chemical activity in the organic phase
   γ_{aq} = chemical activity in the aqueous phase
   T_m = melting point of solid chemical, °C

 The regression analysis was performed with a training set of chemicals of various types. The measured solubility values of the training set ranged over 8 orders of magnitude. For most chemicals, $\gamma_{org} \approx 1$.

 Another method of correcting for bias in the UNIFAC estimates for chemicals with very low solubility was discussed by Al-Sahhaf (1989), but the method was demonstrated and tested with a limited variety of chemicals.

The method underestimates the solubilities of large, highly hydrophobic priority pollutants (Arbuckle, 1983). Solubility estimates for polychlorobiphenyls are also too low (Arbuckle, 1986). Chen et al. (1993) demonstrate that their revised interaction parameters significantly improve the accuracy of the UNIFAC model in estimating activity coefficients of highly hydrophobic chemicals.

7.3.4 Quantitative Structure–Solubility Relationships

Molecular connectivity–solubility relationships are reported for various classes of chemicals. Several relationships are described by Nirmalakhandan and Speece (1988, 1989) that apply to haloalkanes, haloalkenes, alkylbenzenes, halobenzenes, and alkanols. For chloro-, bromo-, and alkyl-substituted benzenes with mixed substitution, an applicable relationship is

$$\log S = 1.790 - 0.934 {}^1\chi^v - 1.01\Phi \tag{7.28}$$

n = 38, s = 0.20, r = 0.960

where S = aqueous solubility, g/g%
${}^1\chi^v$ = first-order valence connectivity
Φ = polarizability factor
= $0.1 N_C + 0.227 N_{Cl}$
N_C = number of carbon atoms
N_{Cl} = number of chlorine atoms
n = number of chemicals in the training set
s = standard deviation, g/g%
r = correlation coefficient

A relationship that applies to chloro- and alkyl-substituted alkanes and alkynes with mixed substitution is

$$\log S = 0.795 + 1.325 {}^3\chi_c - 1.192 {}^3\chi_c^v + 1.934\Phi \tag{7.29}$$

n = 38, s = 0.22, r = 0.964

where S = aqueous solubility, g/g%
${}^3\chi_C^v$ = third-order valence connectivity
Φ = polarizability factor
= $0.17 N_H - 0.57 N_C$
N_H = number of hydrogen atoms
N_C = number of carbon atoms
n = number of chemicals in the training set
s = standard deviation, g/g%
r = correlation coefficient

For alcohols, an applicable relationship is

$$\log S = 4.52 - 4.018 {}^1\chi + 2.905 {}^1\chi^v + 0.185 {}^3\chi_p^v \qquad (7.30)$$

n = 50, s = 0.11, r = 0.980

where S   = aqueous solubility, g/g%
       ${}^1\chi$ = first-order connectivity
       ${}^1\chi^v$ = first-order valence connectivity
       ${}^3\chi_p^v$ = third-order valence connectivity
       n   = number of chemicals in the training set
       s   = standard deviation, g/g%
       r   = correlation coefficient

A general relationship applicable to all chemical classes is

$$\log S = 2.209 + 1.653 {}^0\chi - 1.312 {}^0\chi^v + 1.00\Phi \qquad (7.31)$$

n = 145, s = 0.318, r = 0.962

where S   = aqueous solubility, g/g%
       ${}^0\chi$ = zero-order connectivity
       ${}^0\chi^v$ = zero-order valence connectivity
       Φ = polarizability factor
            = $-0.361 N_H - 0.963 N_{Cl} + 0.767 N_=$
       N_H = number of hydrogen atoms
       N_{Cl} = number of chlorine atoms
       $N_=$ = number of double bonds
       n   = number of chemicals in the training set
       s   = standard deviation, g/g%
       r   = correlation coefficient

    The solubility values of the chemicals in the training set range from miscible to 1×10^{-5} g/g%. The model was verified with a test set of 55 alcohols, aldehydes, alkyl ethers, alkyl esters, alkynes, and chloro- and alkyl-substituted aromatic hydrocarbons. Good agreement was obtained. The model does not apply to ketones, amines, polycyclic aromatic hydrocarbons, and polychlorobiphenyls.
    A general model applicable to polycyclic aromatic hydrocarbons, polychlorobiphenyls, and polychlorodibenzodioxins is reported by Nirmalakhandan and Speece (1989). In response to criticism concerning incorrect and omitted data in the training set (Yalkowsky and Mishra, 1990), the model was revised. The revised model is (Speece, 1990)

$$\log C_W^S = 1.564 + 1.627 {}^0\chi - 1.372 {}^0\chi^v + 1.000\Phi' \qquad (7.32)$$

n = 470, s = 0.355, r = 0.990

where S   = aqueous solubility, mol/L
    $^0\chi$ = zero-order connectivity
    $^0\chi^v$ = zero-order valence connectivity
    Φ' = polarizability factor
    n   = number of chemicals in the training set
    s   = standard deviation, mol/L
    r   = correlation coefficient

The polarizability factor is calculated as follows:

$$\Phi' = -0.936N_{Cl} - 0.361N_H - 2.620N_F + 1.474N_I + 0.636N_{NH_2} + 0.833N_{NH}$$
$$- 1.695N_{NO_2} - 0.767N_= - 1.24I_A + 1.014I_K - 3.332I_D$$

(7.33)

where N_i = number of groups i in molecular structure; the groups are hydrogen (H),
    fluorine (F), chlorine (Cl), iodine (I), NH_2^- (NH_2), –NH– (NH), –NO_2
    (NO_2), and double bond (=)
    I_A = indicator function for alkanes and alkenes
    I_K = indicator function for ketones and aldehydes
    I_D = indicator function for dibenzodioxins

The solubility values of chemicals in the training set range from miscible to 4 ×
10^{-11} mol/L.

The model fails to account for the dependence of the solubility of crystalline
chemicals on the melting point. As a result, the solubilities of high-melting isomers
are systematically underestimated (Yalkowsky and Mishra, 1990).

Abramowitz and Yalkowsky (1990) describe a correlation between total molecular surface area, melting point, and aqueous solubility that they use to estimate the
solubility of 209 polychlorobiphenyl congeners. Patil (1991) describes a correlation
between molecular connectivity and aqueous solubility applicable to chlorobenzenes
and polychlorobiphenyls. It is

$$\log S_W^S = -0.122 - 0.907^1\chi^v - 0.0299\left(^1\chi^v\right)^2$$

(7.34)

n = 71, s = 9.4, r = 0.98

where S_W^S = aqueous solubility, g/L
    $^1\chi^v$ = first-order valence connectivity
    n   = number of chemicals in the training set
    s   = standard deviation, mol/L
    r   = correlation coefficient

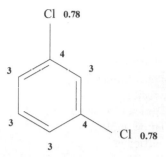

Figure 7.1   The hydrogen-suppressed structure of 1,3-dichlorobenzene. δ^v values are shown next to each atom.

The solubility values of chemicals in the training set range from 1×10^{-2} to 1×10^{-11} mol/L. The model was verified with a test set of 140 polychlorobiphenyls. The variance between estimated and measured values are reported to lie within the measurement error of the data used. An aqueous solubility–molecular connectivity correlation incorporating the melting point of the solid solute is also reported, but only marginal improvement in accuracy is obtained.

Example

Estimate the aqueous solubility of 1,3-dichlorobenzene, $C_6H_4Cl_2$, using Equation 7.34.

1. Calculate the first-order connectivity index for dichlorobenzene. The hydrogen-suppressed structure is shown along with valence delta values in Figure 7.1.

$$^1\chi^v = \Sigma_{bonds} \left(\delta_i^v \delta_j^v \right)^{-0.5}$$

$$= (2)(3 \cdot 3)^{-0.5} + (4)(3 \cdot 4)^{-0.5} + (2)(4 \cdot 0.78)^{-0.5}$$

$$= 2.469$$

2. Estimate C_W^S using Equation 7.34.

$$\log S_W^S = -0.122 - 0.907 \ ^1\chi^v - 0.0299 \left(^1\chi^v \right)^2$$

$$= -0.122 - (0.907)(2.469) - 0.0299 (2.469)^2$$

$$= -2.435$$

The estimated aqueous solubility of 1,3-dichlorobenzene is 3.7×10^{-3} g/L. The reported aqueous solubility of 1,3-dichlorobenzene is 1.3×10^{-3} g/L (GCA, 1985).

Multivariate regression models employing many structural descriptors selected with the aid of neural networks have been described with which to estimate the aqueous solubility (Sutter and Jurs, 1996) and activity coefficient (Chow et al., 1995) of organic chemicals. Increasingly accurate and complex structure-solubility relationships continue to be proposed. As powerful desktop computers are now widely available, complex models such as these are becoming widely used for property estimation.

7.3.5 Hybrid Models

Meylan et al. (1996) describe several estimation methods that employ a log C_W^S–log K_{OW} correlation along with molecular descriptors. The models are

$$\log C_W^S = 0.342 - 1.0374 \log K_{OW} - 0.0108 \left(T_m - 25\right) + \sum f_i \qquad (7.35)$$

n = 1450, s = 0.452, r^2 = 0.960, ME = 0.348

$$\log C_W^S = 0.796 - 0.854 \log K_{OW} - 0.00728(MW) + \sum f_i \qquad (7.36)$$

n = 1450, s = 0.585, r^2 = 0.934, ME = 0.442

$$\log C_W^S = 0.693 - 0.96 \log K_{OW} - 0.0092 \left(T_m - 25\right) - 0.00314(MW) + \sum f_i \quad (7.37)$$

n = 1450, s = 0.409, r^2 = 0.970, ME = 0.313

where C_W^S = aqueous solubility, mol/L
   K_{OW} = octanol–water partition coefficient
   T_m = melting point, °C (= 25 for liquids)
   MW = molecular weight, amu
   f_i = chemical class correction factor
   n = number of chemicals in training set
   s = standard deviation, log units
   r^2 = correlation coefficient
   ME = mean error, log units

Values of the correction factors, f_i, are given in Table 7.9. The log C_W^S values of chemicals in the training set range from –12 to 1.5. The models were validated with a test set of 817 diverse chemicals. Values of log K_{OW} were estimated with LOGKOW (see Appendix A8). Measured melting point values were used. The validation study of Equation 7.35 produced a standard deviation of 0.672, a correlation coefficient (r^2) of 0.882, and a mean error of 0.523. The validation study of Equation 7.36 produced a standard deviation of 0.723, a correlation coefficient (r^2) of 0.865, and a mean error of 0.560. The validation study of Equation 7.37 produced a standard deviation of 0.615, a correlation coefficient (r^2) of 0.902, and a mean error of 0.480.

Table 7.9 Correction Factors for Water Solubility Estimates

Factor	Application Rules	n[a]	Correction Factor for Equation[b] 7.35	7.36	7.37
Aliphatic alcohol	Compounds with one OH attached to aliphatic carbon; excluded: compounds with multiple OH, acetamide, amino, azo, or –(S = 0)	18	0.466	0.510	0.424
Aliphatic acid	Compounds with acid attached to aliphatic carbon; excluded: amino acids and compounds with C(=O)–N–C–COOH	70	0.689	0.395	0.650
Aliphatic amine	Applies to primary, secondary, or tertiary liquid amines; amine must have only aliphatic carbons attached	37	0.883	1.008[c]	0.834
Aromatic acid	Compounds with acid attached to aromatic carbon; excluded: compounds with any amino-type substituent [e.g., NH$_2$: –NH–C(=O)]	46	1.104	—[d]	0.898
Phenol	Compounds with OH attached to aromatic carbon; excluded: compounds with any amino-type substituent [e.g., NH$_2$; –NH–C(=O)]; compounds with nitro or alkyloxy ortho to OH	91	1.092	0.580	0.961
Alkylpyridine	Applies to pyridine and compounds with pyridine ring; only alkyl substituents permitted on ring; carbon attached to ring must be –CH$_3$, CH$_2$, or –CH<	11	1.293	1.300	1.243
Azo	Applies to all compounds with –N=N–; both Ns must be attached to C	12	−0.638	−0.432	−0.341
Nitrile	Applies to all compounds with nitrile (–CN) except N–C–CN	23	−0.381	−0.265	−0.362
Hydrocarbon	Applies to aliphatics with only C and H	33	−0.112	−0.537	−0.441
Nitro	Applies to aliphatic and aromatic compounds with –NO$_2$; excluded: N attachment (i.e., N–NO$_2$; aromatic ring attachment if ring has OH or amino)	75	−0.555	−0.390	−0.505
–SO$_2$	Applies to any aromatic compound with sulfonamide on ring plus any other substituent that is ketone, sulfone, or sulfanamide; also applies to any aliphatic compound with –S(=O)–C–C(=O)–C	36	−1.187	−1.051	−0.865
Fluoroalkane	Applies to any alkane with two or more fluorines	8	−0.832	−0.742	−0.945
PAH[e]	Apples to polycyclic aromatic hydrocarbons; compounds must contain at least three rings, at least two of which are aromatic; aromatic rings need not be fused to one another	58	—	−1.110	—

Table 7.9 (continued) Correction Factors for Water Solubility Estimates

Factor	Application Rules	n^a	Correction Factor for Equation[b] 7.35	7.36	7.37
Multi-N[e]	Applies to compounds with two or more aliphatic Ns, one attached to $C(=O)$, $S(=O)$, or $C(=S)$; compounds with four or more aromatic Ns; compounds with two or more aromatic Ns and one or more aliphatic Ns attached to $C(=O)$, $S(=O)$, or $C(=S)$; excluded: nitrogen in nitrile, nitro, azo; barbiturate and metal compounds	44	—	−1.310	
Amino acid[e]	Applies to all amino acids	11	—	−2.070	

[a] n is the number of chemicals in the training set.
[b] Derived as the regression coefficients for the indicated equation.
[c] For Equation 7.36 only, any compound with acetamide, acid, or imide is excluded.
[d] Not statistically significant.
[e] Factor is used only in Equation 7.36 which does not include T_m as input.

Reprinted from Meylan, W.M., Howard, P.H., and Boethling, R.S., 1996. *Environ. Toxicol. Chem.*, **15**, 100–106. With permission.

The use of the molecular weight and correction factors in Equations 7.35 to 7.37 significantly enhances the accuracy of the log C_W^S–log K_{OW} correlation. The three models should provide reliable estimates of C_W^S, particularly when used with measured values of K_{OW} from a reliable data set. Equation 7.36 provides a useful method of estimating C_W^S values of chemicals for which measured T_m values are not available.

Example

Estimate the aqueous solubility of nitrobenzene using Equation 7.37.

1. Identify any molecular features listed in Table 7.9 that require correction factors. Table 7.9 shows that the correction factor for nitro compounds is −0.505 when using Equation 7.37.
2. Estimate aqueous solubility using Equation 7.37. Nitrobenzene, $C_6H_5NO_2$, has a molecular weight of 123.11 g/mol and a log K_{OW} value of 1.85 (Sangster, 1989). Since it is a liquid at 25°C, $T_m = 25$.

$$\log C_W^S = 0.693 - 0.96 \ \log K_{OW} - 0.0092 \left(T_m - 25\right) - 0.00314 \ (MW) + \Sigma \ f_i$$

$$\log C_W^S = 0.693 - 0.96 \ (1.85) - 0.0092 \ (25 - 25) - 0.00314 \ (123.11) - 0.505$$

$$= -1.98$$

The measured value of log C_W^S is −1.80 (Yalkowsky and Banerjee, 1992). The estimate of log C_W^S is in error by 9.7%. The estimate of C_W^S is in error by 33.9%.

Table 7.10 Range of Reported Measured Aqueous Solubilities (mol/L) of Some Hydrocarbons

Chemical	$-\log C_W^S$ range	Ref.
Benzene	1.64–1.98	a,b,c,d
Toluene	1.77–2.25	c,d,e
Tetrachloroethylene	2.53–2.92	g,n
1,4-Dichlorobenzene	3.21–3.68	c,d,h,i
1,2,4-Trichlorobenzene	3.57–5.78	a,d,j,k
1,2,3-Trichlorobenzene	3.76–4.17	d,h
1,2,3,4-Tetrachlorobenzene	4.25–5.31	d,h,i
Lindane	4.50–6.29	c,f,j
Pentachlorobenzene	5.27–6.27	c,d,i,j
Pyrene	6.18–6.80	d,l
Hexachlorobenzene	6.78–7.78	d,h,i,j,m
2,3,4,2′,3′,4′-Hexachlorobiphenyl	7.56–9.11	h,n
p,p′-DDT	7.95–11.65	d,g,j,m
2,4,5,2′,4′,5′-Hexachlorobiphenyl	8.47–11.44	d,o

[a] Chiou et al., 1983
[b] Wijayaratne and Means, 1984
[c] Garten and Trabalke, 1983
[d] Yalkowski et al., 1983
[e] Chiou and Schmedding, 1982
[f] Chiou et al., 1987
[g] Davies and Dobbs, 1984
[h] Miller et al., 1984
[i] Yalkowski and Valvani, 1980
[j] Kenaga and Goring, 1980
[k] Bharath et al., 1984
[l] Pearlman et al., 1984
[m] Miller et al., 1985
[n] Bruggeman et al., 1982
[o] Muir et al., 1985

After Sabljic, A., 1987. *Environ. Sci. Technol.,* **21**, 358–366.

7.4 SENSITIVITY TO ENVIRONMENTAL PARAMETERS AND METHOD ERROR

The estimates made by the models described in this chapter can be no more accurate than the data used to derive and validate them. Solubility can be measured accurately, in general, but values below 1.0 mg/L are difficult to measure and may be unreliable. Interlaboratory comparisons show that the discrepancy between reported log C_W^S values of individual chemicals can be as much as 3.0 log units (Sabljic, 1987). This is illustrated by the data in Table 7.10.

The temperature dependence of the mole fraction solubility of a chemical in water is described by the van't Hoff equation.

$$\frac{d \ln x_W^S}{dT} = \frac{\Delta H_S}{RT^2} \tag{7.38}$$

where ΔH_S is heat of solution. Ideally, aqueous solubility increases exponentially with temperature, but this is not always observed in real systems. For a liquid, $\Delta H_S = \Delta H_S^e$, where ΔH_S^e is the excess enthalpy of solution. The value of ΔH_S^e is usually small and can be negative. The value can even change from positive to negative with change in temperature. Since solubility is not sensitive to temperature changes near 25°C and varies comparatively little in the temperature range from 0 to 40°C, the temperature dependence is often ignored.

For gases and solids, $\Delta H_S = \Delta H_{tr} + \Delta H_S^e$, where ΔH_{tr} is the heat of transition to supercooled liquid in the case of solids, and to superheated liquids in the case of vapors. For solids, ΔH_{tr} is the heat of fusion. Its value is usually large and positive and, as a result, the solubility of a solid increases as the temperature approaches the melting point. Indeed, the solubility of solids is very sensitive to temperature. The solubility of a rigid organic molecule should increase exponentially, doubling for each 20 to 30° rise in temperature (Yalkowsky and Banerjee, 1992).

For gases, $\Delta H_{tr} = -\Delta H_{vap}$, where ΔH_{vap} is the heat of vaporization. The solubility of a gas decreases as temperature approaches the boiling point.

The aqueous solubility of a gas or a vapor is linearly related to the vapor pressure by Henry's law: $P_g = x_W/H$, where P_g (Pa) is the gas pressure, x_W is the mole fraction of solute, and H (Pa·L/mol) is Henry's law constant. This topic is discussed in Chapter 8.

The presence of dissolved salts and minerals decreases the solubility of gases, nonpolar solutes, and nonionic solutes. The effect of salt concentration on the solubility of nonpolar and nonionic chemicals may be estimated using (Eganhouse and Calder, 1976)

$$\log\left(\frac{C_W^S}{C}\right) = K_s C_s \tag{7.39}$$

where C_W^S = molar solubility in pure water, mol/L
$\quad C$ = molar solubility in solution, mol/L
$\quad K_S$ = Setschenow or salting-out constant, L/mol
$\quad C_S$ = molar concentration of salt, mol/L

The salting-out effect is not large. The solubilities of many hydrocarbons in seawater are reduced by between 10 and 50% of their solubilities in distilled water. The salting-out effect is not as pronounced for polar chemicals (Schwarzenbach et al., 1993).

Example

Compare the solubility of benzene in seawater to the solubility of benzene in fresh water. For benzene, C_6H_6, $K_S \approx 0.2$ (Schwarzenbach et al., 1993). In seawater, $C_S \approx 0.5\ M$.

1. Estimate C_W^S/C using Equation 7.39.

$$\log \left(C_W^S/C \right) = K_S[\text{salt}]$$

$$= (0.2 \ \text{L/mol}) (0.5 \ \text{mol/L}$$

$$= 0.1$$

$$C_W^S/C = 1.26$$

The salting-out effect reduces the solubility of benzene by about 26%.

The presence of dissolved organic matter (humic and fulvic acids and other nonionic solutes) increases the solubility of nonionic organic chemicals.

The presence of cosolvents increases the solubility of organic chemicals. For instance, methyl t-butyl ether greatly increases the solubility of hydrocarbons in gasolines to which it is added. The magnitude of the cosolvent effect can be estimated with the UNIFAC method (Groves, 1988).

Li and Andren (1995) evaluated five methods of estimating ln x_W^S of polychlorinated biphenyl congeners in mixtures of water and n-alcohols. Twelve congeners, ranging from mono-substituted to hexa-substituted, and 77 different mixtures were considered. The UNIFAC method produced an overall estimation error of 35%. This was about as accurate as any of the other methods considered. A model that estimates the solubility of aromatic hydrocarbons in the presence of cosolvents was implemented in the program ARSOL (The Aromatic Solute Solubility in Solvent–Water Mixtures) (Fu et al., 1986). It also employs UNIFAC-derived activity coefficients.

A quick estimate of the effect of cosolvents can be made with the following relationship (Lyman, Reidy, and Levy, 1992):

$$\ln \left(\frac{x^M}{x_W^S} \right) = \sigma f \tag{7.40}$$

where x_W^S = mole fraction solubility in water
x^m = mole fraction solubility in mixed solvents
σ = solute surface area and interfacial free energy parameter
f = volume fraction of cosolvent

The solubility of mixtures of organic solutes is difficult to accurately estimate but may be calculated as a mole-fraction weighted average. The mole-fraction weighted average solubility of the components of gasoline is 0.130 g/L. The solubility of additive-free gasoline ranges from 0.130 to about 0.200 g/L at 20°C (Lyman, Reidy, and Levy, 1992).

The solubility of weak organic bases will increase and the solubility of weak organic acids will decrease with increasing pH. The magnitude of the effect is determined by the chemical's acid or base dissociation constant.

The effect of pressure on solubility is negligible under ambient conditions. At pressures of 10^6 Pa, solubility can be expected to increase to a maximum before decreasing again.

REFERENCES

Abramowitz, R. and Yalkowsky, S.H., 1990. Estimation of aqueous solubility and melting point of PCB congeners. *Chemosphere*, **21**, 1221–1229.

Al-Sahhaf, T.A., 1989. Prediction of solubility of hydrocarbons in water using UNIFAC. *Chromatographia*, **23**, 243–346.

Amidon, G.L. and Anik, S.T., 1976. Comparison of several molecular topological indexes with molecular surface area in aqueous solubility estimation. *J. Pharm. Sci.*, **65**, 801–805.

Arbuckle, W.B., 1983. Estimating activity coefficients for use in calculating environmental parameters. *Environ. Sci. Technol.*, **17**, 537–542.

Arbuckle, W.B., 1986. Using UNIFAC to calculate aqueous solubilities. *Environ. Sci. Technol.*, **20**, 1060–1064.

Banerjee, S., 1985. Calculation of water solubility of organic compounds with UNIFAC-derived parameters. *Environ. Sci. Technol.*, **19**, 369–370.

Banerjee, S., Yalkowsky, S.H., and Valvani, S.C., 1980. Water solubility and octanol/water partition coefficients of organics. Limitations of the solubility-partition coefficient correlation. *Environ. Sci. Technol.*, **14**, 1227–1229.

Baughman, G.L. and Perenich, T.A., 1989. Measuring the solubilities of disperse dyes. *Textile Chemist and Colorist*, **21**, 33–37.

Bowman, B.T. and Sans, W.W., 1983. Determination of octanol–water partitioning coefficients (K_{OW}) of 61 organophosphorus and carbamate insecticides and their relationship to respective water solubility (S) values. *J. Environ. Health*, **B18(6)**, 667–683.

bri. 1996. P.O. Box 7834, Atlanta, GA 30357-0834.

Briggs, G.G., 1981. Theoretical and experimental relationships between soil adsorption, octanol–water partition coefficients, water solubilities, bioconcentration factors, and the parachor. *J. Agric. Food Chem.*, **29**, 1050–1059.

Bruggemann, R. and Altschuh, J., 1991. A validation study for the estimation of aqueous solubility from m-octanol/water partition coefficients. *Sci. Total Environ.*, **109/110**, 41–57.

Bruggemann, R. and Munzer, B., 1988. Physico-chemical data estimation for environmental chemicals. In *Physical Property Prediction in Organic Chemistry*. Proceedings of the Beilstein Workshop 16–20th May, 1988, Schloss Korb, Italy. Jochum, C., Hicks, M.G., and Sunkel, J., Eds. Springer-Verlag, Berlin, pp. 303–334.

Bruggemann, W.A., van der Steen, J., and Hutzinger, O., 1982. Reversed-phase thin-layer chromatography of polynuclear aromatic hydrocarbons and chlorinated biphenyls: relationship with hydrophobicity as measured by aqueous solubility and octanol–water partition coefficient. *J. Chromatogr.*, **238**, 335–346.

Campbell, J.R. and Luthy, R.G., 1985. Prediction of aromatic solute partition coefficients using the UNIFAC group contribution model. *Environ. Sci. Technol.*, **19**, 980–985.

Chen, F., Holten-Anderson, J., and Tyle, H., 1993. New developments of the UNIFAC model for environmental application. *Chemosphere*, **26**, 1325–1354.

Chiou, C.T. and Freed, V.H., 1977. Chemodynamic Studies on Benchmark Industrial Chemicals. Report No. NSF/RA-770286. National Science Foundation, Washington, D.C.

Chiou, C.T., Freed, V.H., Schmedding, D.W., and Kohnert, R.L., 1977. Partition coefficients and bioaccumulation of selected organic compounds. *Environ. Sci. Technol.*, **11**, 475–478.

Chiou, C.T., Schmedding, D.W., and Manes, M., 1982. Partitioning of organic compounds in octanol–water systems. *Environ. Sci. Technol.*, **16**, 4–10.

Chow, H., Chen, H., Ng, T., Myrdal, P., and Yalkowsky, S.H., 1995. Using backpropagation networks for the estimation of aqueous activity coefficients of aromatic organic compounds. *J. Chem. Inf. Comput. Sci.*, **35**, 723–728.

Dickhut, R.M., Miller, K.E., and Andren, A.W., 1994. Evaluation of total molecular surface area for predicting air–water partitioning properties of hydrophobic aromatic chemicals. *Chemosphere*, **29**, 283–297.

Doucette, W.J. and Andren, A.W., 1988. Aqueous solubility of selected biphenyl, furan, and dioxin congeners. *Chemosphere*, **17**, 243–252.

Eganhouse, R.P. and Calder, J.A., 1976. The solubility of medium molecular weight aromatic hydrocarbons and the effects of hydrocarbon co-solutes and salinity. *Geochim. Cosmochim. Acta*, **40**, 555–561.

Fredenslund, A., Gmehling, J., and Rasmusen, P., 1977. *Vapor-Liquid Equilibria Using UNIFAC*. Elsevier, Amsterdam.

Fredenslund, A., Jones, R.L., and Prausnitz, J.M., 1975. Group contribution estimation of activity coefficients in nonideal liquid mixtures. *AIChE. J.*, **21**, 1086–1099.

Fu, J.K., Brooks, C., and Luthy, R.G., 1986. ARSOL, Aromatic Solute Solubility in Solvent-Water Mixtures. Departments of Chemistry and Civil Engineering. Carnegie Mellon University, Pittsburg, PA; Fu, J.K. and Luthy, R.G., 1986. Aromatic compound solubility in solvent-water mixtures. *J. Env. Eng.*, **112**, 328–345; Fu, J.K. and Luthy, R.G., 1986. Effect of organic solvent on sorption of aromatic solutes in soils. *J. Env. Eng.*, **112**, 346–366.

GEMS. 1986. Graphical Exposure Modeling System (GEMS) User's Guide, Vol. 5. Estimation Appendix E. AUTOCHEM, Task 3-2 under U.S. EPA Contract 68-02-3970, U.S. Environmental Protection Agency, Washington, D.C.

Gmehling, J., Rasmussen, P., and Fredenslund, A., 1982. Vapor-liquid equilibria by UNIFAC group contribution. Revision and extension. II. *Ind. Eng. Chem. Process. Des. Dev.*, **21**, 118–127.

Groves, F.R., Jr., 1988. Effect of cosolvents on the solubility of hydrocarbons in water. *Environ. Sci. Technol.*, **22**, 282–286.

Hafkenscheid, T.L. and Tomlinson, E., 1981. Estimation of aqueous solubilities of organic non-electrolytes using liquid chromatographic retention data. *J. Chromatogr.*, **218**, 409–425.

Hafkenscheid, T.L. and Tomlinson, E., 1983. Isocratic chromatographic retention data for estimating aqueous solubilities of acidic, basic, and neutral drugs. *Int. J. Pharm.*, **16**, 1–20.

Hansch, C., Quinlan, J.E., and Lawrence, G.L., 1968. The linear free-energy relationship between partition coefficients and aqueous solubility of organic liquids. *J. Org. Chem.*, **33**, 347–350.

Howard, P.H., 1989. *Handbook of Environmental Fate and Exposure Data for Organic Chemicals*, Volumes 1–3. Lewis Publishers, Chelsea, MI.

Irmann, F., 1965. Eine einfache Korrelation zwichen Wasserloslichkeit und Struktur von Kohlenwasserstoffen und Halogenkohlenwasserstoffen. *Chemie. Ing. Techn.*, **37**, 789–798.

Isnard, P. and Lambert, S., 1989. Aqueous solubility and n-octanol/water partition coefficient correlations. *Chemosphere*, **18**, 1837–1853.

Israelchvili, J.N., 1992. *Intermolecular and Surface Forces*. Second Edition. Academic Press, New York.

International Union of Pure and Applied Chemistry. 1987. *Solubility Data Series*. Pergamon Press, London.

Kamlet, M.J., Doherty, R.M., Carr, P.W., MacKay, D., Abraham, M.H., and Taft, R.W., 1988a. Linear solvation energy relationships. XLIV. Parameter estimation rules that allow accurate prediction of octanol/water partition coefficients and other solubility and toxicity properties of polychlorinated biphenyls and polycyclic aromatic hydrocarbons. *Environ. Sci. Technol.*, **22**, 503–509.

Kamlet, M.J., Doherty, R.M., Abraham, M.H., Marcus, Y., and Taft, R.W., 1988b. Linear solvation energy relationships. XLVI. An improved equation for correlation and prediction of octanol/water partition coefficients of organic nonelectrolytes (including strong hydrogen bond donor solutes). *J. Phys. Chem.*, **92**, 5244–5255.

Kan, A.T. and Tomson, M.B., 1996. UNIFAC prediction of aqueous and nonaqueous solubilities with environmental interest. *Environ. Sci. Technol.*, **30**, 1369–1376.

Kenaga, E.E. and Goring, C.A.I., 1980. Relationships between water solubility, soil sorption, octanol–water partitioning, and concentration of soil biota. Proc. 3rd Symp. Aquatic Tox., Special Technical Publication 707. ASTM, Philadelphia, PA, pp. 78–107.

Klopman, G., Wang, S., and Balthasar, D.M., 1992. Estimation of aqueous solubility of organic molecules by the group contribution approach. Application to the study of biodegradation. *J. Chem. Inf. Comput. Sci.*, **32**, 474–482.

Kohlenbrander, J.P., Drehahl, A., and Reinhard, M., 1995. *DESOC User's Guide.* Stanford Bookstore, Stanford, CA 94305-3079.

Kuhne, R., Ebert, R.-U., Kleint, F., Schmidt, G., and Schuurmann, G., 1995. Group contribution methods to estimate water solubility of organic chemicals. *Chemosphere*, **30**, 2061–2077.

Lee, Y.-C., Myrdal, P.B., and Yalkowsky, S.H., 1996. Aqueous functional group activity coefficients (AQUAFAC). IV. Applications to complex organic compounds. *Chemosphere*, **33**, 2129–2144.

Leo, A., 1993. Calculating log P_{OCT} from structures. *Chem. Rev.*, **93**, 1281–1306.

Li, A. and Andren, A.W., 1995. Solubility of polychlorinated biphenyls in water/alcohol mixtures. II. Predictive methods. *Environ. Sci. Technol.*, **29**, 3001–3006.

Lide, D.R., 1994. *CRC Handbook of Chemistry and Physics.* 74th ed., CRC Press, Boca Raton, FL.

Lyman, W.J., 1990. Solubility in Water. Chapter 2. In *Handbook of Chemical Property Estimation Methods. Environmental Behavior of Organic Compounds.* Lyman, W.J., Reehl, W.F., and Rosenblatt, D.H., Eds. American Chemical Society, Washington, D.C.

Lyman, W. J. and Potts, R.G., 1987. *CHEMEST: User's Guide — A Program for Chemical Property Estimation*, Version 2.1. A. D. Little, Cambridge, MA; Boethling, R. S., Campbell, S. E., Lynch, D.G., and LaVeck, G.D., 1988. Validation of CHEMEST, an on-line system for the estimation of chemical properties, *Ecotoxicol. Environ. Safety*, **15**, 21–30; Lynch, D.G., Tirado, N.F., Boethling, R.S., Huse, G.R., and Thom, G.C., 1991. Performance of on-line chemical property estimation methods with TSCA premanufacture notice chemicals. *Sci. Total Environ.*, **109/110**, 643–648.

Lyman, W.J., Reidy, P.J., and Levy, B., 1992. *Mobility and Degradation of Organic Contaminants in Subsurface Environments*, C.K. Smoley, Chelsea, MI.

Macedo, E.A., Weidlich, U., Gmehling, J., and Rasmussen, P., 1983. Vapor-liquid equilibria by UNIFAC group contribution. Revision and extension. III. *Ind. Eng. Chem. Process. Des. Dev.*, **22**, 676–678.

Mackay, D., Bobra, A., Shiu, W.Y., and Yalkowsky, S.H., 1980. Relationships between aqueous solubility and octanol–water partition coefficients. *Chemosphere*, **9**, 701–711.

Mackay, D. and Shiu, W.Y., 1977. Aqueous solubility of polynuclear aromatic hydrocarbons. *J. Chem. Eng. Data*, **22**, 399–402.

Mackay, D., Shiu, W.Y., and Ma, K.C., 1995. *Illustrated Handbook of Physical-Chemical Properties and Environmental Fate for Organic Chemicals*, Volumes 1–4. Lewis Publishers, Chelsea, MI.

McAuliffe, C., 1966. Solubility in water of paraffin, cycloparaffin, olefin, acetylene, cycloolefin, and aromatic hydrocarbons. *J. Phys. Chem.*, **70**, 1267–1275.

Meylan, W.M., Howard, P.H., and Boethling, R.S., 1996. Improved method for estimating water solubility from octanol/water partition coefficient. *Environ. Toxicol. Chem.*, **15**, 100–106.

Miller, M.M., Wasik, S.P., Huang, G.L., Shiu, W.Y., and Mackay, D., 1985. Relationships between octanol–water partition coefficients and aqueous solubility. *Environ. Sci. Technol.*, **19**, 522–529.

Muller, M. and Klein, W. 1992. Comparative evaluation of methods predicting water solubility. *Chemosphere*, **25**, 769–782.

Myrdal, P., Ward, G.H., Dannenfelser, R.M., Mishra, D., and Yalkowsky, S.H., 1992. AQUAFAC 1: aqueous function of group activity coefficients: application to hydrocarbons. *Chemosphere*, **24**, 1047–1061.

Myrdal, P., Ward, G.H., Simamora, P., and Yalkowsky, S.H., 1993. AQUAFAC: aqueous functional group activity coefficients. SAR and QSAR. *Environ. Res.*, **1**, 53–61.

Myrdal, P.B., Manka, A.M., and Yalkowsky, S.H., 1995. Aquafac 3: aqueous functional group activity coefficients; application to the estimation of aqueous solubility. *Chemosphere*, **30**, 1619–1637.

Nirmalakhandan, N.N. and Speece, R.E., 1988. Prediction of aqueous solubility of organic chemicals based on molecular structure. *Environ. Sci. Technol.*, **22**, 328–338.

Nirmalakhandan, N.N. and Speece, R.E., 1988. Structure-activity relationships. *Environ. Sci. Technol.*, **22**, 610.

Nirmalakhandan, N.N. and Speece, R.E., 1989. Prediction of aqueous solubility of organic chemicals based on molecular structure. II. Application to PNAs, PCBs, PCDDs, etc. *Environ. Sci. Technol.*, **23**, 708–713.

Patil, G.S., 1991. Correlation of aqueous solubility and octanol–water partition coefficient based on molecular structure. *Chemosphere*, **22**, 723–738.

Penning, W., Roi, R., and Boni, M., 1990. ECDIN — the European data bank on environmental chemicals. *Toxicol. Environ. Chem.*, **25**, 251–264.

Prausnitz, J.M., 1969. *Molecular Thermodynamics of Fluid-Phase Equilibria*. Prentice-Hall, Englewood Cliffs, NJ.

Reid, R.C., Prausnitz, J.M., and Poling, B.E., 1987. *The Properties of Gases and Liquids*. McGraw-Hill, New York.

Sabljic, A., 1991. Chemical topology and ecotoxicology. *Sci. Total Environ.*, **109/110**, 197–220.

Sabljic, A., 1987. On the prediction of soil sorption coefficients of organic pollutants from molecular structure: application of molecular topology model. *Environ. Sci. Technol.*, **21**, 358–366.

Sangster, J., 1989. Octanol-water partition coefficients of simple organic compounds. *J. Phys. Chem. Ref. Data*, **18**, 1111–1229.

Sangster, J., 1993. LOGKOW DATABANK. Sangster Research Laboratories, Montreal, Quebec, Canada.

Schwarzenbach, R.P., Gschwend, P.M., and Imboden, D.M., 1993. *Environmental Organic Chemistry*, John Wiley & Sons, New York.

Speece, R.E., 1990. Comment on "Prediction of aqueous solubility of organic chemicals based on molecular structure. II. Application to PNAs, PCBs, PCDDs, etc." *Environ. Sci. Technol.*, **24**, 929–930.

Sutter, J.M. and Jurs, P.C., 1996. Prediction of aqueous solubility for a diverse set of hetero-atom-containing organic compounds using a quantitative structure–property relationship. *J. Chem. Inf. Comput. Sci.,* **36**, 100–107.

Suzuki, T., 1991. Development of an automatic estimation system for both the partition coefficient and aqueous solubility. *J. Comput. Aided Mol. Design,* **5**, 149–166.

Tewari, Y.B., Miller, M.M., Wasik, S.P., and Martire, D.E., 1982. Aqueous solubility and octanol/water partition coefficients of organic compounds at 25.0°C. *J. Solution Chem.,* **11**, 435–445.

Tiegs, D., Gmehling, J., Rasmussen, P., and Fredenslund, A., 1987. Vapor-liquid equilibria by UNIFAC group contribution. Revision and extension. IV. *Ind. Eng. Chem. Process. Des. Dev.,* **26**, 159–161.

Valsaraj, K.T., 1995. *Elements of Environmental Engineering.* CRC/Lewis Publishers, Boca Raton, FL.

Valvani, S.C., Yalkowsky, S.H., and Roseman, T.J., 1981. Solubility and partitioning. IV. Aqueous solubility and octanol–water partition coefficients of liquid nonelectrolytes. *J. Pharm. Sci.,* **70**, 502–507.

Verschueren, K., 1983. *Handbook of Environmental Data on Organic Chemicals.* Second Edition. Van Nostrand Reinhold, New York.

Wakita, K., Yasimoto, M., Miyamoto, S., and Watanabe, H., 1986. A method for calculation of the aqueous solubility of organic compounds by using new fragment solubility con-stants. *Chem. Pharm. Bull. (Tokyo),* **34**, 4663–4681.

Yalkowsky, S.H., 1979. Estimation of entropies of fusion of organic compounds. *Ind. Eng. Chem. Fundam.,* **18**, 108–111.

Yalkowsky, S.H., 1993. Estimation of the aqueous solubility of complex organic compounds. *Chemosphere,* **26**, 1239–1261.

Yalkowsky, S.H. and Banerjee, S., 1992. *Aqueous Solubility. Methods of Estimation for Organic Compounds.* Marcel Dekker, New York.

Yalkowsky, S.H. and Dannenfelser, R.M., 1990. *Arizona Database.* Fifth Edition. College of Pharmacy, University of Arizona, Tucson; Dannenfelser, R.-M. and Yalkowsky, S.H., 1988. The Arizona database for nonelectrolytes. In *Physical Property Prediction in Organic Chemistry.* Proceedings of the Beilstein Workshop 16–20th May, 1988, Schloss Korb, Italy. Jochum, C., Hicks, M.G., and Sunkel, J., Eds. Springer-Verlag, Berlin, pp. 499–508.

Yalkowsky, S.H., Orr, R.J., and Valvani, S.C., 1979. Solubility and partitioning. III. The solubility of halobenzenes in water. *I&EC Fundam.,* **18**, 351–353.

Yalkowsky, S.H. and Mishra, D.S., 1990. Comment on "Prediction of aqueous solubility of organic chemicals based on molecular structure. II. Application to PNAs, PCBs, PCDDs, etc." *Environ. Sci. Technol.,* **24**, 927–929.

Yalkowsky, S.H., Pinal, R., and Banerjee, S., 1988. Water solubility: a critique of the solva-tochromic approach. *J. Pharm. Sci.,* **77**, 74–144.

Yalkowsky, S.H. and Valvani, S.C., 1979. Solubility and partitioning. II. Relationships between aqueous solubility, partition coefficients, and molecular surface areas of rigid aromatic hydrocarbons. *J. Chem. Eng. Data,* **24**, 127–129.

Yalkowsky, S.H. and Valvani, S.C., 1980. Solubility and partitioning. I. Solubility of nonelec-trolytes in water. *J. Pharm. Sci.,* **69**, 912–922.

Yalkowsky, S.H., Valvani, S.C., and Mackay, D., 1983. Estimation of the aqueous solubility of some aromatic compounds. *Residue Rev.,* **85**, 43–55.

Yalkowsky, S.H., Valvani, S.C., and Roseman, T.J., 1983. Solubility and partitioning. IV. Octanol solubility and octanol–water partition coefficient. *J. Pharm. Sci.,* **72**, 866–870.

Yen, C-P.C., Perenich, T.A., and Baughman, G.L., 1989. Fate of dyes in aquatic systems II. Solubility and octanol/water partition coefficients of disperse dyes. *Environ. Toxicol. Chem.,* **8**, 981–986.

CHAPTER 8

Air–Water Partition Coefficient

8.1 INTRODUCTION

This chapter describes methods of estimating air–water partition coefficients and Henry's law constants for hydrophobic chemicals, primarily. The **air–water partition coefficient**, also called the dimensionless Henry's law coefficient, is defined as:

$$K_{AW} = \frac{C_g}{C_w} \tag{8.1}$$

where C_g (mol/m^3) is the vapor concentration of chemical and C_w (mol/m^3) is the aqueous solution concentration of chemical in an air/water system at equilibrium. The air–water partition coefficient may also be given in terms of mole fraction of the volatile chemical; $K'_{AW} = y_i/x_i$, where y_i is the mole fraction of chemical vapor in air and x_i is the mole fraction of chemical in water. The two partition coefficients are related; $K_{AW} = K'_{AW} V_{ml}/V_{mg}$, where V_{ml} (mol/m^3) is the molar volume of liquid water and V_{mg} (mol/m^3) is the molar volume of air.

Chemical partitioning between air and water is often described with the **Henry's law constant**, the ratio of the chemical's partial pressure in solution to its concentration in solution at equilibrium:

$$H = \frac{P_g}{C_W} \tag{8.2}$$

where P_g (Pa) is the equilibrium partial pressure of chemical vapor in air.

The Henry's law constant describes the pressure dependence of the solubility of a chemical vapor. It is reported in terms of any convenient pressure and concentration unit. For instance, aqueous concentrations are frequently reported in terms of solute mole fraction, x_{Wi}, and the corresponding Henry's law constant is $p/x_{Wi} = H'(\text{Pa})$. At low concentrations, $H' \approx H/V_{Wm}$, where V_{Wm} (L/mol) is the partial molar volume of water.

The air–water partition coefficient and the Henry's law constant are related to each other by $H = K_{AW}RT$, where R (8.314 Pa m³/K mol) is the gas constant and T (K) is the ambient temperature.

Henry's law constants of chemicals of environmental interest range from about 1×10^3 kPa·m³/mol for highly volatile hydrophobic chemicals to roughly 1×10^{-7} kPa·m³/mol for hydrophilic chemicals with low vapor pressures. Chemicals that are not highly volatile may also have large Henry's law constants if their aqueous solubility is low. Several compilations of Henry's law constants have been published (Mackay and Shiu, 1981; Shiu and Mackay, 1986; Mackay et al., 1992). Measurements are difficult to make, and many of the values given in these compilations are estimated, particularly when H is reported to be below 2×10^{-5} kPa·m³/mol (Nielsen et al., 1994).

8.2 BACKGROUND

Imagine that a solute in an aqueous solution is in equilibrium with its vapor in the air above the solution. The condition for equilibrium is that $f_g = f_l$, where f_g (Pa) is the fugacity of the vapor and f_l (Pa) is the fugacity of the solute in aqueous solution. Now, $f_g = \phi_i P_i$, where ϕ_i is the vapor's fugacity coefficient and P_i (bar) is the partial pressure of the chemical vapor, and $f_l = \gamma_i x_i P_i^S$, where γ_i is the chemical's activity coefficient in solution, x_i is the mole fraction of solute in solution, and P_i^S (bar) is the saturation vapor pressure of the solute in solution. Equating the two fugacity expressions, $\phi_i P_i = \gamma_i x_i P_i^S$ or

$$\frac{p_i}{x_i} = \frac{\gamma_i P_i^S}{\phi_i} = H' \tag{8.3}$$

where H' is the Henry's law constant. Additional background material on the solubility of gases and vapors in water is presented in Chapter 7.

Group contribution methods of estimating the air–water partition coefficient are based on the assumption that it is an additive property. Individual molecular fragments contribute quantifiable amounts to the size of a molecule and to the polar interactions between solute and water molecules, and so they make specific contributions to log K_{AW}. To some degree, air–water partitioning is also a constitutive property. Intramolecular interactions of polar groups in the molecule minimize molecular size and reduce overall molecular polarity. In this case, the presence of topological features that enhance intramolecular interactions requires that group-contribution estimates of log K_{AW} be corrected. Additional background material on group contribution methods is presented in Chapter 2.

8.3 ESTIMATION METHODS

Several different approaches can be taken in estimating Henry's law constants. Equation 8.3 shows that H can be estimated from the ratio of the saturation vapor pressure to the activity coefficient of organic chemical in water. Arbuckle (1983) uses this approach after first estimating γ_i from molecular structure by the UNIFAC method. A structure–Henry's law correlation based on molecular connectivity was described by Nirmalakhandan and Speece (1988). Hine and Mookerjee (1975) proposed a typical group contribution method to estimate H. Meylan and Howard (1991, 1992) revised this method and implemented it in a computer program. Suzuki et al. (1992) developed a group contribution method that also incorporates molecular connectivity indices. Suzuki's method is implemented in a computer program (Drefahl and Reinhard, 1993). The various approaches are discussed below.

8.3.1 Quantitative Property-K_{AW} Relationships

Mackay and Shiu (1981) first proposed estimating the Henry's law constant of a slightly or moderately soluble chemical ($C_W^S < 1.5$ mol/L) from the ratio of its saturation vapor pressure to its aqueous solubility. The method is based upon a simplified version of Equation 8.3. First, $\phi_i \approx 1$ for most organic vapors. Carboxylic acids and other chemicals that associate in the gas phase are exceptions. Second, if γ_i^S is the solute's activity coefficient and x_i^S its concentration in a saturated water solution, $1/\gamma_i \approx 1/\gamma_i^S = x_i^S$ for organic chemicals that are not highly soluble in water, and

$$H' = \frac{P_i^S}{x_i^S} \tag{8.4}$$

where H' = Henry's law constant, Pa
P_i^S = saturation vapor pressure of chemical, Pa
x_i^S = aqueous solubility of chemical

If C_W^S (mol/L) is the solute's molar solubility, $C_W^S \approx x_W^S/V_{ml}$ for hydrophobic chemicals. Substituting $C_W^S \cdot V_{ml}$ for x_W^S in Equation 8.4 and rearranging, we obtain

$$\frac{P_i^S}{C_W^S} = V_{ml} \cdot H' = H \tag{8.5}$$

where H   = Henry's law constant, Pa m³/mol
P_i^S = saturation vapor pressure, Pa
C_W^S = aqueous solubility, mol/m³

An important point is that we use the vapor pressure of the pure chemical but the solubility of the chemical saturated with water, in our derivation of Equation 8.5. This is strictly valid only if water is not very soluble in the organic chemical.

The air–water partition coefficient is estimated as

$$K_{AW} = \frac{1}{RT} \frac{P^S}{C_W^S} \tag{8.6}$$

where K_{AW} = air–water partition coefficient
        P^{is} = saturation vapor pressure, Pa
        C_W^S = aqueous solubility, mol/m³
        R   = 8.3145 Pa m³/mol K
        T   = temperature, K

The values of both P^S and C_W^S must apply to the same temperature and physical state (liquid or solid) of the chemical.

The estimation method of Mackay and Shiu has been used for polychlorinated biphenyls (Shiu and Mackay, 1986), pesticides (Mackay and Shiu, 1981; Suntio et al., 1988), and various other hydrocarbons in water (Mackay and Shiu, 1981). The error of the estimate depends upon the accuracy of the input data. Mackay and Shiu (1981) demonstrate an estimation error of 10% or less for chemicals with easily measured and accurate vapor pressure and aqueous solubility values. Solubilities below 1×10^{-3} g/l and vapor pressures below 10 torr are difficult to measure and may be no better than order of magnitude estimates, however. The method should not be used to estimate Henry's law constants of chemicals with such low vapor pressures and solubilities.

Most toxic chemicals on federal and state agency priority lists are nonionic organic compounds, and environmental specialists are often required to estimate their volatilization rates from dilute aqueous solution. Equations 8.5 and 8.6 are applicable to this problem, and most values of Henry's law constant reported in the literature are computed using them.

Example

Estimate the values of H and K_{AW} for DDT, $C_{14}H_9Cl_5$, at 20°C using Equations 8.5 and 8.6. For DDT, the reported vapor pressure at 20°C is 2.53×10^{-5} Pa and the reported aqueous solubility is $2.7 \pm 2.1 \times 10^{-3}$ g/m³ (Mackay and Shiu, 1981). The molar mass is 352.46 g/mol. $C_W^S = (2.7 \times 10^{-3}$ g/m³)/(352.46 g/mol) = 7.7×10^{-6} mol/m³.

1. Estimate H using Equation 8.5.

$$H = P^S / C_W^S$$

$$= \left(2.53 \times 10^{-5}\ \text{Pa}\right) / 7.7 \times 10^{-6}\ \text{mol}/\text{m}^3$$

$$= 3.3\ \text{Pa} \cdot \text{m}^3 / \text{mol}$$

2. Estimate K_{AW} using Equation 8.6.

$$K_{AW} = (1/RT)\left(P^S / S_W\right) = H/RT$$

$$= \left(3.3\ \text{Pa} \cdot \text{m}^3 / \text{mol}\right) / \left(8.3145\ \text{Pa}\ \text{m}^3 / \text{mol}\ \text{K}\right)(293\ \text{K})$$

$$= 1.4 \times 10^{-3}$$

Reported values of K_{AW} range from 5.2×10^{-4} to 2.4×10^{-3}. Values reported most recently range from 5.2×10^{-4} to 9.5×10^{-4} (Fendinger et al., 1989). The discrepancy between estimated and measured values of K_{AW} is due, in large measure, to error in the solubility value of DDT used.

8.3.2 Quantitative Structure–K_{AW} Relationships

A relationship between K_{AW} and molecular connectivity is reported by Nirmalakhandan and Speece (1988). The model is

$$\log K_{AW} = 1.29 + 1.005\phi - 0.468\,^1\chi^v - 1.258I \qquad (8.7)$$

n = 180, s = 0.262, r = 0.99

where K_{AW} = air–water partition coefficient
  ϕ = chemical polarizability
     $= \Sigma_i n_i G_i$
  n_i = number of groups i in molecular structure
  G_i = contribution of group i to polarizability
  $^1\chi^v$ = the first-order valence connectivity index
  I = indicator variable
  n = number of chemicals in training set
  s = standard deviation
  r = correlation coefficient

The group contributions to molecular polarizability, G_i, are listed in Table 8.1. The indicator variable accounts for the effect of hydrogen bonding. It is assigned a value of 1 for all chemicals containing an electronegative element (oxygen, nitrogen, halogen, etc.) attached directly to >CH–, –CH$_2$–, or –CH$_3$. It is also assigned a value of 1 for acetylinic and aromatic chemicals in which the hydrogen atoms are partially substituted. Otherwise, $I = 0$.

Table 8.1 Optimized Group Contributions to Polarizability, G_i

Atom/Bond	G_i	Atom/Bond	G_i
Carbon	0.577	Iodine	0.407
Hydrogen[a]	−0.120	Fluorine	−0.570
Oxygen	−0.825	Cycle	−0.952
Hydroxyl	−3.701	Double bond	−0.859
Chlorine	−0.187	Triple bond	−0.109
Bromine	−0.222		

[a] Attached to carbon atom only.

Reprinted from Nirmalakhandan, N.N. and Speece, R.E., 1988. *Environ. Sci. Technol.*, **22**, 1349–1357. With the permission of the American Chemical Society.

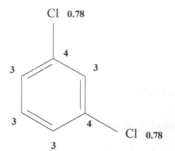

Figure 8.1 The hydrogen-suppressed graph of 1,3-dichlorobenzene. δ^v values are given next to each atom.

The model applies to halocarbons, hydrocarbons, alcohols, and acid esters. It does not apply to amines, ethers, aldehydes, and ketones. The log K_{AW} values of the chemicals in the training set range from -5.21 to $+2.12$. The model was validated with a test set of 20 chemicals including polyhalogenated alcohols, aromatics, and PAH.

Example

Estimate the value of K_{AW} for 1,3-dichlorobenzene, $C_6H_4Cl_2$, using Equation 8.7.

1. Calculate the first-order connectivity index for dichlorobenzene. The hydrogen-suppressed structure is shown along with valence delta values in Figure 8.1.

$$^1\chi^v = \Sigma_{bonds} \left(\delta_i^v \delta_j^v \right)^{-0.5}$$

$$= (2)(3 \cdot 3)^{-0.5} + (4)(3 \cdot 4)^{-0.5} + (2)(4 \cdot 0.78)^{-0.5}$$

$$= 2.469$$

2. Estimate ϕ. The G_i values are given in Table 8.1.

$$\phi = \Sigma_i n_i G_i$$

$$= 6 \cdot G_c + 4 \cdot G_H + 2 \cdot G_{Cl}$$

$$= (6)(0.577) + (4)(-0.120) + (2)(-0.187)$$

$$= 2.608$$

3. Estimate log K_{AW} using Equation 8.7. The chemical contains chlorine, and $I = 1$.

$$\log K_{AW} = 1.29 + 1.005 \ \phi - 0.468 \ ^1\chi^v - 1.258 \ I$$

$$= 1.29 + (1.005)(2.608) - (0.468)(2.469) - 1.258$$

$$= 1.498$$

The reported value of log K_{AW} of 1.3-dichlorobenzene is -0.72 (Nirmalakhandan and Speece, 1988).

Much recent work has been devoted to finding structure–K_{AW} relationships for polychlorobiphenyls (Hawker, 1989; Brunner et al., 1990; Dunnivant et al., 1992). Sabjlic and Gusten (1989) investigated the correlation of the K_{AW} values of polychlorinated biphenyl compounds (PCBs) with simple and valence connectivity indices from zero to sixth order. Unexpectedly, K_{AW} was found to be only weakly correlated with connectivity indices that code for molecular size. They report a satisfactory correlation between K_{AW} and two fourth-order connectivity indices, ${}^4\chi_p$ and ${}^4\chi_{pc}$, which encode information such as the degree of substitution and the proximity of substituents. For instance, the ${}^4\chi_{pc}$ index is well correlated with both the number of chlorine atoms and the number of ortho-chlorine atoms in a PCB molecule (Brunner et al., 1990). Sabjlic and Gusten's model is

$$K_{AW} = -6.56 + 4.01 {}^4\chi - 2.93 {}^4\chi_{pc} \qquad (8.8)$$

n = 18, s = 0.311, r = 0.982, F(2,5) = 199

where K_{AW} = air–water partition coefficient
${}^4\chi$ = fourth-order molecular connectivity
${}^4\chi_{PC}$ = fourth-order path/cluster molecular connectivity

The model applies to PCBs with two to six chlorine substituents having K_{AW} values in the range from 0 to 8.

Brunner et al. (1990) describe a similar correlation that was developed with a training set of 58 polychlorinated biphenyls:

$$\log K_{AW} = -1.15 - 0.42 {}^4\chi_{pc} \qquad (8.9)$$

n = 58, s = 0.173, r = 0.929

The K_{AW} values of the chemicals in the training set range from 0.38 to 10.

Multivariate structure–K_{AW} correlations involving many molecular descriptors selected with neural networks have been described recently (Russell et al., 1992). Estimation methods based on such correlations appear to be highly accurate (Katritzky et al., 1996). As powerful desktop computers are now widely available, these methods are growing increasingly popular.

8.3.3 Group Contribution Methods

Hine and Mookerjee (1975) describe a bond contribution method of estimating K_{AW} values. For simple monofunctional chemicals, the estimates agree with measured values within a factor of 3. The method is less accurate when multiple polar groups are present in a molecule, presumably because intramolecular interactions between polar groups are not accounted for. Hine and Mookerjee's method was revised by Meylan and Howard (1991, 1992), and accuracy was significantly enhanced. Meylan and Howard's model is

$$\log K_{AW} = \sum_i a_i g_i + \sum_j b_j F_j \qquad (8.10)$$

n = 345, s = 0.45, r^2 = 0.94

where K_{AW} = air–water partition coefficient at 25°C
a_i = number of bonds of type i in molecular structure
g_i = contribution of bond i to log K_{AW}
b_j = number of features j in molecular structure
F_j = correction factor for feature j
n   = number of chemicals in training set
s   = standard deviation
r   = correlation coefficient

The bond contribution values, g_i, are given in Table 8.2. The correction factors, F_j, are given in Table 8.3. The training set contains a wide variety of chemical classes including aliphatic and aromatic hydro- and halocarbons, alcohols, aldehydes, ketones, ethers, epoxides, aliphatic acids, esters, phenols, amines, nitriles, pyridines, anilines, PAH, and pesticides.

The K_{AW} values of the chemicals in the training set range from −2.5 to +8. The values are either measured or estimated with Henry's law. Meylan and Howard's model was validated with a set of 74 diverse and structurally complex chemicals not included in the training set. When applied to the test set, the model produced a standard deviation of 0.457 and a correlation coefficient of 0.96.

Example

Estimate K_{AW} of 1-propanol using the method of Meylan and Howard.

1. Identify all bond types and features in the molecular structure. The contributions are given in Tables 8.2 and 8.3. The molecule has seven C–H, two C–C, one C–O, and one O–H bond. It is a linear alcohol.
2. Estimate K_{AW} using Equation 8.10.

$$\log K_{AW} = \Sigma_i a_i g_i + \Sigma_j b_i F_i$$

$$= (7)(-0.1197) + (2)(0.2326) + (1)(1.0855) + (1)(3.2301) + (1)(-0.20)$$

$$= 3.5112$$

The measured value is 3.55 (Meylan and Howard, 1991). The error is 0.04 log units.

Table 8.2   Bond Contribution Values to Log K_{AW} at 25°C

Bond[a]	g_i	Bond[a]	g_i	Bond[a]	g_i
C–H	−0.1197	C_d–Cl	0.0426	C_a–CN	1.8606
C–C	0.1163	C_d–CN	2.5514	C_a–CO	1.2387
C–C_a	0.1619	C_d–O	0.2051	C_a–Br	0.2454
C–C_d	0.0635	C_d–F	−0.3824	C_a–NO_2	2.2496
C–C_t	0.5375	C_t–H	0.0040	CO–H	1.2101
C–CO	1.7057	$C_t \equiv C_t$	0.0000[c]	CO–O	0.0714
C–N	1.3001	C_a–H	−0.1543	CO–N	2.4261
C–O	1.0855	C_a–C_a	0.2638[e]	CO–CO	2.4000
C–S	1.1056	C_a–C_a	0.1490[f]	O–H	3.2318
C–Cl	0.3335	C_a–Cl	−0.0241	O–P	0.3930
C–Br	0.8187	C_a–OH	0.5967[b]	O–O	−0.4036
C–F	−0.4184	C_a–O	0.3473[b]	O=P	1.6334
C–I	1.0074	C_a–N_a	1.6282	N–H	1.2835
C–NO_2	3.1231	C_a–S_a	0.3739	N–N	1.0956[d]
C–CN	3.2624	C_a–O_a	0.2419	N=O	1.0956[d]
C–P	0.7786	C_a–S	0.6345	N=N	0.1374
C=S	−0.0460	C_a–N	0.7304	S–H	0.2247
C_d–H	−0.1005	C_a–I	0.4806	S–S	−0.1891
C_d=C_d	0.0000[c]	C_a–F	−0.2214	S–P	0.6334
C_d–C_d	0.0997	C_a–C_d	0.4391	S=P	−1.0317
C_d–CO	1.9260				

[a] C: single-bonded aliphatic carbon; C_d: olefinic carbon; C_t: triple-bonded carbon; C_a: aromatic carbon; N_a: aromatic nitrogen; S_a: aromatic sulfur; O_a: aromatic oxygen; CO: carbonyl (C=O); CN: cyano (C≡N). Note: the carbonyl, cyano, and nitro groups are treated as single atoms.

[b] Two separate types of aromatic carbon-to-oxygen bonds have been derived: (a) the oxygen is part of an −OH group, and (b) the oxygen is not connected to hydrogen.

[c] The C=C and C≡C bonds are assigned a value of zero by definition.

[d] Value specific for nitrosamines.

[e] Intraring aromatic carbon to aromatic carbon.

[f] External aromatic carbon to aromatic carbon (e.g., biphenyl).

8.4 SENSITIVITY TO ENVIRONMENTAL PARAMETERS AND METHOD ERROR

The estimation errors of the models described in this chapter are comparable to solubility measurement errors. Several different methods of measuring Henry's law constants are in use (Fendinger et al., 1989; Brunner et al., 1990; Tse et al., 1992; Nielsen et al., 1994). The variance between K_{AW} values measured by the different methods ranges from 2 to 70%. The measurement is most difficult to make accurately for hydrophobic chemicals of low volatility.

Henry's law constants are sensitive to temperature, the presence of electrolytes, the presence of cosolvents, and the presence of sorbents. The temperature dependence is the same as that for the saturation vapor pressure, doubling for every 10°C increase in temperature. The effect is discussed in detail by Valsaraj (1995).

Table 8.3   Correction Factors for Log K_{AW} at 25°C

Feature	F_j
Linear or branched alkane[a]	−0.75
Cyclic alkane[a]	−0.28
Monoolefin[a,b]	−0.20
Cyclic monoolefin[a,b]	+0.25
Linear or branched aliphatic alcohol[a]	−0.20
Adjacent aliphatic ether groups (–C–O–C–O–C–)	−0.70
Cyclic monoether	+0.90
Epoxide	+0.50
Each additional aliphatic alcohol function (–OH) above one	−3.00
Each additional aromatic nitrogen within a single ring above one	−2.50
A fluoroalkane with only one fluorine	+0.95
A chloroalkane with only one chlorine	+0.50
A totally chlorinated chloroalkane	−1.35
A totally fluorinated fluoroalkane	−0.60
A totally halogenated halofluoroalkane	−0.90

[a] Can have no substituents except alkyl groups.
[b] Can have only one olefinic double bond.

The presence of electrolytes leads to the "salting out" effect discussed in Chapter 7. Salting out leads to an increase in Henry's law constant of about 20 to 80% in seawater (Schwarzenbach et al., 1993).

Gas-liquid partitioning of nonpolar organics in moderately concentrated solutions and of polar organics under most conditions does not conform well to Henry's law. As we saw in Section 8.2, the Henry's law constant is proportional to the activity coefficient and the saturation vapor pressure of the chemical. For dilute solutions of nonelectrolytes, the activity coefficient is not very different from the infinite-dilution activity coefficient, and the Henry's law constant changes by only a few percent from infinite dilution to saturation.

The value of the Henry's law constant can be estimated from the value of γ_W. The method is based upon a simplified version of Equation 8.3. First, $\phi_i \approx 1$ for most organic vapors. Carboxylic acids and other chemicals that associate in the gas phase are exceptions. Equation 8.3 becomes $P_i/x_i = \gamma_i P_i^S$ or

$$H' = \gamma_i P_i^S \tag{8.11}$$

where H' = Henry's law constant, Pa
       P_i^S = saturation vapor pressure of chemical, Pa
       γ_i = activity coefficient of chemical in water

Since the solute concentration is low and the solute is a hydrophobic chemical, $\gamma_i \approx \gamma_i^\infty$, where γ_i^∞ is the solute's activity coefficient at infinite dilution. Substituting the infinite-dilution activity coefficient in Equation 8.11, we obtain:

$$H \approx \gamma_W^\infty V_{ml} P^S \tag{8.12}$$

where   H   = Henry's law constant, Pa·m³/mol
  γ_W^∞ = activity coefficient at infinite dilution
  V_{ml} = molar volume of water, m³/mol
  P^S = saturation vapor pressure, Pa

The infinite-dilution activity coefficient is estimated using the UNIFAC method (Arbuckle, 1983) described in Chapter 7. This method is useful for estimating K_{AW} values of solute alone (Tse et al., 1992; Nielsen et al., 1994) and in the presence of cosolvents.

The presence of a cosolvent increases the solubility of a hydrophobic solute and lowers the value of the Henry's law constant. Munz and Roberts (1987) show that cosolvents significantly reduce the value of K_{AW} at cosolvent levels above 10 g/L. This would be a very high contaminant concentration in most environmental systems. Since the UNIFAC method estimates activity coefficients of mixtures, Arbuckle's method (1983) can be used to estimate the effect of cosolvents on air–water partitioning.

Hydrophobic chemicals accumulate at the air-water interface. At the high area/volume ratios observed in turbulent systems and in atmospheric aerosols, the value of K_{AW} may be much higher than that estimated using the methods described in this chapter. For instance, Glotfelty et al. (1987) report that pesticides are enriched in fog droplets at concentrations several thousand times that expected from K_{AW} values of the chemicals. The effect has been discussed by Valsaraj (1995) and by Hoff et al. (1993). It is the basis of the "solvent sublation" separation process (Valsaraj et al., 1986).

Surfactant films greatly enhance the solubility of hydrophobic organic chemicals in fog droplets. The degree of enrichment depends upon droplet size (Lo and Lee, 1996).

The value of K_{AW} is not very sensitive to changes in pressure. K_{AW} can be considered a constant within the normal range of atmospheric pressure.

REFERENCES

Arbuckle, W.E., 1983. Estimating activity coefficients for use in calculating environmental parameters. *Environ. Sci. Technol.*, **17**, 537–542.

Brunner, S., Hornung, E., Santl, H., Wolff, E., Piringer, O.G., Altschuh, J., and Bruggemann, R., 1990. Henry's law constants for polychlorinated biphenyls: experimental determination and structure–property relationships. *Environ. Sci. Technol.*, **24**, 1751–1754.

Dean, J.A., 1996. *Lange's Handbook of Chemistry*. 14th Edition. McGraw-Hill, New York.

Drefahl, A. and Reinhard, M., 1993. Similarity-based search and evaluation of environmentally relevant properties for organic compounds in combination with the group contribution approach. *J. Chem. Inf. Comput. Sci.*, **33**, 886–895.

Dunnivant, F.M., Eizerman, A.W., Jurs, P.C., and Hasan, M.N., 1992. Quantitative structure–property relationships for aqueous solubilities and Henry's law constants of polychlorobiphenyls. *Environ. Sci. Technol.*, **26**, 1567–1573.

Fendinger, N.J., Glotfelty, D.E., and Freeman, H.P., 1989. Comparison of two experimental techniques for determining air/water Henry's law constants. *Environ. Sci. Technol.*, **23**, 1528–1531.

Glotfelty, D.E., Seiber, J.N., and Liljedahl, L.A., 1987. Pesticides in fog. *Nature,* **325**, 602–608.

Grain, C.F., 1990. Vapor Pressure. Chapter 14. In *Handbook of Chemical Property Estimation Methods. Environmental Behavior of Organic Compounds.* Lyman, W.J., Reehl, W.F., and Rosenblatt, D.H., Eds. American Chemical Society, Washington, D.C.

Hawker, D.W., 1989. Vapor pressures and Henry's law constants of polychlorobiphenyls. *Environ. Sci. Technol.,* **23**, 1250–1253.

Hine, J. and Mookerjee, P.K., 1975. The intrinsic hydrophilic character of organic compounds. Correlations in terms of structural contributions. *J. Org. Chem.,* **40**, 292–298.

Hoff, J.T., Mackay, D., Gillham, R., and Shiu, W.Y., 1993. Partitioning of organic chemicals at the air-water interface in environmental systems. *Environ. Sci. Technol.,* **27**, 2174–2180.

Katritzky, A.R., Mu, L., and Karelson, M., 1996. A QSPR study of the solubility of gases and vapors in water. *J. Chem. Inf. Sci.,* **36**, 1162–1168.

Lide, D.R., 1994. *CRC Handbook of Chemistry and Physics.* 74th Edition. CRC Press, Boca Raton, FL.

Mackay, D. and Shiu, W.Y., 1981. A critical review of Henry's law constants for chemicals of environmental interest. *J. Phys. Chem. Ref. Data,* **10**, 1175–1199.

Mackay, D., Shiu, W.Y., and Ma, K.C., 1992. *Illustrated Handbook of Physical-Chemical Properties and Environmental Fate for Organic Chemicals. Volumes 1–4.* Lewis Publishers, Boca Raton, FL.

Meylan, W.M. and Howard, P.H., 1991. Bond contribution method for estimating Henry's law constants. *Environ. Toxicol. Chem.,* **10**, 1283–1293.

Meylan, W.M. and Howard, P.H., 1992. *Henry's Law Constant Program.* Lewis Publishers, Boca Raton, FL.

Meylan, W.M., Howard, P.H., and Boethling, R.S., 1992. Molecular topology/fragment contribution method for predicting soil sorption coefficients. *Env. Sci. Technol.,* **26**, 1560–1567.

Munz, C. and Roberts, P.V., 1987. Air-water phase equilibria of volatile organic solutes. *J. Am. Water Works Assoc.,* **79**, 62–70.

Nielsen, F., Olsen, E., and Fredenslund, A., 1994. Henry's law constants and infinite dilution activity coefficients for volatile organic compounds in water by a validated batch air stripping method. *Environ. Sci. Technol.,* **28**, 2133–2138.

Nirmalakhandan, N.N. and Speece, R.E., 1988. QSAR model for predicting Henry's constant. *Environ. Sci. Technol.,* **22**, 1349–1357.

Russell, C.J., Dixon, S.L., and Jurs, P.C., 1992. Computer-assisted study of the relationship between molecular structure and Henry's law constant. *Anal. Chem.,* **64**, 1350–1355.

Sabjlic, A. and Gusten, H., 1989. Predicting Henry's law constants for polychlorinated biphenyls. *Chemosphere,* **19**, 1503–1511.

Schwarzenbach, R.P., Gschwend, P.M., and Imboden, D.M., 1993. *Environmental Organic Chemistry.* John Wiley & Sons, New York.

Shiu, W.Y. and Mackay, D., 1986. A critical review of aqueous solubilities, vapor pressures, Henry's law constants, and octanol-water partition coefficients of the polychlorinated biphenyls. *J. Phys. Chem. Ref. Data,* **15**, 911–929.

Suzuki, T., Ohtaguchi, K., and Koide, K., 1992. Application of principal components analysis to calculate Henry's law constants from molecular structure. *Computers Chem.,* **16**, 41–52.

Suntio, L.R., Shiu, W.Y., and Mackay, D., 1988. Critical review of Henry's law constants for pesticides. *Rev. Environ. Contam. Toxicol.,* **103**, 1–59.

Tse, G., Orbey, H., and Sandler, S.I., 1992. Infinite dilution activity coefficients and Henry's law coefficients of some priority water pollutants determined by a relative gas chromatographic method. *Environ. Sci. Technol.,* **26**, 2017–2022.

Valsaraj, K.T., 1995. *Elements of Environmental Engineering.* Lewis Publishers, Boca Raton, FL.

Valsaraj, K.T., Porter, J.L., Lilienfeldt, E.K., and Springer, C., 1986. Solvent sublation for the removal of hydrophobic chlorinated compounds from aqueous solutions. *Water Research,* **20**, 1161–1175.

CHAPTER 9

Octanol–Water Partition Coefficient

9.1 INTRODUCTION

The **octanol–water partition coefficient** is *the ratio of a chemical's molar concentration in the 1-octanol phase to its molar concentration in the water phase of an octanol–water system at equilibrium*:

$$K_{OW} = \frac{C_{Oi}}{C_{Wi}} \qquad (9.1)$$

where C_{Oi} (mol/L) is the solute concentration in the octanol phase, and C_{Wi} (mol/L) is the solute concentration in the water phase. The value of K_{OW} is measured at "room temperature" ($25 \pm 5°C$) and at a total solute concentration of less than 0.01 mol/L. Under these conditions, there is only a slight dependence of the K_{OW} value on temperature and solute concentration. Since the octanol phase is saturated with water ($2.3\ M\ H_2O$ in octanol) and the water phase is saturated with octanol ($4.5 \times 10^{-3}\ M$ octanol in water), K_{OW} is not equal to the ratio of the solubility of a chemical in pure octanol to its solubility in pure water.

The value of $\log_{10} K_{OW}$ (also written log P, log P_{oct}, log P_{ow}) is usually reported. Measured values of log K_{OW} range from about -4 for hydrophilic chemicals to $+8.5$ for hydrophobic chemicals. It is relatively easy to measure log K_{OW} values below 5 accurately, and large compilations of such data are available in the literature (Hansch and Leo, 1979) and in computer data bases (Leo, 1993, Sangster, 1993). Accurate methods of measuring octanol–water partition coefficients of hydrophobic chemicals ($5 \geq \log K_{OW} \geq 8$) were developed only after 1982. Discrepancies between published log K_{OW} values of a particular chemical can be as large as 0.4 to 3.5 log units (Sabljic, 1987), and comparatively little reliable data are available. The most accurate data compilation available contains measured K_{OW} values for 10,000 chemicals (Leo and Weininger, 1989). A tabulation of much of the data is available in print (Sangster, 1989; Hansch et al., 1995).

Due to the work of Hansch, Leo, and co-workers, the correlation between K_{OW} and the partitioning of chemicals in environmental systems is widely studied (Leo et al., 1971). For this reason, methods of estimating the soil sorption coefficient, bioconcentration factor, and biodegradation rate constant of a chemical often begin with the value of K_{OW}. In Chapter 7, we saw that K_{OW} values are also a good starting point for estimating aqueous solubility.

9.2 BACKGROUND

Imagine that chemical i is partitioned between 1-octanol and water. At equilibrium, the chemical potential is the same in both phases: $u_{Oi} = u_{Wi}$, where u_{Oi} is the chemical potential in the octanol phase and u_{Wi} is the chemical potential in the water phase. Now, $u_{Oi} = u_i^0 + RT\ln a_{Oi}$ and $u_{Wi} = u_i^0 + Rt\ln a_{Wi}$, where u_i^0 is the chemical potential of the pure liquid chemical at temperature T, a_{Oi} is the chemical's activity in octanol, and a_{Wi} is the chemical's activity in water. Since $u_{Oi} = u_{Wi}$, $u_i^0 + RT\ln a_{Oi} = u_i^0 + Rt\ln a_{Wi}$, and $a_{Oi} = a_{Wi}$ at equilibrium. Since activity is the product of concentration and activity coefficient, $\gamma_{Oi}x_{Oi} = \gamma_{Wi}x_{Wi}$ at equilibrium, where γ_{Oi} is the activity coefficient of chemical i in octanol, γ_{Wi} is the activity coefficient of chemical i in water, x_{Oi} is the mole fraction of chemical i in octanol, and x_{Wi} is the mole fraction of chemical i in water. In dilute solution, $x_{Oi} \approx n_{Oi}/n_O = C_{Oi} \cdot V_{Om}$, where n_{Oi}/n_O is the mole ratio of chemical to octanol in the octanol phase and V_{Om} is the partial molar volume of octanol. Similarly for the water phase, $x_{Wi} \approx C_{Wi} \cdot V_{Wm}$, where V_{Wm} is the partial molar volume of the water phase. Combining these relationships, we find that

$$\gamma_{Oi} \cdot C_{Oi} \cdot V_{Om} = \gamma_{Wi} \cdot C_{Wi} \cdot V_{Wm} \qquad (9.2)$$

The octanol–water partition coefficient, K_{OW}, is defined in Equation 9.1 as the equilibrium concentration ratio of a chemical i in the octanol phase to that in the water phase. Rearranging Equation 9.2, we obtain an expression for K_{OW} in terms of the solute activity coefficients and the solvent partial molar volumes:

$$K_{OW} = \frac{\gamma_{Wi}}{\gamma_{Oi}} \frac{V_{Wm}}{V_{Om}} \qquad (9.3)$$

Since K_{OW} values are always reported for dilute solutions at 25°C, the partial molar volumes of the octanol and water phases are roughly the same for every measurement. A 2.3 M solution of H_2O in octanol has $V_{Om} = 0.12$ L/mol, and a 4.5×10^{-3} M solution of octanol in water has $V_{Wm} = 0.018$ L/mol. Assuming that the presence of low levels of solute does not significantly alter the molar volumes, $V_{Wm}/V_{Om} \approx 0.15$ for all K_{OW} measurements and $K_{OW} \approx 0.15 \gamma_{Wi}/\gamma_{Oi}$. We conclude that the value of K_{OW} depends primarily on the ratio of the activity coefficients γ_{Oi} and γ_{Wi}.

Chemicals with similar chemical nature readily mix together: "like dissolves like." Most organic chemicals form nearly ideal solutions with 1-octanol and γ_{Oi}

values lie in the range of 1 to 10 (Schwarzenbach et al., 1993). On the other hand, aqueous solutions of organic chemicals are not ideal, and γ_{Wi} values of organic chemicals span the range from 10^{-1} to 10^{11}. So, $K_{OW} = 0.15\ \gamma_{Wi}/\gamma_{Oi} \approx A \cdot \gamma_{Wi}$, where A is a constant, and the difference in K_{OW} value from chemical to chemical is almost entirely due to variation in γ_{Wi} with chemical structure. In Section 7.2, we saw that γ_{Wi} is related to the aqueous solubility of a chemical. Therefore, K_{OW} is determined almost entirely by a chemical's solubility in water.

Another way of thinking about this is to view the octanol–water partition coefficient as being related to the ratio of the solubility of chemical in water-saturated octanol to the solubility of chemical in octanol-saturated water. Mackay (1991) argues that the solubility of organic chemicals in octanol is fairly constant, ranging between 0.2 and 2 M, while the aqueous solubility values range over many orders of magnitude. Therefore, the value of K_{OW} depends primarily upon the chemical's solubility in water.

The value of K_{OW} is related to molecular structure in a complex way. It depends on the size of the solute molecule, on the strength of polar interactions between the solute and water, and on the ability of the solute to form hydrogen bonds with water (Taft et al., 1985; Hawker, 1989). K_{OW} must depend on the size of a solute molecule as this relates to the size of the solvent cage or cavity needed to hold the molecule. This can be understood in terms of the maximization of entropy in the process of partitioning solute between the two solvents. Mixing is spontaneous, and the overall entropy must increase to a maximum in the process. The increase is moderated by a decrease in entropy of the water in forming a hydration shell of solvent molecules around each solute molecule in aqueous solution (Israelchvili, 1992). The larger the solute molecule, the larger the hydration shell, and the less entropically favorable the process of mixing with water, in general. Therefore, aqueous solubility decreases (McAuliffe, 1966; Amidon and Anik, 1976) and K_{OW} increases with molecular size (Leo et al., 1976). Molecular features such as unsaturation, branching, and molecular flexibility help to minimize molecular size, increasing solubility and decreasing K_{OW}.

Water is a polar solvent, and interactions with the polar portion of a solute molecule serve to increase the entropy of forming hydration shells around the solute. We expect that maximizing the ratio of polar to nonpolar surface area of a solute molecule maximizes its aqueous solubility and minimizes its K_{OW}. Hydrogen bonding between solute molecules and water also enhances solubility. The hydrogen bond is a moderately attractive force occurring between hydrogen atoms on one small, highly electronegative atom such as N, O, or F and the lone pair electrons on another small, highly electronegative atom such as N, O, or F. Either the solute or water can donate the hydrogen atom.

These qualitative principles are formalized in the quantitative linear solvation energy relationships introduced by Kamlet and co-workers (Kamlet et al., 1988a; Yalkowsky et al., 1988). The values of K_{OW} and other solubility-related properties are estimated with group contributions to molecular volume, solute polarity/polarizability, hydrogen-bond donor acidity, and hydrogen-bond acceptor basicity. The rules governing group contributions to solute properties are given in two papers by Kamlet et al. (1988a, 1988b).

Group contribution methods of estimating K_{OW} values are based on the assumption that aqueous solubility is an additive property. (See Section 2.3.2.) Individual molecular fragments contribute quantifiable amounts to the size of a molecule and to the polar interactions between solute and solvent molecules, and they make quantifiable contributions to the value of log K_{OW}. To a large degree, octanol–water partitioning is also a constitutive property. Intramolecular interactions of polar groups in the molecule minimize molecular size and reduce overall molecular polarity. In this case, the presence of topological features that enhance intramolecular interactions requires a correction to the group-contribution estimates of log K_{OW} values.

9.3 ESTIMATION METHODS

Methods of estimating K_{OW} values developed before 1981 are described by Lyman (1990). Many new methods have since been proposed, and older methods have been revised. New methods have been proposed by Broto et al. (1984), Viswanadhan et al. (1989), Suzuki and Kudo (1990), and Klopman et al. (1994). All are reasonably accurate and broadly applicable. The first three methods have been implemented in computer programs SmilogP (Convard et al., 1994), DESOC (Kolenbrander et al., 1995), and CHEMCALC2 (Suzuki, 1991), respectively. The method of Klopman et al. is applicable to a wide variety of chemicals, is as accurate as the others, and is easy to apply by hand. It is described in Section 9.3.1.

The widely used group contribution method of Hansch and Leo has been substantially revised and expanded. It has also grown difficult to apply correctly without the use of a computer (Leo, 1993). The method has been implemented in the computer program CLOGP3 (Leo and Weininger, 1989). Meylan and Howard (1995) also developed a group-contribution method of estimating log K_{OW} values which has been implemented in the computer program LOGKOW. Both are described in Section 9.3.

Quantitative aqueous solubility–K_{OW} relationships are reviewed by Lyman (1990), Isnard and Lambert (1989), and Yalkowsky and Banerjee (1992). Work on methods of estimating K_{OW} values using activity coefficients calculated by the UNIFAC method are described by Arbuckle (1983) and Chen et al. (1993). This is discussed in Chapter 7.

Useful quantitative structure–K_{OW} relationships involving structural descriptors have been reported. Correlations between molecular connectivity and K_{OW} are described by Kier and Hall (1976), Doucette and Andren (1988), and Basak et al. (1990). They are described in Section 9.3.4.

9.3.1 Klopman's Method

The octanol–water partition function is a complex parameter, and most methods of estimating its value are complex, as well. A group contribution method which is comparatively simple and straightforward was derived using the Computer Automated Structure Evaluation (CASE) approach (Klopman and Wang, 1991; Klopman et al., 1994). Only 64 molecular fragments (groups 1 to 64 in Table 9.1) and 25 correction factors (groups 69 to 93 listed in Table 9.1) were identified by stepwise multiple regression analysis as contributing significantly to the value of log K_{OW} of a set of 1663 chemicals of various types. Also, five factors (groups 94 to 98 in Table 9.1) were added to correct large estimation errors observed due to molecular folding that enhances intramolecular interactions between polar groups and aromatic rings (groups 94 to 96), alkanes (group 97), and unsaturated hydrocarbons (group 98). Klopman's model is

$$\log K_{OW} = a + \sum_i b_i B_i + \sum_j c_j C_j \qquad (9.4)$$

n = 1663, SD = 0.3817, r^2 = 0.928, F(94, 1568) = 217.77

where K_{OW} = octanol–water partition coefficient
     a   = –0.703
     b_i = number of fragments i in molecular structure
     B_i = contribution of fragment i to K_{OW}
     c_j = number of correction fragments j
     C_j = contribution of fragment j to K_{OW}
     n   = number of chemicals in the training set
     SD = standard deviation, log units
     r   = correlation coefficient
     F   = F ratio

Values of the fragment contributions, B_i and C_j, are given in Table 9.1. The training set consists of aliphatic and aromatic hydrocarbons, alcohols, ethers, phenols, ketones, aldehydes, carboxylic acids, esters, amines, nitriles, amides, anilides, sulfur- and nitro-containing hydrocarbons, amino acids, halocarbons, and multifunctional chemicals. The measured values of log K_{OW} of chemicals in the set span the range from –3 to + 6.5. The performance of Klopman's method by chemical class is shown in Table 9.2. Comparisons with other methods are discussed in Section 9.4 and given in Table 9.3.

Example

Estimate the K_{OW} of 2,6-di-*sec*-butylphenol, $(C_4H_9)_2C_6H_3OH$, using the method of Klopman et al.

1. Identify the significant fragments found in the molecular structure.
   The structure contains four CH_3, two CH_2, two –CH<, three =C*H–, one >C*<, and one phenolic –OH group.
2. Estimate log K_{OW} using Equation 9.4. The fragment constants are given in Table 9.1.

Table 9.1   Group Contribution Values and Correction Factors to Log K_{OW} of the Basic Group Set[a]

Group	n[b]	Freq[c]	Contribution	Remarks
		Group Contributions		
1. $-CH_3$	855	1391	0.661	
2. $-CH_2-$	610	1180	0.415	
3. $-CH<$	138	169	0.104	
4. $>C<$	99	107	-0.107	
5. $=CH_2$	31	37	0.553	
6. $=CH-$	70	108	0.315	Not in $-CHO$
7. $=C<$	39	39	0.470	Not in $-CO-$, $-CS-$
8. $=C=$	21	22	1.748	
9. $-C\equiv CH$	4	4	0.262	Including $HC\equiv CH$
10. $-C\equiv$	2	4	0.131	Not in $-C\equiv N$, $-C\equiv CH$
11. $-C_rH_2-$	148	420	0.360	
12. $-C_rH<$	107	287	0.104	
13. $>C_r<$	27	31	0.064	
14. $=C_rH-$	1331	5946	0.380	
15. $=C_r<$	1322	3065	0.129	Not in $-C_rO-$
16. $-F$	36	45	0.468	Connected to C_a
17. $-F$	81	212	0.487	Not connected to C_a
18. $-Cl$	147	214	0.905	Connected to C_a
19. $-Cl$	47	108	0.713	Not connected to C_a
20. $-Br$	58	79	1.088	Connected to C_a
21. $-Br$	23	25	1.021	Not connected to C_a
22. $-I$	25	26	1.442	Connected to C_a
23. $-I$	6	6	1.209	Not connected to C_a
24. $-OH$	118	120	-0.681	Primary alcohol
25. $-OH$	65	78	-0.575	Secondary alcohol
26. $-OH$	8	9	-0.415	Tertiary alcohol
27. $-OH$	185	196	0.135	Phenol
28. $-OH$	16	16	-0.190	The others
29. $-O_r-$	67	73	0.103	Not in ester
30. $-O-$	235	270	-0.402	Not in ester
31. $-CHO$	15	15	0.009	Aldehyde
32. $-COOH$	72	73	0.467	Connected to C_a
33. $-COOH$	141	148	-0.263	Not connected to C_a
34. $-COO-$	195	199	-0.414	
35. $-C_rOO-$	7	8	-0.874	
36. $-CONH_2$	86	92	-0.795	
37. $-CONH-$	239	325	-1.006	
38. $-CON<$	94	109	-1.283	
39. $-CON=$	17	18	-1.661	
40. $-CO-$	68	70	-0.493	Not in $-COOH$ $-COO-$, $-CONH_{2(1,0)}-$
41. $-C_rO-$	28	41	-0.187	Not in $-C_rOO-$
42. $-NO$	36	39	-0.469	Not in $-NO_2$
43. $-PO$			nd	Not in $-PO_4$
44. $-SO-$	5	5	-1.320	Not in $-SO_2$
45. $-NH_2$	36	36	-0.894	Primary
46. $-NH_2$	21	21	-0.759	Secondary, tertiary
47. $-NH_2$	132	139	-0.402	Aniline
48. $-NH_2$	70	73	0.050	The other, not in $-CONH_2$

Table 9.1 (continued)   Group Contribution Values and Correction Factors to Log K_{ow} of the Basic Group Set[a]

Group	n^b	Freq^c	Contribution	Remarks
49. –NH–	94	95	0.021	Not in –CONH–
50. –N<	42	44	–0.937	Not in –CON<
51. –N$_r$H–	73	77	–0.160	Not in –CON$_r$H–
52. –N$_r$<	111	122	–1.027	Not in –CON$_r$<
53. –C≡N	28	28	–0.067	Connected to C_a
54. –C≡N	43	45	0.072	Not connected to C_a
55. =NH			nd	
56. =N–	71	74	0.739	
57. =N$_r$–	63	102	–0.034	Including aromatic N
58. –NO$_2$	140	160	0.220	Connected to C_a
59. –NO$_2$	9	9	0.079	Not connected to C_a
60. –SH	5	5	0.875	
61. –S–	37	38	0.485	
62. –S$_r$–	24	24	0.812	
63. =S			nd	Not in –N=C=S
64. –CS–	9	9	–0.042	
65. –SO$_2$	50	50	–0.818	
66. –S$_r$O$_2$	3	3	–0.984	
67. –P=			nd	Not in –P=O
68. –P=(<)	4	4	–0.450	

Correction Factors

Group	n^b	Freq^c	Contribution	Remarks
69. HO–C=N–	5	5	–1.133	
70. HO–CO–C=N–	3	3	–3.578	
71. –NH–N=CH–X	7	7	0.363	X not N=
72. HCO–X	12	12	0.736	X not C
73. NH$_{2(1,0)}$–CO–NH$_{2(1,0)}$	89	182	0.510	
74. OH$_{(0)}$–CO–NH$_{2(1,0)}$	82	83	0.652	
75. –CO–NH$_{(0)}$–CO–	46	116	0.541	
76. –CH$_{2(1,0)}$–NH–CH$_{2(1,0)}$–	35	72	–0.367	
77. –CH$_{2(1,0)}$–O–CH$_{2(1,0)}$–	69	148	–0.121	
78. –N=C(NH$_{2(1)}$)–N=	16	24	–0.185	
79. HO–C=C–CO–OH	9	10	0.419	
80. HO–C=C–CO–	27	27	0.730	
81. HO–CO–CH$_{2(1)}$–NH$_{2(1)}$	13	13	–1.846	
82. =NH$_{(0)}$–N=N–N–	14	15	0.320	Including =NH$_{(0)}$–N=N–CH$_{(0)}$–, =NH$_{(0)}$–N=CH$_{(0)}$–N=
83. –C–N(–NH$_2$)–C=N	7	7	0.178	
84. HO–CO–CH$_2$–O–	67	67	0.261	
85. NH$_{2(1)}$–CH$_2$–CH$_2$–OH	16	17	–0.324	
86. NH$_{2(1)}$–CO–N–NO	8	8	0.704	
87. –N=CH$_{(0)}$–CH$_{(0)}$=C–OH	10	12	–0.494	
88. NO$_2$–C=CH$_{(0)}$–CH$_{(0)}$=C–OH	3	4	0.302	
89. NO$_2$–C=CH$_{(0)}$–CH$_{(0)}$=C–NH$_2$	8	14	0.185	
90. NH$_2$–C=CH$_{(0)}$–CH$_{(0)}$ =C–CO–OH	23	47	–0.530	Including NH$_2$–C=CH(0)–C–CO–OH
91. NH$_{2(1)}$–C=CH$_{(0)}$ –CH$_{(0)}$=C–SO$_2$–NH$_{2(l)}$)	14	28	–0.466	
92. –CO–NH–C=CH$_{(0)}$ –CH$_{(0)}$=C–OH	16	32	–0.187	

Table 9.1 (continued)   Group Contribution Values and Correction Factors to Log K_{ow} of the Basic Group Set[a]

Group	n^b	Freq[c]	Contribution	Remarks
93. $CH_3-CH_2-CH_2-CH_2-CH_2$ $-CH_2-$	14	14	0.824	Not in the alkane
94. $OH-(CH_2)_n-$pyridine	9	12	−0.650	$n \geq 3$
95. $NH_2-(CH_2)_n-$pyridine	9	12	−0.545	$n \geq 3$
96. $NH_2-CO-(CH_2)n-$pyridine	12	16	−0.903	$n \geq 2$
97. No. of atoms in alkane	13	13	0.095	Including cycloalkane
98. Unsaturated hydrocarbon	67	67	0.872	

[a] r denotes an atom in a ring system. a denotes an aromatic carbon atom. A symbol in parenthesis indicates the open valence not filled by hydrogen. A number in parenthesis indicates the allowed number of hydrogen atoms. nd indicates the value is not determined.
[b] n is the number of chemicals in the training set.
[c] Freq is the number of times the group appears in the training set.

From Klopman, G., Li, Ju-Y., Wang, S., and Dimayuga, M., 1994. *J. Chem. Inf. Comput. Sci.*, **34**, 752–781. With permission.

Table 9.2   Estimation Results of Klopman's Method by Class of Chemical[a]

Chemical Class	n	SD, B_i	SD, Model
Aliphatic hydrocarbon	28	1.14	0.28
Aromatic hydrocarbon	52	0.77	0.20
Alcohol, ether, phenol	74	0.36	0.31
Ketone, aldehyde	27	0.33	0.31
Acid, ester	43	0.32	0.30
Amine, nitrile	66	0.62	0.38
Amide, anilide	23	0.59	0.39
Sulfur-containing hydrocarbon	9	0.54	0.36
Nitro-containing hydrocarbon	18	0.35	0.35
Amino acid	10	1.03	0.24
Halogenated hydrocarbon	54	0.39	0.43
Multifunctional chemicals	1259	0.58	0.40
All	1663	0.58	0.38

[a] n denotes the number of chemicals in the training set. SD, B_i denotes the standard deviation of the group contribution. SD, model denotes the standard deviation of the model estimate.

From Klopman, G., Li, Ju-Y., Wang, S., and Dimayuga, M., 1994. *J. Chem. Inf. Comput. Sci.*, **34**, 752–781. With permission.

$$\log KOW = a + \Sigma_i b_i B_i + \Sigma_j c_j C_j$$

$$= -0.703 + (4)(0.661) + (2)(0.415) + (2)(0.104) + (3)(0.380)$$

$$+ (1)(0.129) + (1)(0.135)$$

$$= 4.25$$

The measured value of log K_{OW} is 4.36. The estimate error is 2.6%. Interestingly, the CLOGP estimate is 5.39 (Leo, 1993). This is one of about 100 chemicals (out of 7996 in the STARLIST data base) that CLOGP doesn't handle well.

Table 9.3 Correlation Coefficient, r, and Standard Deviation, SD, Using Various Estimation Methods[a]

Method[b]	All chemicals[c]	Nucleosides only	Bases only
KLOGP	0.89(0.46)	0.91(0.47)	0.77(0.46)
BLOGP	0.40(1.20)	0.43(1.37)	0.81(0.53)
CLOGP	0.71(0.93)	0.75(1.02)	0.60(0.63)
ALOGP	0.84(0.51)	0.87(0.55)	0.80(0.42)

[a] Standard deviation given in parentheses.
[b] ALOGP: method of Viswanadhan et al. (1989); BLOGP: method of Bodor et al. (1989); CLOGP: method of Leo (1991); KLOGP: method of Klopman et al. (1994).
[c] Test set consists of 47 nucleosides and bases.

From Klopman, G., Li, Ju-Y., Wang, S., and Dimayuga, M., 1994. *J. Chem. Inf. Comput. Sci.*, **34**, 752–781. With permission.

9.3.2 Hansch and Leo's Method and CLOGP3

Hansch, Leo, and co-workers pioneered work on the estimation of octanol–water partition coefficients (Leo et al., 1971). A major improvement in the estimation of K_{OW} was the additive-constitutive method (Nys and Rekker, 1973; Rekker, 1977), which was adopted and substantially improved by Hansch and Leo (1979). The original method of Hansch and Leo has been described in detail (Lyman, 1990). The CLOGP computer program, the first application program based on Hansch and Leo's estimation method, was described by Chou and Jurs (1980).

CLOGP3, a widely used computerized fragment-contribution model, is an improved and updated version described by Leo (1993). It calculates the value of log K_{OW} of neutral solute molecules including zwitterions. Leo (1993) reports an overall correlation of 0.970 and a standard deviation of 0.398 between measured and predicted log K_{OW} when applied to a test set of 7800 diverse chemicals. The data set does not include some chemicals for which CLOGP3 produces large predictive errors: nucleosides, chemicals exhibiting keto/enol tautomerism, etc. Viswanadhan et al. report an overall correlation of 0.711 and a standard deviation of 0.93 between measured and predicted values of 47 nucleosides and nucleobases.

The method tends to overestimate the octanol–water partition coefficient of chemicals with log $K_{OW} > 5.0$ (Doucette and Andren, 1988; Sabljic, 1989). Doucette and Andren report an average percent error of 10.34% in predicting log K_{OW} of 64 highly hydrophobic aromatic and haloaromatic chemicals. The errors ranged from less than 1% for benzene to about 36% for decachlorobiphenyl. Some other problems with the method, such as inability to properly account for long-range hydrogen-bonding interactions among fragments, have been discussed by Leo (1993).

The method is adaptive in that it is continually being modified to improve its accuracy and extend its applicability as chemicals are added to the training set. Adaptivity has reached the point that even the fundamental rule by which fragments are defined, a strong point of this method, is broken for α-amino acids (Leo, 1991). Over the years, the method has grown so complex that software should be used to estimate K_{OW} values of chemicals with complex molecular structures. Hand calculations are not for the faint-hearted or for those who are poorly versed in chemistry. Even the designers of the method use the computer version in order to uniformly and correctly apply the rules to estimate K_{OW} of complex molecules (Leo, 1991). An indication of the difficulty in applying Leo's method uniformly by hand is that some published estimates of log K_{OW} can't be duplicated (Leo, 1993). For more details, see Hansch and Leo (1979) and the *CLOGP3 User's Guide* (Leo and Weininger, 1989).

Hansch and Leo's model is

$$\log K_{OW} = \sum_i a_i \cdot f_i + \sum_j b_j \cdot F_j \tag{9.5}$$

where K_{OW} = octanol–water partition coefficient
     a_i = number of fragments i in molecular structure
     f_i = contribution of fragment i to log K_{OW}
     b_j = number of structural features j in structure
     F_j = contribution of structural feature j to K_{OW}

A partial list of fragment values, a_i, is given in Table 9.4. Unlike most other group contribution methods, Hansch and Leo's fragment contributions are not derived by multiple regression analysis of K_{OW} data taken from a large training set. Instead, each value is derived from data on a small set of simple molecules containing the fragment of interest. The method cannot be applied to a chemical that contains a fragment that has not been evaluated.

The term "fragment" is carefully defined by Leo and Hansch in order to eliminate any ambiguity in the process of defining fragments so that it can be done repeatably by a computer. As a compromise between making fragments as large as possible, which would aid in accurately accounting for intramolecular interactions, and making all fragments atoms, promoting broad applicability of the method, atomic fragments are limited to isolating carbon atoms, hydrogen atoms bound to isolating carbon atoms, and heteroatoms bound to isolating carbon atoms. All other fragments are multiple-atom fragments.

An isolating carbon is a carbon atom that is not doubly or triply bonded to a heteroatom. The isolating carbon and its attached hydrogens are considered hydrophobic fragments (Leo, 1993). However, an isolating carbon atom can be aromatically bonded to a heteroatom.

A polar fragment is any atom or group of atoms in a molecule bounded by isolating carbon atoms. Polar fragments contain no isolating carbon atoms. They may have one or more bonds to isolating carbon atoms designated by A for aliphatic and a for aromatic. The bonds connecting a fragment to an isolating carbon atom are its "valence" bonds.

Polar fragments make the most negative contribution to K_{OW} when bonded to sp^3 isolating carbon atoms. Fragment values become more positive as the isolating carbon atoms to which they are attached allow for charge delocalization. Several types of isolating carbon atoms are identified. They are aliphatic (A), benzyl- (Z = $C_6H_5CH_2-$),

Table 9.4 Fragment Contributions to Log K_{ow}

Fragment	f_i	$f_i^{\phi a}$	$f_i^{\phi\phi}$	f_i, Other Types
Hydrocarbon Increments				
–H	0.23	0.23		
>C<	0.20	0.20		
=C< aromatic	0.13[b]			
=CH– aromatic	0.355			
–CH$_3$	0.89	0.89		
–C$_6$H$_5$	1.90			
Oxygen Increments				
–O–	–1.82	–0.61	0.53	Vinyl,[c] –1.21
				X1[d], –0.22
				X2, 0.17
–O– aromatic	–0.08			
–OH	–1.64	–0.44		X1, 0.32
				Benzyl[e], –1.34
–O$^-$		–3.64		
Carbonyl Increments				
–C(O)–	–1.90	–1.09	–0.50	X1, –0.83
				Benzyl, –1.77
–C(O)– aromatic	–0.59			
–C(O)H	–1.10	–0.42		
–C(O)O–	–1.49	–0.56	–0.09	X1, –0.36
				Vinyl, –1.18
				Benzyl, –1.38
–OC(O)– aromatic	–1.40			
–C(O)OH	–1.11	–0.03		Benzyl, –1.03
–C(O)O^{-2}	–5.19	–4.13		
–C(O)NH–	–2.17	–1.81	–1.06	Vinyl, –1.51
–C(O)NH$_2$	–2.18	–1.26		X1, –0.82
				Benzyl, –1.99
–OC(O)H	–1.14	–0.64		
–OC(O)NH–	–1.79	–1.45		Vinyl, –0.91
–OC(O)NH$_2$	–1.58	–0.82		Benzyl, –1.24
–C(O)C(O)–	–3.00		–0.30	
–NHC(O)NH–	–2.18	–1.57	–0.82	
–NHC(O)NH$_2$	–2.18	–1.07		
>NC(O)NH$_2$		–2.25	–2.15	
>NC(O)H	–2.67	–1.59		
–NHC(O)N<		–2.29		–NHC(O)Na–, –2.42
–NHC(O)H		–0.64		
Nitrogen Increments				
–N<	–2.18	–0.93	–0.50	
–NH–	–2.15	–1.03	–0.09	X1, –0.37
–NH– aromatic	–0.65			
–NH$_2$	–1.54	–1.00		X1, –0.23
				Benzyl, –1.35
–NH(OH)		–1.11		

Table 9.4 (continued)   **Fragment Contributions to Log K_{OW}**

Fragment	f_i	$f_i^{\phi a}$	$f_i^{\phi\phi}$	f_i, Other Types
$-NHNH-$			-0.74	Benzyl, -2.84
$-NH(NH_2)$		-0.65		
$-N=$aromatic	-1.12			
$-N<$ aromatic	-1.10			
$>N-\phi$ aromatic	-0.56			
$-N=N-$			0.14	
$-N=N-$ aromatic	-2.14			
$-NHN=N-$ aromatic	-0.86			
$-NNN-$		0.69		
$-N=NN<$		-0.85		X1, -0.67
$-NO$		0.11		
$-NO_2$	-1.16	-0.03		X2, 0.09
$-ONO_2$	-0.36			
$>NNO$	-2.40	-0.84		
$-CN$	-1.27	-0.34		Benzyl, -0.88
$-CH=N-$		-1.03	0.08	
$-CH=NOH$	-1.02	-0.15		
$>C=NH$			-1.29	
$-NHCN$		-0.03		
$-CH=NN<$		-1.71		
$-CH=NNH-$ aromatic	-0.47			
$-N=CHNH-$ aromatic	-0.79			
$-N=CH-O-$ aromatic	-0.71			
$-CH=N-O-$ aromatic	-0.63			
$-N=CHN=$aromatic	-1.46			

Sulfur Increments

Fragment	f_i	$f_i^{\phi a}$	$f_i^{\phi\phi}$	f_i, Other Types
$-S-$	-0.79	-0.03	0.77	
$-S-$ aromatic	0.36			
$-SH$	-0.23	0.62		
$-S(O)-$	-3.01	-2.12	-1.62	
$-SO_2-$	-2.67	-2.17	-1.28	
$-SO_2O-$	-2.11	-2.06	-0.62	$-SO_2O-a,^f$ -1.42
$-SO_2O^{-2}$	-5.87	-4.53		
$-OSO_3^{-2}$	-5.23			
$-SO_2N<$		-2.09		
$-SO_2NH-$		-1.75	-1.10	$-SO_2NH-a$, -1.72
$-SO_2(NH_2)$		-1.59		X1, -1.04
$-SO_2NH(NH_2)$	-2.04			
$-NHSO_2(NH_2)$		-1.50		
$-SCN$	-0.48	0.64		Benzyl, -0.45
$-C(=S)O-$	-1.11			
$-N=CH-S-$ aromatic	-0.29			
$-C(=S)NH-$	-2.00			$-C(=S)NHa$, -0.96
$-C(=S)NH_2$		-0.41		
$-NHC(=S)NH-$		-1.79		
$-NHC(=S)NH_2$	-1.29	-1.17		

Halogen Increments

Fragment	f_i	$f_i^{\phi a}$	$f_i^{\phi\phi}$	f_i, Other Types
$-F$	0.38	0.37		Vinyl, 0.00
$-Cl$	0.06	0.94		Vinyl, 0.50

Table 9.4 (continued)   Fragment Contributions to Log K$_{OW}$

Fragment	f$_i$	f$_i^{\phi a}$	f$_i^{\phi\phi}$	f$_i$, Other Types
–Br	0.20	1.09		Vinyl, 0.64
				Benzyl, 0.48
–I	0.59	1.35		Vinyl, 0.97
–CF$_3$		1.11		
–SF$_5$		1.45		
–SO$_2$F		0.30		
–IO$_2$		–3.23		
–N=CCl$_2$		0.64		

Phosphoropus Increments

–P(O)<				aP(O)aa, –2.45
–P(O)O$_2$<		–2.33		
–OP(O)O$_2$<	–2.29	–1.71		X1, –1.50
>NP(S)(N<)$_2$	–3.37			
–SP(S)O$_2$<	–2.89			
–SP(O)(O–)NH–	–2.18			
–SP(O)(NH$_2$)O–	–2.50			

Other Increments

–As(OH)$_2$O–		–1.84		
–As(O)(OH)$_2$		–1.90		
–B(OH)$_2$		–0.32		
–Se–	0.45			
>Si<	–0.09	0.65		Benzyl, –0.38

[a] Superscript ϕ denotes contribution of fragment attached to an aromatic ring, e.g., Ar–O–. Superscript $\phi\phi$ denotes contribution of fragment attached to two aromatic rings, e.g., Ar–O–Ar.

[b] Contribution of carbon atom shared by aromatic rings = 0.225. Contribution of carbon atom shared by aromatic rings and bonded to a hetero atom or a nonisolating C atom = 0.44.

[c] Vinyl means an isolated double bond, >C=C<. Contribution of halogens = $(1/2)(\delta + \delta^\phi)$. Contribution of fragments containing N and O atoms = $(2/3)\delta + (1/3)\delta^\phi$.

[d] X denotes attachment to aromatic ring with a second electron-withdrawing substitutent. X1 used when Hammett σ_i of second substituent exceeds 0.50 (e.g., –F, –CN, and –NH$_3$+) and when two halogen atoms are also attached to ring. X2 used when Hammett σ_i > 0.75 (e.g., –NO$_2$ and –OCN).

[e] Benzyl means a C$_6$I I$_5$Cl I$_2$– group.

[f] a denotes group attached to an aromatic ring as shown.

From Hansch, C. and Leo, A., 1979. *Substituent Constants for Correlation Analysis in Chemistry and Biology.* John Wiley & Sons, New York; Leo, A.J., 1993. *Chem. Rev.,* **93**, 1281–1306; Lyman, W.J., 1990. Chapter 1. In *Handbook of Chemical Property Estimation Methods. Environmental Behavior of Organic Compounds.* Lyman, W.J., Reehl, W.F., and Rosenblatt, D.H., Eds. American Chemical Society, Washington, D.C.

vinyl- (V = isolated >C=C<), styryl- (Y = C$_6$H$_5$CH=C<), and aromatic (a). The contribution of a polar fragment to log K$_{OW}$ depends on the type of the isolating carbon atoms to which it is attached, i.e., on the nature of its valence bonds. In general, the contribution increases in this order: aliphatic, benzyl, vinyl, styryl, aromatic. If measured values are not available for Z-, V-, and Y-type isolating carbon atoms, the following estimates are made: $Z = A + 0.02$, $V = [(a - A)/2] + A$, $Y = [3(a - A)/4] + A$ (Leo, 1991).

More or less accurate contributions of the fragments of the hydrocarbon portion of the solute can be derived directly from values given in Table 9.4. The contribution of a $-CH_3$ group to log K_{OW} can be derived by adding the contributions of the hydrogen atoms and the isolating carbon atom: $3 (0.23) + 0.20 = 0.89$. However, a polar fragment cannot be broken up into smaller fragments or constructed by rearranging fragments. For instance, a new fragment cannot be constructed by replacing a hydrogen in a listed fragment with another atom or group of atoms. Since the group contribution of hydrogen, $f_i = 0.23$, applies only to the nonpolar portion of the molecule, no polar fragment contribution can be derived by adding and subtracting component values in such a case.

Hydrophobic atom fragments are defined as isolating carbon atoms and the hydrogen attached to isolating carbon atoms. All other fragments are considered polar even if they make positive contributions to log K_{OW}. To account for the various types of intermolecular interactions (London forces, dipolar forces, and hydrogen bonding), two types of polar fragments are classified: halogen atoms (no H-bonding ability) designated as type X, and other fragments (containing N, O, S, and P) designated as H-polar groups (type Y). The halogen class is itself subdivided: the presence of fluorine decreases log K_{OW} when bound to aliphatic carbon, and the presence of Br, Cl, and I increases log K_{OW}. The nonhalogen class is also subdivided into groups according to the strength of interactions with α-halogens: $-S(=O)-$ and $-SO2-$ (Y-3 group) the most sensitive to the presence of halogens, $-CONH-R$, $-O-R$, $-S-R$, and $-NHR$ (Y-2 group) less sensitive, and all others (Y-1 group) least sensitive. The varying sensitivity of the Y groups is accounted for as follows: add $+0.9$ to Y-3 value for each α-halogen, $+0.9$ to Y-2 value for first α-halogen and roughly half that for each of the next two α-halogens, and $+0.9$ to Y-1 value for the first α-halogen only.

The value of log K_{OW} is reduced when compact polar groups are present. Conjugation of the polar fragments with double bonds and aromatic rings delocalizes electric charge, increasing K_{OW}. K_{OW} is increased in the order attachment to aliphatic, benzyl- ($C_6H_5CH_2-$), vinyl- ($>C=C<$), styryl- ($C_6H_5CH=CH-$), and aromatic group. This is accounted for in Table 9.4.

Isolated double bonds are slightly hydrophylic and make a slight negative contribution to log K_{OW}. Triple bonds are hydrophylic and make a large negative contribution. Conjugation overcomes this effect so that 1,2-diphenylethane, trans-stilbene, and diphenylacetylene have log K_{OW} values of 4.79, 4.81, and 4.78, respectively.

Some contributions of structural features to log K_{OW}, F_j, are listed in Table 9.5. Geometric features (multiple bonding, chain flexing, and chain branching) reduce the size of the solute molecule, making the solute less hydrophobic and decreasing log K_{OW}. Double and triple carbon-carbon bonds are shorter and more polarizable than single carbon-carbon bonds and contribute to increased aqueous solubility and reduced K_{OW}. The correction is made for multiple bonds in aliphatic hydrocarbons but not in aromatic systems because it is already incorporated in the aromatic fragment contributions.

"Flexing" of hydrocarbon chains and alicyclic rings reduces the solute surface area and required cavity size, among other things, increasing aqueous solubility and

Table 9.5   Some Common Intramolecular Interaction Factors Useful for Log K_{ow} Estimation

Feature	F_j		
Geometric Features			
Multiple bonding (unsaturation)			
Double bond	-0.09[a]		
Triple bond	-0.50[a]		
Skeletal flexing			
Hydrocarbon chains	$(n-1)(-0.12)$[b]		
Alicyclic rings	$(n-1)(-0.09)$[b]		
Chain branching			
Nonpolar chain	-0.13		
Polar chain	-0.22		
Electronic Features			
Polyhalogenation			
2 on same carbon atom	0.60		
3 on same carbon atom	1.59		
4 on same carbon atom	2.88		
2 on adjacent sp^3 carbon atom	0.28		
3 on adjacent sp^3 carbon atom	0.56		
4 on adjacent sp^3 carbon atom	0.84		
5 on adjacent sp^3 carbon atom	1.12		
6 on adjacent sp^3 carbon atom	1.40		
Polar fragments[c]	In chain	In alicyclic ring	In aromatic ring
On same C atom	$-0.42\ (f_1 + f_2)$		
On adjacent C atom	$-0.26\ (f_1 + f_2)$	$-0.32\ (f_1 + f_2)$	$-0.16\ (f_1 + f_2)$
On C atoms separated by one C atom	$-0.10\ (f_1 + f_2)$	$-0.20\ (f_1 + f_2)$	$-0.08\ (f_1 + f_2)$
Intramolecular hydrogen bonding			
With $-OH$	1.0		
With $-NH$	0.6		

[a] The factor value includes deductions for removing hydrogen atoms to produce the unsaturation (after Schwarzenbach et al., 1993).
[b] n = number of bonds in chain or ring.
c. Nonhalogen polar fragments. f_i values are given in Table 9.4.

From Hansch, C. and Leo, A., 1979. *Substituent Constants for Correlation Analysis in Chemistry and Biology.* John Wiley & Sons, New York; Schwarzenbach, R.P., Gschwend, P.M., and Imboden, D.M., 1993. *Environmental Organic Chemistry.* John Wiley & Sons, New York, p. 133.

reducing K_{OW}. The size of the flexing factor correction is proportional to the number of bonds, n, between carbon atoms in the chain outside of polar fragments and aromatic rings. The correction is $-0.12(n-1)$ for aliphatic chains and $-0.9n$ for less-flexible aliphatic rings. A molecule with one carbon-carbon bond is rigid, and so the first bond in a chain is not counted. No correction is made for the flexibility of a polar fragment, since it is accounted for in the fragment constant.

    Branching also reduces solute surface area, increasing aqueous solubility and to a lesser degree solubility in octanol, reducing K_{OW}. Polar and hydrogen-bonding groups located at branching points further enhance aqueous solubility and reduce K_{OW}. The correction for a polar group branch $F_{br\,polar}$ and for a chain branch $F_{br\,nonpolar}$ correct for this effect. Alicyclic fusions are considered chain branches and an alicyclic H-polar fragment gets a group branch correction. Branching factors are not given for halogen substitution, however, because this is accounted for in another way. A branching node can also be a tertiary amine [correction = -0.08(n - 1)] and a phosphate ester [correction = -0.19(n - 1)].

Electronic features (shielding, conjugation, intramolecular hydrogen bonding) account for the intramolecular interactions between electron-attracting substituents which compete with solute–solvent interactions. They reduce the net solute polarity, making the solute more hydrophobic and increasing log K_{OW}. There are two types of polar fragments, halogen atoms ("S-polar" fragments) and hydrogen-bonding groups ("H-polar" fragments). There are four types of interactions among these groups: (1) dipole-dipole interactions among halogens, (2) interactions of halogens with hydrogen-bonding substituents, (3) interactions between hydrogen-bonding substituents, and (4) intramolecular hydrogen bonding. The closer the fragments are to each other, the larger the correction.

The polyhalogenation correction increases with the number of halogen atoms and is greatest when the halogens are bonded to the same carbon atom. Polyhalogenation increases the value of K_{OW} when the halogens are on adjacent singly bonded carbon atoms but not on adjacent multiply bonded carbon atoms. The carbon-halogen bonds contribute to the chain "flexing" factor but not to the branching factor, since this is incorporated into the polyhalogenation factor. Similar considerations explain the trend of F-values of interactions among hydrogen-bonding substituents.

Example

Estimate the K_{OW} of benzyl bromide, C_7H_7Br, using the method of Leo.

1. Identify the fragments and features found in the molecular structure $C_6H_5-CH_2-Br$. The fragment and feature contributions are given in Tables 9.4 and 9.5, respectively.

Fragment/Feature	Contribution
6 aromatic isolating carbons	0.780
1 aliphatic isolating carbon	0.195
7 hydrogens on isolating carbons	1.589
Bromine atom	0.480
1 chain	-0.120

2. Estimate log K_{OW} using Equation 9.5.

$$\log K_{OW} = a + \Sigma_i a_i f_i + \Sigma_j b_j F_j$$

$$= 0.780 + 0.195 + 1.589 + 0.480 - 0.120$$

$$= 2.924$$

The measured value of log K_{OW} is 2.92 as listed in the STARLIST data base (Leo and Weininger, 1989).

9.3.3 Method of Meylan and Howard

Meylan and Howard (1995) developed an estimation method that can be used
to estimate log K_{OW} values for any chemical with accuracy comparable to that offered
by the method of Hansch and Leo. It has been implemented in the computer program
LOGKOW. The fragment contributions and correction factors were evaluated with
a multiple regression analysis of data taken primarily from the STARLIST data set.
The model is

$$\log K_{OW} = \sum_i n_i f_i + \sum_j m_j c_j + 0.229 \tag{9.6}$$

n = 2351, SD = 0.216, r² = 0.982, ME = 0.161

where K_{OW} = octanol–water partition coefficient
n_i = number of fragments i in molecular structure
f_i = contribution of fragment i to log K_{OW}
m_j = number of correction factors j
c_j = value of correction j
n   = number of chemicals in training set
SD   = standard deviation
r²   = correlation coefficient
ME   = mean error

Some values of the fragment contributions, f_i, are given in Table 9.6. The fragments are listed in order of precedence. If there is more than one way of classifying
an atom or fragment in the chemical structure, fragments listed first in Table 9.6 are
assigned before any fragments that are listed below them. For example, an aliphatic
oxygen attached to both nitrogen and carbonyl is assigned to the "–ON– (nitrogen
attachment)" fragment which is listed above the "–O– (carbonyl attachment)" fragment. Also, the fragment value is contributed by the core atom or fragment not
including the attached atoms. For example, the contribution of a cyano fragment
"N=CS (sulfur attachment)" is that of the cyano group, N=C–. The attached sulfur
is considered to be a separate core atom.

Table 9.7 lists correction factors, c_j, involving aromatic ring substitution patterns.
Table 9.8 lists some miscellaneous correction factors. In general, every ring substitution factor that can be applied to a structure should be applied, even when there
appears to be some overlap. On the other hand, subgroups of fragments listed in the
table of miscellaneous factors are ignored. For example, two miscellaneous factors
are "–C(=O)NC(=O)NC(=O)–" and "–NC(=O)NC(=O)–". The second is a sub-
group of the first, and if both are present in the structure of a molecule, the second
should be ignored.

Table 9.6   Some Fragment Descriptions and Group Contribution Values, f_i

Fragment	f_i (freq)[a]
Aromatic atoms	
Carbon	0.2940 (6735)
Oxygen	−0.0423 (164)
Sulfur	0.4082 (306)
Aromatic nitrogen	
Nitrogen [N==O, oxide type]	−2.4729 (91)
Nitrogen [+5 valence type]	−6.6500 (51)
Nitrogen at a fused ring location	−0.0001 (12)
Nitrogen in a five–member ring	−0.5262 (851)
Nitrogen in a six–member ring	−0.7324 (1050)
Aliphatic carbon	
$-CH_3$ (methyl)	0.5473 (5257)
$-CH_2-$	0.4911 (4630)
$-CH$	
>C< (no hydrogens, single bonds, three or more carbons attached)	0.2676 (591)
C (no hydrogens)	0.9723 (907)
Olefinic and acetylenic carbon	
=C< (two aromatic attachments)	−0.4186 (11)
$=CH_2$	0.5184 (162)
=CH– or =C<	0.3836 (1044)
≡CH or ≡C–	0.1334 (46)
Carbonyls and thiocarbonyls[b]	
CHO– (aldehyde, aliphatic attachment)	−0.9422 (23)
CHO– (aldehyde, aromatic attachment)	−0.2828 (49)
–C(=O)OH (acid, aliphatic attachment)	−0.6895 (547)
–C(=O)OH (acid, aromatic attachment)	−0.1186 (180)
–NC(=O)N– (urea-type)	1.0453 (612)
NC(=O)O (carbamate-type)	0.1283 (340)
NC(=O)S (thiocarbamate type)	0.5240 (20)
–C(=O)O (ester, aliphatic attachment)	−0.9505 (599)
–C(=O)O (ester, aromatic attachment)	−0.7121 (335)
–C(=O)N (aliphatic attachment)	−0.5236 (1474)
–C(=O)N (aromatic attachment)	0.1599 (490)
–C(=O)S (thioester, aliphatic attachment)	−1.100 (5)
–C(=O)– (noncyclic, two aromatic attachments)	−0.6099 (38)
–C(=O)– (cyclic, two aromatic attachments)	−0.2063 (16)
–C(=O)– (cyclic, aromatic, olefinic attachment)	−0.5497 (56)
–C(=O)– (olefinic attachment)	−1.2700 (191)
–C(=O)– (aliphatic attachment)	−1.5586 (152)
–C(=O)– (one aromatic attachment)	−0.8666 (189)
NC(=S)N (thiourea-type)	1.2905 (45)
Cyano (–C≡N)	
N≡CS (sulfur attachment)	0.3540 (7)
N≡CN (nitrogen attachment)	0.3731 (36)
C≡NC=N	0.0562 (53)
–C=–N (other aliphatic attachment)	−0.9218 (94)
–C≡N (aromatic attachment)	−0.4530 (153)

Table 9.6 (continued)   Some Fragment Descriptions and Group Contribution Values, f_i

Fragment	f_i (freq)[a]
Aliphatic nitrogen	
$-NO_2$ (aliphatic attachment)	−0.8132 (51)
$-NO_2$ (aromatic attachment)	−0.1823 (728)
>N< (+5 valence, single bonds)	−6.6000 (85)
−N=C=S (aliphatic attachment)	0.5236 (13)
−N=C=S (aromatic attachment)	1.3369 (24)
−NP (phosphorus attachment)	−0.4367 (84)
−N (two aromatic attachments)	−0.4657 (88)
−N (one aromatic attachment)	−0.9170 (2209)
−N(O) (nitroso, +5 valence)	−1.0000 (33)
−N=C (aliphatic attachment)	−0.0010 (543)
$-NH_2$ (aliphatic attachment)	−1.4148 (1027)
−NH− (aliphatic attachment)	−1.4962 (1597)
−N< (aliphatic attachment)	−1.8323 (1274)
−N(O) (nitroso)	−0.1299 (149)
−N=N− (azo, includes both N)	0.3541 (118)
Aliphatic oxygen	
−OH (nitrogen attachment)	−0.0427 (52)
−OH (phosphorus attachment)	0.4750 (8)
−OH (olefinic attachment)	−0.8855 (23)
−OH (carbonyl attachment)	0.0[c]
−OH (aliphatic attachment)	−1.4086 (861)
−OH (aromatic attachment)	−0.4802 (558)
=O	0.0[c]
−O− (two aromatic attachments)	0.2923 (77)
−OP (aromafic attachment)	0.5345 (88)
−OP (aliphatic attachment)	−0.0162 (180)
−ON− (nitrogen attachment)	0.2352 (115)
−O− (carbonyl attachment)	0.0[c]
−O− (one aromatic attachment)	−0.4664 (1156)
−O− (aliphatic attachment)	−1.2566 (584)
Phosphorus	
−P=O	−2.4239 (153)
−P=S	−0.6587 (74)
Aliphatic sulfur	
$-SO_2N$ (aromatic attachment)	−0.2079 (417)
$-SO_2N$ (aliphatic attachment)	−0.4351 (45)
−SOOH (sulfonic acid)	−3.1580 (11)
$-SO_2O$ (aliphatic attachment)	−0.7250 (28)
−S(=O)− (one aromatic attachment)	−2.1103 (57)
$-SO_2-$ (one aromatic attachment)	−1.9775 (76)
$-SO_2-$ (two aromatic attachments)	−1.1500 (42)
$-SO_2-$ (aliphatic attachment)	−2.4292 (23)
−S(=O)− (aliphatic attachment)	−2.5458 (11)
−S−S (disulfide)	0.5497 (133)
−S− (one aromatic attachment)	0.0535 (133)
−S− (two aromatic attachments)	0.5335 (41)
−SP (phosphorus attachment)	0.6270 (45)

Table 9.6 (continued)   Some Fragment Descriptions and Group
                        Contribution Values, f_i

Fragment	f_i (freq)[a]
−S− (two nitrogen attachments)	1.200 (40)
−SC=(aliphatic C=)	−0.1000 (118)
−S− (aliphatic attachment)	−0.4045 (106)
=S	0.0[c]
Halogens	
Halogen {all} (nitrogen attachment)	0.0001 (138)
−F (aliphatic attachment)	−0.0031 (377)
−F (aromatic attachment)	0.2004 (248)
−Cl (aliphatic attachment)	0.3102 (274)
−Cl (aromatic attachment)	0.6445 (1124)
−Br (aliphatic attachment)	0.3997 (58)
−Br (aromatic attachment)	0.8900 (284)
−I (aliphatic attachment)	0.8146 (16)
−I (aromatic attachment)	1.1672 (94)
Silicon	
−Si− (aromatic or oxygen attachment)	0.6800 (15)
−Si− (aliphatic attachment)	0.3004 (5)

[a] Freq is the number of times the fragment occurs in the training set.
[b] Carbonyl is always considered aliphatic in this method.
[c] By definition; included as part of other fragments.

After Meylan, W.M. and Howard, P.H., 1995. *J. Pharm. Sci.*, **84**, 83–92.
With permission from the American Chemical Society.

The method was validated with a test set of 6055 diverse chemicals, including compounds that differ from and are more structurally complex than chemicals in the training set. When applied to the test set, SD = 0.408, r^2 = 0.943, and ME = 0.31.

While the method can be applied by hand to simple molecules, it is best to use the computer program. The complete method employs 130 fragment contributions and 235 correction factors. Applying these correctly and uniformly to a complex molecular structure requires some experience. Also, the accuracy of this method is enhanced with the group exchange method described in Chapter 2. The computer program includes a data base from which reference compounds, chemicals with structures similar to the chemical of interest, and their measured values of log K_{OW} are selected.

9.3.4   Structure-K_{OW} Relationships

Several structure property studies of the relationship between molecular connectivity index and K_{OW} have been reported. Kier and Hall (1976) developed independent correlations between molecular connectivity indices and K_{OW} values of hydrocarbons, monofunctional alcohols, ethers, ketones, acids, esters, and amines. These are given in Table 9.9.

Class-specific relationships, such as those shown in Table 9.9, are not easy to apply to polyfunctional chemicals. If the chemical falls under more than one class, there are no rules to help decide which regression formula to use.

Table 9.7   Correction Factors Involving Aromatic Ring Substitution Patterns

Fragment	c_i (freq)[a]
Ortho interactions	
–COOH/–OH	1.1930 (19)
–OH/ester	1.2556 (31)
Amino (at 2-position) on pyridine	0.6421 (53)
Alkyloxy (or alkylthio) ortho to one aromatic nitrogen	0.4549 (67)
Alkoxy ortho to two aromatic nitrogens (or pyrazine)	0.8955 (43)
Alkylthio ortho to two aromatic nitrogens (or pyrazine)	0.5415 (25)
Carboxamide [–C(=O)N] ortho to an aromatic nitrogen	0.6427 (37)
Any[b]/–NHC(=O)C (e.g., 2-methylacetanilide)	–0.5634 (94)
Any[b] two/–NHC(=O)C (e.g., 2,6-dimethylacetanilide)	–1.1239 (40)
Any[b]/–C(=O)NH (e.g., 2-methylbenzamide)	–0.7352 (26)
Any[b] two/–C(=O)NH (e.g., 2,6-dimethylbenzamide)	–1.1284 (16)
Amino-type[c]/–C(=O)N	0.6194 (17)
Can be either ortho or non-ortho	
–NO$_2$ with –OH, –N<, or –N=N–	0.5770 (104)
–C≡N with –OH or –N (e.g., cyanophenols or amines)	0.5504 (39)
–NO$_2$/–NC(=O) (cyclic type)	0.3994 (12)
–NO$_2$/–NC(=O) (noncyclic type)	0.7181 (30)
Non-ortho reactions	
–N</–OH (e.g., 4-aminophenol)	–0.3510 (50)
–N</ester (e.g., 4-aminobenzoic acid methyl ester)	0.3953 (49)
–OH/ester	0.6487 (9)
Others	
Amino-type[c] (at 2-position) on triazine, pyrimidine, or pyrazine	0.8566 (254)
NC(=O)NS on triazine or pyrimidine (2-position)	–0.7500 (6)
1,2,3-Trialkyloxy	–0.7317 (40)

[a] Freq denotes the number of chemicals in the data base to which the correction applies.
[b] Any means any ortho substituent other than hydrogen (with the exception of –OH or an amino-type).
[c] Can be a primary, secondary, or tertiary amine, including –N–C(=O) types.

From Meylan, W.M. and Howard, P.H., 1995. *J. Pharm. Sci.*, **84**, 83–92. Reprinted with permission from the American Chemical Society.

Doucette and Andren (1988) compared six methods of estimating the K_{OW} values of highly hydrophobic aromatic hydrocarbons, both halogenated and nonhalogenated, exhibiting K_{OW} values as large as 8.58. Compound types include benzenes, biphenyls, dibenzofurans, and dibenzo-*p*-dioxins. The most successful (smallest average error) used a structure–property relationship using first-order connectivity indices. The model is

$$\log K_{OW} = -0.0853 + 1.272 {}^1\chi^v - 0.0499 \left({}^1\chi^v \right)^2 \qquad (9.15)$$

n = 64, r^2 = 0.964

where K_{OW} = octanol–water partition coefficient
         ${}^1\chi^v$ = first-order valence connectivity index

Table 9.8   Some Miscellaneous Correction Factors

Fragment	c_i (freq)[a]
Various carbonyl factors	−0.5865 (22)
More than one aliphatic −C(=O)OH	1.7838 (30)
HO−CC(=O)CO−	0.9739 (24)
−C(=O)−C−C(=O)N	1.0254 (93)
−C(=O)NC(=O)NC(=O)− (e.g., barbiturates)	0.6074 (135)
−NC(=O)NC(=O)− (e.g., uracils)	−1.0577 (69)
Cyclic ester (nonolefin type)	−0.2969 (10)
Cyclic ester (olefin type)	−2.0238 (60)
Amino acid (α-carbon type)	−0.7203 (135)
Di-N urea/acetamide aromatic substituent	−0.3662 (125)
C[C(=O)OH] aromatic (e.g., phenylacetic acid)	0.1984 (115)
di-N-aliphatic substituted carbamate	0.3263 (33)
−NC(=O)CR (R is one halogen)	0.6365 (43)
−NC(@O)CR$_{2+}$ (R is two or more halogens)	0.4193 (35)
CC(=O)NCC(=O)OH	1.5505 (20)
CC(=O)NC(C(=O)OH)S−	0.4874 (38)
(Ar−O or −C−O)−CC(=O)NH−	−1.0000 (48)
>C=NOC(=O)	
Various ring factors	0.7525 (124)
1,2,3-Triazole ring	−0.1621 (392)
Pyridine ring (nonfused)	0.8856 (71)
sym-Triazine ring	−0.3421[b] (295)
Fused aliphatic ring correction	
Various alcohol, ether, and nitrogen factors	0.4064 (271)
More than one aliphatic −OH	0.6365 (44)
−NC(C−OH)C−OH	0.5494 (142)
−NCOC	1.0649 (86)
HO−CHCOCH−OH	0.5944 (63)
HO−CHC(OH)CH−OH	1.1330 (159)
−NH−NH− structure	0.7306 (57)
>N−N< structure	

[a] Freq denotes the number of chemicals in the data base to which the correction applies.

[b] The number of fused ring corrections applied to a structure depends upon the number of fused aliphatic carbons, the atoms in the rings, and the number of cyclic "bridges" (3-D-type ring system); the number of corrections is equivalent to the maximum number of free "fused carbons", carbons that can have, or do have, a linear substituent; the number of corrections is decreased by "bridging" and noncarbon ring members.

From Meylan, W.M. and Howard, P.H., 1995. *J. Pharm. Sci.*, **84**, 83–92. Reprinted with permission from the American Chemical Society.

The average error is 4.19%, and the range of error is −16.8 to 11.9% for the 64 compounds in the training set.

Basak et al. (1990) describe a connectivity–K_{OW} model that applies to aliphatic and aromatic hydrocarbons and halocarbons that do not form hydrogen bonds such as alkanes, alkylbenzenes, polycyclic aromatic hydrocarbons, chlorinated alkanes, chlorobenzenes, and PCBs. The model is

Table 9.9   Molecular Connectivity/K_{OW} Correlations for Various Classes of Chemicals

Hydrocarbons (including cyclic and noncyclic alkanes, alkenes, alkynes, substituted benzenes, naphthalene, and phenanthrene)

$$\log K_{OW} = 0.406 + 0.884 \; ^1\chi^v \tag{9.7}$$

$$n = 45, \; s = 0.160, \; r = 0.975$$

Aliphatic alcohols[a]

$$\log K_{OW} = -0.985 + 0.860 \; ^1\chi^v \tag{9.8}$$

$$n = 42, \; s = 0.195, \; r = 0.9612$$

Aliphatic ethers[a,b]

$$\log K_{OW} = -1.411 + 0.988 \; ^1\chi \tag{9.9}$$

$$n = 12, \; s = 0.083, \; r = 0.9762$$

Aliphatic ketones[a]

$$\log K_{OW} = -1.468 + 0.985 \; ^1\chi^v \tag{9.10}$$

$$n = 16, \; s = 0.098, \; r = 0.9929$$

Aliphatic carboxylic acids[c]

$$\log K_{OW} = -0.859 + 1.615 \; ^1\chi^v - 0.550 \; ^1\chi \tag{9.11}$$

$$n = 9, \; s = 0.099, \; r = 0.9979$$

Esters[c]

$$\log K_{OW} = -1.778 + 0.256 \; ^1\chi^v + 0.541 \; ^0\chi^v - 0.535 \; ^4\chi_{PC} \tag{9.12}$$

$$n = 24, \; s = 0.064, \; r = 0.9988$$

Aliphatic amines[c,d]

$$\log K_{OW} = -1.001 + 0.696 \; ^0\chi^v - 0.566 \; ^6\chi_C^v - 0.248 \; \delta_N^v \tag{9.13}$$

$$n = 28, \; s = 0.184, \; r = 0.9797$$

Alcohols and ethers[c,e]

$$\log K_{OW} = -2.529 + 1.042 \; ^1\chi^v - 0.219 \; ^3\chi_C + 0.274 \; \delta_O^v \tag{9.14}$$

$$n = 61, \; s = 0.154, \; r = 0.9725$$

[a] Modest improvement is reported with multiple regression approach.
[b] The best correlation is obtained with connectivity rather than valence connectivity index as shown.
[c] The correlation with $^1\chi^v$ alone produces a large standard error.
[d] δ_N^v is the valence delta value.
[e] δ_O^v is the valence delta value.

From Kier, L.B. and Hall, L.H., 1976. *Molecular Connectivity in Chemistry and Drug Research.* Academic Press, New York.

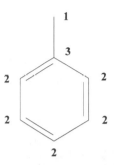

Figure 9.1   The hydrogen-suppressed graph of toluene. δ values are given next to each atom.

$$\log K_{OW} = -3.127 - 1.644 IC_0 + 2.120^5\chi_C - 2.9140^6\chi_{CH} + 4.208^0\chi^v$$
$$+ 1.060^4\chi^v - 1.020^4\chi_{PC}^v$$

(9.16)

n = 137, s = 0.26, r² = 0.97

Example

Estimate the K_{OW} of toluene using Equation 9.15.

1. The hydrogen-suppressed graph is shown in Figure 9.1. The delta values for CH₃C₆H₅ are $\delta_1 = 1$, $\delta_2 = 4$, $\delta_3 = 3$, $\delta_4 = 3$, $\delta_5 = 3$, $\delta_6 = 3$, and $\delta_7 = 3$. The first-order connectivity index is calculated:

$$^1\chi_P = \Sigma_{bonds} \left(\delta_i\delta_j\right)^{-0.5}$$

$$= (4)(3\cdot3)^{-0.5} + (2)(4\cdot3)^{-0.5} + (1)(4\cdot1)^{-0.5}$$

$$= 2.41$$

2. The value of K_{OW} is estimated using Equation 9.15.

$$\log K_{OW} = -0.0853 + 1.272 \, ^1\chi^v - 0.0499 \left(^1\chi^v\right)^2$$

$$= -0.0853 + 1.272(2.41) - 0.0499(2.41)$$

$$= 2.86$$

The measured value of log K_{OW} is 2.65 as listed in the STARLIST data base (Leo and Weininger, 1989). The estimate of log K_{OW} is in error by 8%. The estimate of K_{OW} is in error by 62%.

9.4 SENSITIVITY TO ENVIRONMENTAL PARAMETERS AND METHOD ERROR

The reported K_{OW} value of a chemical may vary greatly depending on the experimental method employed. For instance, measured values of log K_{OW} for benzene range from 1.56 to 2.34 log units. For hexachlorobenzene, reported values range from 4.13 to 7.42 log units (Sabljic, 1991). Additional examples are given in Table 9.10. Data of this sort are used to develop and validate the models described in this chapter, and we expected the model's estimates to be no more accurate.

The standard deviation of Klopman's model, 0.38 log units, lies within the measurement error given for the data set, 0.4 log units. The standard deviation by class of chemical is given in Table 9.2. Clearly, the model is remarkably good regardless of chemical type. The model does produce estimation errors of 0.8 log units or more for 71 chemicals in the training set. Each contained a unique fragment in its structure which may be accounted for in future as more chemicals with the fragment are added to the data set.

Klopman's method was compared with three other group contribution methods in its ability to estimate the K_{OW} values of 35 nucleosides and 12 nucleobases. The results are given in Table 9.3. The methods ALOGP (Viswanadhan et al., 1989), BLOGP (Bodor et al., 1989), and CLOGP (Leo, 1991) are described in the literature. KLOGP, the method of Klopman et al., is the best at estimating K_{OW} values of nucleosides. The relatively poor results Klopman's method gives for nucleobases are caused by a large error in the K_{OW} estimate of one chemical, purine.

The predictive accuracy of the CLOGP3 program is better than that of KLOGP. As was noted above, Leo (1993) reports an overall correlation of 0.970 and a standard deviation of 0.398 between measured and predicted log K_{OW} values of a training set of 7800 diverse chemicals. Meylan and Howard (1995) report that, with a test set of 7250 chemicals, CLOGP produces a standard deviation of 0.3 and a correlation coefficient (r^2) of 0.96. This is comparable to the precision of K_{OW} measurements. It has been pointed out that the predictive accuracy of CLOGP3 may not be as good with chemicals that are not members of the STARLIST data set, however (Bodor et al., 1989). This problem is not unique to CLOGP3. It is a concern with estimation methods for most chemical properties.

The ambient temperature and solute concentration influence both γ_{Wi} and V_{Om}, and so affect K_{OW}. Below 0.01 M, K_{OW} depends only slightly on solute concentration (Lyman, 1990). We can assume that K_{OW} is independent of solute concentration in this range but not at higher levels. Likewise, the temperature dependence is slight; log K_{OW} usually varies by less than 0.03 log units per degree, while K_{OW} varies by about 1 log unit per 5°C (Valsaraj, 1995).

Ionic strength affects γ_{Wi} and, as a result, it affects the partitioning of organic chemicals between octanol and water. The presence of salts and minerals reduces the aqueous solubility of nonionic chemicals. This is called the **salting-out effect**. Seawater has an electrolyte concentration of 0.51 M, and this causes a decrease in the aqueous solubility of neutral, nonpolar chemicals by as much as 50% over that in pure water. The decrease in solubility of polar organic chemicals may not be so

Table 9.10 Range of Reported Experimental K_{OW} Values of Some Hydrocarbons

Chemical	Range of log K_{OW}	Ref.
1,2-Dichloroethane	1.45–1.79	a,b,c,
Benzene	1.56–2.15	a,d
Toluene	2.11–2.73	a,d,e
Chlorobenzene	2.18–3.79	d,f,g
Trichloroethylene	2.29–3.30	c,h,i
Naphthalene	3.01–4.70	d,f
1.3-Dichlorobenzene	3.24–3.60	a,d,j,k
Pentachlorophenol	3.32–5.86	l,m
3,4,5-Trichlorophenol	3.69–4.41	b,w
p,p'-DDT	3.98–6.36	d,n
2,2'-Dichlorobiphenyl	4.04–5.00	g,n,o,p
Hexachlorobenzene	4.13–7.42	c,l,n,o,q
1,2,3,5-Tetrachlorobenzene	4.46–4.94	a,g,r
Pentachlorobenzene	4.88–5.69	a,k,r,s,t
Aldrin	5.52–7.40	o,q
2,4,5,2',4',5'-Hexachlorobiphenyl	6.34–8.18	s,u,v,x

[a] Hansch and Leo, 1979. *Substituent Constants for Correlation Analysis in Chemistry and Biology.* John Wiley & Sons, New York.
[b] Koch, 1984. *Environ. Toxicol. Chem.,* **7**, 331–346.
[c] Davies and Dobbs, 1984. *Water Res.,* **18**, 1253–1262.
[d] Garst and Wilson, 1984. *J. Pharm. Sci.,* **73**, 1616–1623.
[e] Ogata et al., 1984. *Bull. Environ. Contam. Toxicol.,* **33**, 561–567.
[f] Veith et al., 1979. *Fish Res. Board Can.,* **36**, 1040–1048.
[g] Bruggeman et al., 1982. *J. Chromatogr.,* **238**, 335–346.
[h] Arbuckle, 1983. *Environ. Sci. Technol.,* **17**, 537–542.
[i] Wasik et al., 1983. *Residue Rev.,* **85**, 29–42.
[j] Chiou et al., 1983. *Environ. Sci. Technol.,* **17**, 227–231.
[k] Konnemann et al., 1979. *J. Chromatogr.,* **178**, 559–565.
[l] Esser and Moser, 1982. *Ecotoxicol. Environ. Saf.,* **6**, 131–148.
[m] McKim et al., 1985. *Toxicol. Appl. Pharmacol.,* **77**, 1–10.
[n] Chiou and Schmedding, 1982. *Environ. Sci. Technol.,* **16**, 4–10.
[o] Garten and Trabalka, 1983. *Environ. Sci. Technol.,* **17**, 590–595.
[p] Sugiura et al., 1978. *Chemosphere,* **7**, 731–736.
[q] Briggs, 1981. *J. Agric. Food Chem.,* **29**, 1050–1059.
[r] Konemann and van Leeuwen, 1980. *Chemosphere,* **9**, 3–19.
[s] Kenaga and Goring, 1980. In *Aquatic Toxicology,* Eaton, Parrish, and Hendricks, Eds. American Society for Testing and Materials, Philadelphia, PA, pp. 78–115.
[t] Watari et al., 1982. *Anal. Chem.,* **54**, 702–705.
[u] Karickhoff et al., 1979. *Water Res.,* **13**, 241–248.
[v] Yalkowski et al., 1983. *Residue Rev.,* **85**, 43–55.
[w] Schellenberg et al., 1984. *Environ. Sci. Technol.,* **18**, 652–657.
[x] Rapaport and Eisenrich, 1984. *Environ. Sci. Technol.,* **18**, 163–170.

After Sabljic, A., 1987. *Environ. Sci. Technol.,* **21**, 358–366.

pronounced, however. The partitioning of neutral organic chemicals between octanol and water should increase with increasing ionic strength, although electrolytes are distributed to some degree between both phases, and the exact magnitude of the effect is uncertain.

The degree of ionization and the distribution ratio of organic acids and bases is sensitive to the ionic strength of solution.

Depending on the value of the dissociation constants, the degree of ionization of an organic acid or base depends strongly on the pH of the solution. Increasing pH may cause an increase in the aqueous solubility of organic acids and a decrease in the aqueous solubility of bases. The distribution ratio will also be affected by pH.

REFERENCES

Amidon, G.L. and Anik, S.T., 1976. Comparison of several molecular topological indexes with molecular surface area in aqueous solubility estimation. *J. Pharm. Sci.*, **65**, 801–805.

Arbuckle, W.B., 1983. Estimating activity coefficients for use in calculating environmental parameters. *Environ. Sci. Technol.*, **17**, 537–542.

Basak, S.C., Niemi, G.J., and Veith, G.D., 1990. Optimal characterization of structure for prediction of properties. *J. Math. Chem.*, **4**, 185–205.

Broto, P., Moreau, G., and Vandycke, C., 1984. Molecular structures: perception, autocorrelation descriptor and sar studies. *Eur. J. Med. Chem. Chim. Ther.*, **19**, 71–78.

Bodor, N., Gabanyi, Z., and Wong, C.-K., 1989. A new method for the estimation of partition coefficient. *J. Am. Chem. Soc.*, 111, 3783–3786.

Chen, F., Holten-Anderson, J., and Tyle, H., 1993. New developments of the UNIFAC model for environmental application. *Chemosphere*, **26**, 1325–1354.

Chou, J.T. and Jurs, P.C., 1979. Computer-assisted computation of partition coefficients from molecular structures using fragment constants. *J. Chem. Inf. Comput. Sci.*, **19**, 172–178.

Convard, T., Dubost, J.P., le Solleu, H., and Kummer, E., 1994. SmilogP: a program for a fast evaluation of theoretical log P from the SMILES code of a molecule. *Quant. Struct. Act. Relat.*, **13**, 34–37.

Doucette, W.J. and Andren, A.W., 1988. Estimation of octanol/water partition coefficients: evaluation of six methods for highly hydrophobic aromatic hydrocarbons. *Chemosphere*, **17**, 345–359.

Hansch, C. and Leo, A., 1979. *Substituent Constants for Correlation Analysis in Chemistry and Biology*. John Wiley & Sons, New York.

Hansch, C., Leo, A., and Hoekman, D., 1995. *Exploring QSAR. Volume 2: Hydrophobic, Electronic, and Steric Constants*. American Chemical Society, Washington, D.C.

Hawker, D., 1989. The relationship between octan-1-ol/water partition coefficient and aqueous solubility in terms of solvatochromic parameters. *Chemosphere*, **19**, 1585–1593.

Isnard, P. and Lambert, S., 1989. Aqueous solubility and n-octanol/water partition coefficient correlations. *Chemosphere*, **18**, 1837–1853.

Israelchvili, J.N., 1992. *Intermolecular and Surface Forces*, 2nd ed., Academic Press, New York.

Kamlet, M. J., Doherty, R. M., Carr, P. W., MacKay, D., Abraham, M. H., and Taft, R. W., 1988a. Linear solvation energy relationships. XLIV. Parameter estimation rules that allow accurate prediction of octanol/water partition coefficients and other solubility and toxicity properties of polychlorinated biphenyls and polycyclic aromatic hydrocarbons. *Environ. Sci. Technol.*, **22**, 503–509.

Kamlet, M.J., Doherty, R.M., Abraham, M.H., Marcus, Y., and Taft, R.W., 1988b. Linear solvation energy relationships. XLVI. An improved equation for correlation and prediction of octanol/water partition coefficients of organic nonelectrolytes (including strong hydrogen bond donor solutes). *J. Phys. Chem.*, **92**, 5244–5255.

Kier, L.B. and Hall, L.H., 1976. *Molecular Connectivity in Chemistry and Drug Research*. Academic Press, New York.

Klopman, G. and Wang, S., 1991. A computer automated structure evaluation (CASE) approach to calculated partition functions. *J. Comput. Chem.*, **8**, 1025–1032.

Klopman, G., Li, Ju-Y., Wang, S., and Dimayuga, M., 1994. Computer automated log P calculations based on an extended group contribution approach. *J. Chem. Inf. Comput. Sci.*, **34**, 752–781.

Kolenbrander, J.P., Drefahl, A., and Reinhard, M., 1995. *DESOC User's Guide.* Stanford Bookstore, Stanford, CA 94305-3079.

Leo, A.J., 1991. Hydrophobic parameter: measurement and calculation. *Methods Enzymol.*, **202**, 544–591.

Leo, A.J., 1993. Calculating log P_{oct} from structures. *Chem. Rev.*, **93**, 1281–1306.

Leo, A., Hansch, C., and Elkins, D. 1971. Partition coefficients and their uses. *Chem. Rev.*, **71**, 525–616.

Leo, A., Hansch C., and Jow, Y.C., 1976. Dependence of hydrophobicity of apolar molecules on their molecular volume. *J. Med. Chem.*, **19**, 611–615.

Leo, A. J. and Weininger, D., 1989. *CLOGP, Version 3.54 — User Reference Manual,* Medicinal Chemistry Project. Pomona College, Claremont, CA.

Lyman, W.J., 1990. Octanol/water partition coefficient. Chapter 1. In *Handbook of Chemical Property Estimation Methods. Environmental Behavior of Organic Compounds.* Lyman, W.J., Reehl, W.F., and Rosenblatt, D.H., Eds. American Chemical Society, Washington, D.C.

Mackay, D., 1991. *Multimedia Environmental Models. The Fugacity Approach.* Lewis Publishers, Chelsea, MI, p. 79.

McAuliffe, C., 1966. Solubility in water of paraffin, cycloparaffin, olefin, acetylene, cycloolefin, and aromatic hydrocarbons. *J. Phys. Chem.*, **70**, 1267–1275.

Meylan, W.M. and Howard, P.H., 1995. Atom/fragment contribution method for estimating octanol–water partition coefficients. *J. Pharm. Sci.*, **84**, 83–92.

Nys, G.G. and Rekker, R.F., 1973. statistical analysis of a series of partition coefficients with special reference to the predictability of folding drug molecules. The introduction of hydrophobic fragment constants (f values). *Chim. Therap.*, 8, 521–535.

Rekker, R.F., 1977. *The Hydrophobic Fragment Constant.* Elsevier, New York.

Sabljic, A., 1987. On the prediction of soil sorption coefficients of organic pollutants from molecular structure: application of molecular topology model. *Environ. Sci. Technol.*, **21**, 358–366.

Sabljic, A., 1989. Quantitative modeling of soil sorption for xenobiotic chemicals. *Environ. Health Perspect.*, **83**, 179–190.

Sabljic, A. 1991. Chemical topology and ecotoxicology. *Sci. Total Environ.*, **109/110**, 197–220.

Sangster, J., 1989. Octanol–water partition coefficients of simple organic compounds. *J. Phys. Chem. Ref. Data*, **18**, 1111–1229.

Sangster, J., 1993. LOGKOW DATABANK. Sangster Research Laboratories, Montreal, Quebec, Canada.

Schwarzenbach, R.P., Gschwend, P.M., and Imboden, D.M., 1993. *Environmental Organic Chemistry.* John Wiley & Sons, New York, p. 133.

Suzuki, T. and Kudo, Y., 1990. Automatic log P estimation based on combined additive modeling methods. *J. Comput. Aided Mol. Design*, **4**, 155–198.

Suzuki, T., 1991. Development of an automatic estimation system for both the partition coefficient and aqueous solubility. *J. Comput. Aided Mol. Design*, **5**, 149–166.

Taft, R.W., Abraham, M.H., Famini, G.R., Doherty, R.M., Abboud, J.L.M., and Kamlet, M.J., 1985. Solubility properties in polymers and biological media. V. An analysis of the physicochemical properties which influence octanol–water partition coefficients of aliphatic and aromatic solutes. *J. Pharm. Sci.*, **74**, 807–814.

Valsaraj, K.T., 1995. *Elements of Environmental Engineering*. Lewis Publishers, Boca Raton, FL.

Viswanadhan, V.N., Ghose, A.K., Revankar, G.R., and Robins, R.K., 1989. Atomic physico-chemical parameters for three dimensional structure directed quantitative structure-activity relationships. IV. Additional parameters for hydrophobic and dispersive interaction and their application for an automated superposition of certain naturally occurring nucleoside antibiotics. *J. Chem. Inf. Comput. Sci.,* **29**, 163–172.

Yalkowsky, S.H., Pinal, R., and Banerjee, S., 1988. Water solubility: a critique of the solvatochromic approach. *J. Pharm. Sci.* **77**, 74–144.

Yalkowsky, S.H. and Banerjee, S., 1992. *Aqueous Solubility. Methods of Estimation for Organic Compounds*. Marcel Dekker, New York.

CHAPTER **10**

Soil and Sediment Sorption Coefficient

10.1 INTRODUCTION

This chapter describes methods of estimating soil–water partition coefficients, sorption coefficients, organic matter–water partition coefficients, and organic carbon–water partition coefficients. By **sorption** we mean *the association of a chemical with the solid phase by adsorption to surfaces and by absorption into the solid's matrix*. Sorption retards chemical transport in the soil column and alters chemical reactivity. Interaction with soil surfaces increases the chemical reactivity of many chemical compounds, including water.

The **soil–water partition coefficient**, K_{SW}, is *the ratio of the chemical's total molar concentration in the sorbed phase to its molar concentration in soil solution at equilibrium*. The concentration of sorbed chemical usually is not reported in terms of quantity of sorbate per unit volume of solid as in K_{SW}, but in terms of quantity of sorbate per unit mass of sorbent, and the **soil-water distribution ratio**, K_d (L kg^{-1}), in a two-phase soil–water system at equilibrium is almost universally used. It is

$$K_d = \frac{C_S}{C_W} \tag{10.1}$$

where C_S (mol/kg) is the equilibrium concentration of sorbed chemical and C_W (mol/L) is the equilibrium molar concentration of chemical in solution. The soil–water partition coefficient and the soil–water distribution ratio are related by $K_{SW} = \rho_s K_d$, where ρ_S (kg/l) is the soil density.

The value of K_d often varies with solute concentration. Several different models of the correlation have been proposed (Travis and Etnier, 1981; Calvet, 1989). The Freundlich isotherm is most widely used to describe sorption on heterogeneous soil and sediment surfaces. It is

$$C_S = K_P \cdot C_W^n \tag{10.2}$$

165

where K_p, the **sorption coefficient**, and n are the Freundlich parameters. The value of n is observed to lie in the range from 0.7 to 1.2 (von Open et al., 1991). If n > 1, sorption of a surface-active chemical enhances further sorption. This type of behavior has been reported for chemicals such as metolachlor (Kozac, 1983). If n < 1, sorption deactivates the surface, inhibiting further sorption. Such behavior has been reported for chemicals such as the s-triazines (Hamaker and Thompson, 1972). Comparison of Equations 10.1 and 10.2 shows that $K_p = K_d \cdot C_W^{1-n}$.

If n = 1 in Equation 10.2, sorption is a linear function of solute concentration. In other words, the value of K_p is then independent of solute concentration, and $K_p = K_d$. This is reported for chemicals such as bromacil on peat soil (Angemar et al., 1984). The sorption of all "hydrophobic" chemicals is approximately linear at solute concentrations below about 1×10^{-5} mol/L. Linear sorption isotherms are frequently observed for "nonhydrophobic" chemicals at very low solute concentrations, also.

Soil is a mixture of mineral components and organic matter. Most minerals have polar surfaces which favor the sorption of ionic and polar chemicals and chemicals that form hydrogen bonds. The sorption isotherms of nonpolar organic chemicals on mineral surfaces are linear at low concentrations and normal soil moisture levels (Schwarzenbach and Westall, 1981).

Except under conditions in which mineral surfaces are dry, sorption of neutral, nonpolar organic chemicals occurs primarily on the organic matter in soils and sediments (Karickhoff et al., 1979, 1981; Tanford, 1980). If the amount of organic matter is significant, the quantity of sorbate associated with the mineral fraction can be ignored, $C_S \approx C_{OM} \cdot f_{OM}$, and

$$K_d \approx \frac{C_{OM} \cdot f_{OM}}{C_W} \qquad (10.3)$$

where C_{OM} (mol/kg) is the concentration of sorbate associated with soil organic matter and f_{OM} (kg/kg) is the mass fraction of organic matter in soil. The correlation between K_d and f_{OM}, Equation 10.3, has been demonstrated with various chemicals and pesticides (Calvet, 1989). The mass fraction of mineral matter in soil and sediment cannot be ignored in considering the sorption of polar and ionic chemicals and chemicals that hydrogen-bond with mineral surfaces.

The **organic matter–water partition coefficient**, K_{OM} (L/kg), in a two-phase soil–water system at equilibrium is

$$K_{OM} = \frac{C_{OM}}{C_W} \qquad (10.4)$$

The value of K_{OM} may vary with solute concentration. While the nature of the organic matter varies from soil to soil, K_{OM} is reported to be nearly independent of soil type. The relationship between the nature of the soil organic matter and sorption is discussed by Lambert (1968), Suffett and McCarthy (1989), Grathwohl (1990), and Rutherford et al. (1992). Comparison of Equations 10.3 and 10.4 shows that the soil–water distribution ratio and the soil–water partition coefficient are related by $K_d = f_{OM} \cdot K_{OM}$.

The partitioning of neutral organic chemicals is commonly described with the **organic carbon–water partition coefficient**. It is

$$K_{OC} = \frac{C_{OC}}{C_W} \qquad (10.5)$$

where C_{OC} (mol/kg) is the concentration of sorbate associated with the organic carbon content of soil. The soil–water distribution ratio and the organic carbon–water partition coefficient are related by $K_d = f_{OC} \cdot K_{OC}$, where f_{OC} (kg/kg) is the mass fraction of organic carbon in soil. This linear relationship between K_d and f_{OC} has been observed for various organic solvents and degreasers (Abdul et al., 1987).

The nature of the organic matter varies among soils and sediments. Even so, the organic carbon–water partition coefficient is usually related to the organic-matter partition coefficient with the assumption that the organic matter content of soil or sediment is roughly 1.724 times larger than the organic carbon content (Lyman, 1990). Then, $K_{OC} \approx 1.724 \cdot K_{OM}$ and $\log K_{OC} \approx 0.24 + \log K_{OM}$.

The subject of sorption of organic chemicals on soils and sediments was reviewed by Karickhoff (1984) and by Calvet (1989). Sabljic (1987) and Lohninger (1994) compiled and evaluated data on sorption coefficients of hydrocarbons, halocarbons, and pesticides. Data on sorption coefficients of alkylbenzenes, chlorobenzenes, and polycholorobiphenyls were compiled by Abdul et al. (1987), Vowles and Mantoura (1987), and Paya-Perez et al. (1991). Other good compilations of K_{OC} values are found in publications by Kenaga and Goring (1980), Briggs (1981), and Meylan et al. (1992). Soil characteristics are not standardized in making sorption coefficient measurements. This and other factors cause poor agreement between reported K_p, K_{OM}, and K_{OC} values determined by different laboratories. As a result, the values of sorption coefficients reported in the literature should be viewed as reliable to within a factor of 10, on average.

Although estimation methods that apply to organic acids and bases are described below, most reported methods of estimating sorption coefficients apply to nonionic organic chemicals, and that is the emphasis of this chapter. In the following, some theoretical background is presented in order to show the strengths, weaknesses, and range of applicability of the estimation methods. Then several methods of estimating K_p, K_{OM}, and K_{OC} values are presented, ranging from simple, back-of-the-envelope calculations to what are considered to be the most accurate methods employing quantitative K_{OC}-K_{OW} relationships. Elaborate methods that take soil characteristics into account are also discussed.

10.2 BACKGROUND

The basic premise upon which most models of the sorption of nonionic organic chemicals is based is that the process can be viewed as the partitioning of a chemical between water and an organic phase much like the partitioning of chemical between n-octanol and water (Karickhoff, 1981, 1984; Chiou, 1989). On the basis of a study of the sorption of pesticides on soils, Lohninger (1994) concludes that the sorption of pesticides is controlled by three major factors: (1) sorption is correlated with molecular volume; (2) the presence of nonpolar groups enhances sorption. Structural fragments such as aromatic rings, aliphatic sulfur groups, and chlorine atoms increase the sorption coefficient; (3) the presence of polar fragments decreases sorption. Structural fragments such as sulfonyl-, sulfoxyl-, phosphinoxy-, hydroxy-, and urethane groups decrease the sorption coefficient. The same factors are important controls of octanol–water partitioning, and they are discussed in more detail in Chapter 9.

Group contribution methods of estimating the value of sorption coefficients are based on the assumption that sorption is an additive property. Individual molecular fragments contribute quantifiable amounts to the size of a molecule and to the polar solute–solvent and sorbate–sorbent interactions, and they make quantifiable contributions to the value of sorption coefficients. To a large degree, sorption is also a constitutive property. Intramolecular interactions of polar groups in the molecule minimize molecular size and reduce overall molecular polarity. In this case, the presence of topological features that enhance intramolecular interactions requires a correction to the group-contribution estimates of sorption coefficient values.

Lambert (1967) derived a general expression for the contribution of molecular fragments to the value of the soil sorption coefficient in the following way. Imagine a molecule X–Y–Z composed of structural groups and atoms X, Y, and Z. The standard molar Gibbs energy of X–Y–Z, G^0_{mXYZ}, is related to the group contributions by:

$$G^0_{mXYZ} = g_X + g_Y + g_Z + g_{X,Y} + g_{X,Z} + g_{Y,Z} \qquad (10.6)$$

where the contributions g_X (J), g_Y (J), and g_Z (J) are independent functions of the groups X, Y, and Z, respectively, and the contributions $g_{X,Y}$ (J), $g_{X,Z}$ (J), and $g_{Y,Z}$ (J) result from group interactions of X with Y, X with Z, and Y with Z, respectively. The g_i are additive and constitutive functions of molecular structure.

Suppose the molecule X–Y–Z is partitioned between liquid phase 1 and liquid phase 2; X–Y–Z (l_1) $\rightleftharpoons$ X–Y–Z (l_2). The equilibrium constant, K_{XYZ}, is

$$K_{XYZ} = \frac{C_{XYZ}(l_2)}{C_{XYZ}(l_1)} \qquad (10.7)$$

where $C_{XYZ}(l_1)$ (mol/L) is the equilibrium concentration of X–Y–Z in liquid phase 1 and $C_{XYZ}(l_2)$ (mol/L) is the equilibrium concentration of X–Y–Z in phase 2. The change in standard Gibbs energy of the process is

$$\Delta G^0_{mXYZ} = G^0_{mXYZ}(l_2) - G^0_{mXYZ}(l_1)$$

$$= \left[g_X(l_2) + g_Y(l_2) + g_Z(l_2) + g_{X,Y}(l_2) + g_{X,Z}(l_2) + g_{Y,Z}(l_2) \right]$$

$$- \left[g_X(l_1) + g_Y(l_1) + g_Z(l_1) + g_{X,Y}(l_1) + g_{X,Z}(l_1) + g_{Y,Z}(l_1) \right]$$

$$= -RT \ \ln K_{XYZ}$$

Rearranging and dividing all terms by $-RT$, we obtain:

$$\ln K_{XYZ} = B_X + B_Y + B_Z + C_{X,Y} + C_{X,Z} + C_{Y,Z} = \Sigma_i b_i B_i + \Sigma_j c_j C_j \quad (10.8)$$

where b_i is the number of B_i fragments found in the molecular structure, $B_X = -[g_X(l_2) - g_X(l_1)]/RT$, $B_Y = -[g_Y(l_2) - g_Y(l_1)]/RT$, $B_Z = -[g_Z(l_2) - g_Z(l_1)]/RT$, c_j is the number of C_j interactions which must be accounted for, $C_{X,Y} = -[g_{X,Y}(l_2) - g_{X,Y}(l_1)]/RT$, $\Delta_{X,Z} = -[g_{X,Z}(l_2) - g_{X,Z}(l_1)]/RT$, and $C_{Y,Z} = -[g_{Y,Z}(l_2) - g_{Y,Z}(l_1)]/RT$. In practice, a regression formula of the form $a + \Sigma_i b_i B_i + \Sigma_j c_j C_j$, where a is a constant, will be used. Also, not all group contribution terms may be significant, and only the most important may be retained.

Several group contribution methods of estimating sorption coefficients are described in Section 10.3.

Consider an aqueous solution of chemical in contact with soil or sediment on which the chemical is sorbed. For the sorbate, $f_S = \gamma_S x_S f^0$, where f_S is the sorbate's fugacity, γ_S is the sorbate's activity coefficient, x_S is the mole fraction of sorbate on the sorbent, and f^0 (Pa) is the reference fugacity. At low concentrations, $x_S \approx (C_S/\rho_S)V_{Sm}$, where V_{Sm} is the partial molar volume of sorbed material, and $\gamma_S x_S f^0 = \gamma_S(C_S/\rho_S)V_{Sm}f^0$. For the solute, $f_W = \gamma_W x_W f^0$, where γ_W is the chemical's activity coefficient in solution and x_W is the mole fraction of chemical in solution. At low concentrations, $x_W \approx C_W V_{Wm}$, where V_{Wm} is the partial molar volume of water in soil solution, and $\gamma_W x_W f^0 = \gamma_W C_W V_{Wm}f^0$. At equilibrium, $f_S = f_W$ and $\gamma_S(C_S/\rho_S)V_{Sm}f^0 \approx \gamma_W C_W V_{Wm}f^0$ or $K_d = C_S/C_W \approx (V_{Wm}/V_{Sm}\rho S)(\gamma_W/\gamma_S) \approx (\text{const.})(\gamma_W/\gamma_S)$.*

In general, the magnitudes of γ_W and γ_S vary with the concentrations of solute and sorbate. The sorbed phase can be viewed much like a nearly ideal mixture of organic chemicals. As a result, the variation of γ_S is several orders of magnitude smaller than variation of γ_W for most organic chemicals (Karickhoff, 1984), and variation in K_d is primarily due to variation in γ_W, that is, $K_d \propto \gamma_W$. Since $K_d = f_{OC} K_{OC}$, $K_{OC} \propto \gamma_W/f_{OC}$.

At solute concentrations below about one half the chemical's aqueous solubility or 1×10^{-5} mol/l, whichever is lower, γ_W values of nonionic organic chemicals vary little from the infinite-dilution limiting value. In this case, sorption is a linear function of solute concentration and, as pointed out in Section 10.1, $K_d = K_p$. On the other hand, ionic intermolecular interactions are important and activity coefficients of ions vary significantly even in extremely dilute solutions. Also, γ_W of ionic chemicals is sensitive to the ionic strength of the aqueous solution.** Because of these and other factors such as solute speciation changes, the sorption of ionic chemicals is not well described by linear isotherms.

Chemical partitioning between aqueous solution and organic matter in soils and sediments is similar in some respects to the chemical partitioning of solute between octanol and water (Karickhoff, 1984; Chiou, 1989). In Section 9.2, we saw that $K_{OW} = C_O/C_W \approx 0.15(\gamma_W/\gamma_O)$, where C_O is the molar concentration of chemical in

---

* Soils and sediments are mixtures of phases within which sorption may occur. In this case, the fugacity of chemical in each phase equals its fugacity in aqueous solution.
** The activity coefficient of ionic chemicals can be estimated with the Davies equation up to ionic strengths of 0.5 mol/kg.

n-octanol and γ_O is the activity coefficient of solute in n-octanol. Since K_{OC} is proportional to γ_W, we conclude that $K_{OC} \propto K_{OW}$. Furthermore, the dependence of K_{OW} values on the chemical structure of nonpolar chemicals is related mainly to the value of γ_W and hence to the chemical's aqueous solubility, C_W^S. We conclude that K_{OC} of hydrophobic chemicals is linearly correlated with K_{OW} and roughly inversely correlated with aqueous solubility. This is the basis for the quantitative K_{OC}–K_{OW} and K_{OC}–aqueous solubility correlations described in Section 10.3.4.

10.3 ESTIMATION METHODS

A simple group-contribution method of estimating the K_{OM} values of nonpolar chemicals is described by Okouchi and Saegusa (1989). Simple relationships between the molecular connectivity index and the soil–water partition coefficient are discussed by Sabljic (1984, 1987), by Bahnick and Doucette (1988), and by Sabljic et al. (1989). Sabljic (1987) described a hybrid estimation method that employs a linear relationship between log K_{OM} and the first-order molecular connectivity index, along with a correction term for "solute polarity". Meylan et al. (1992) extend Sabljic's approach by coupling a fragment-contribution method with a linear correlation between log K_{OC} and the first-order molecular connectivity index. A simple, linear correlation between log K_{OC} and the first-order molecular connectivity index is proposed for nonpolar chemicals.

A number of linear K_{OC}–K_{OW} and K_{OC}–C_W^S relationships have been described. Many are discussed by Karickhoff (1981), Sabljic (1989), and Lyman (1990). In addition, Bintein and Devillers (1994) describe a K_p–K_{OW} relationship that takes soil pH and organic carbon fraction explicitly into account.

The methods described in this chapter differ in accuracy, complexity, and range of applicability. The best method for the task is always the simplest one that is applicable to the chemical of interest and that offers satisfactory accuracy. Keep in mind, however, that the measured sorption coefficients that have been used in the model training sets are accurate only to within an order of magnitude, on average, and the methods described in this chapter will not yield any higher accuracy.

10.3.1 Method of Okouchi and Saegusa

Okouchi and Saegusa (1989) propose the following model:

$$\log K_{OM} = 0.16 + 0.62 AI \tag{10.9}$$

n = 72, s = 0.341, r = 0.967

where K_{OM} = organic matter–water partition function, l/kg
      AI   = absorbability index
           = $\sum_i n_i A_i$
      n_i = number of groups i in molecular structure
      A_i = contribution of group i

Table 10.1 Group Contributions to K_{OM}

Group	A_i	Group	A_i
C	0.26	Br	0.86
H	0.12	NO_2	0.21
N	0.26	–C=C–	0.19
O	0.17	Iso	–0.12
S	0.54	Tert	–0.32
Cl	0.59	Cyclo	–0.28

After Okouchi, S. and Saegusa, H., 1989.
Bull. Chem. Soc. Jpn., **62**, 922–924.

The group-contribution values, A_i, are given in Table 10.1. The model applies to alkylbenzenes, chlorobenzenes, chloroalkanes, chloroalkenes, polycyclic aromatic hydrocarbons (PAH), heterocyclic and substituted PAH, and halogenated phenols with log K_{OM} values in the range from 1.5 to 6.4 log units. The training set is listed in Table 10.2. Although contributions for chain branching in aliphatic hydrocarbons ("Iso" and "Tert") are given in Table 10.1, representative chemicals are not found in the training set. Also, only one cycloalkane is included in the training set. The method does not distinguish between isomers.

Example

Estimate log K_{OM} for 2,4,6-trichlorophenol, $C_6H_3Cl_3OH$, using Equation 10.9.

1. Calculate the absorbability index using the values of A_i listed in Table 10.1.

Group	A_i	Contribution
C	0.26	6(0.26) = 1.56
H	0.12	3(0.12) = 0.36
Cl	0.59	3(0.59) = 1.77
O	0.17	1(0.17) = 0.17
–C=C–	0.19	3(0.19) = 0.57
		AI = 4.43

2. Estimate log K_{OM} using Equation 10.9.

$$\log K_{OM} = 0.16 + 0.62 \cdot AI$$

$$= 0.16 + 0.62(4.43)$$

$$= 2.91$$

The measured value of log K_{OC} for 2,4,6-trichlorophenol is 3.02 (Sabljic, 1987). The estimate of log K_{OM} is in error by 4%. The estimate of K_{OM} is in error by 22%.

Table 10.2   Values of the Soil Sorption Coefficient (K_{OM}) and Absorbability Indexes (AI) of the 72 Hydrophobic Organic Compounds

Chemical	log K_{OM}	AI	% Error[a]
Dibenz(a,h)anthracene	6.31	9.49	3.6
3-Methylcholanthrene	6.25	9.09	6.7
Dibenzo(a,i)carbazole	6.14	8.92	6.7
2,3,4,5,6,2′,5′-Heptachlorobiphenyl	5.95	8.75	5.5
Naphthacene	5.81	7.83	13.2
DDT[b]	5.38	8.81	−5.2
7,12-Dimethylbenz(a)athracene	5.37	8.83	−5.6
2,4,5,2′,4′,5 ′-Hexachlorobiphenyl	5.34	8.28	0.2
6-Aminochrysene	5.21	8.21	−1.4
2,3,4,2′,3′,4′-Hexachlorobiphenyl	5.05	8.28	−5.5
Pyrene	4.92	6.88	9.5
9-Methylanthracene	4.81	6.67	10.1
DDE[c]	4.70	8.29	−13.5
2,5,2′,5′-Tetrachlorobiphenyl	4.67	7.34	−1.5
2,4,5,2′,5′-Pentachlorobiphenyl	4.63	7.81	−8.7
2,3,4,2′,5′-Pentachlorobiphenyl	4.50	7.81	−11.9
Aldrin	4.45	8.00	−15.8
2-Aminoanthracene	4.45	6.55	4.6
Anthracene	4.42	6.17	9.3
2,4,4′-Trichlorobiphenyl	4.38	6.87	−1.5
Phenanthrene	4.36	6.22	7.3
Acridine	4.22	6.05	6.7
Dibenzothiophene	4.05	5.76	7.3
2-Methylnaphthalene	3.93	5.01	16.4
2,4′-Dichlorobiphenyl	3.90	6.40	−6.5
Pentachlorophenyl	3.73	5.37	5.9
2,2′-Dichlorobiphenyl	3.68	6.40	−12.9
Hexachlorobenzene	3.59	5.67	−3.0
3 4,5-Trichlorophenyl	3.56	4.43	17.9
Pentachlorobenzene	3.50	5.20	2.7
Butylbenzene	3.39	4.85	6.0
1,2,3-Trichlorobenzene	3.37	4.26	16.4
2 4,5-Trichlorophenol	3.36	4.43	13.0
2,3,4,6-Tetrachlorophenol	3.35	4.90	4.0
I-Naphthol	3.33	4.68	7.5
2-Chlorobiphenyl	3.23	5.93	−19.5
Hexachlorocyclohexane	3.21	6.08	−23.2
1,2,3,5-Tetrachlorobenzene	3.20	4.73	2.8
1 2,4,5-Tetramethylbenzene	3.12	4.85	−2.1
Naphthalene	3.11	4.51	4.4
2,4,6-Trichlorophenol	3.02	4.43	3.2
1,3,5-Trichlorobenzene	2.85	4.26	1.1
1,3,5-Trimethylbenzene	2.82	4.35	−1.9
1,2,3-Trimethylbenzene	2.80	4.35	−2.7
2,4-Dichlorophenol	2.75	3.96	4.3
1,2,4-Trichlorobenzene	2.70	4.26	−4.4
2,3-Dichlorophenol	2.65	3.96	0.7
Bromobenzene	2.60	3.59	7.7
1,4-Dimethylbenzene	2.52	3.85	−1.7
1,4-Dichlorobenzene	2.40	3.79	−5.2

Table 10.2 (continued)   Values of the Soil Sorption Coefficient
(K_{OM}) and Absorbability Indexes (AI) of the 72
Hydrophobic Organic Compounds

Chemical	log K_{OM}	AI	% Error[a]
Toluene	2.39	3.35	5.8
Tetrachloroethene	2.32	3.07	10.5
1, 2-Dichlorobenzene	2.26	3.79	−11.7
1,3-Dimethylbenzene	2.26	3.85	−13.4
1,3-Dichlorobenzene	2.23	3.79	−13.2
4-Bromophenol	2.17	3.76	−15.5
1,2-Dibromo-3-chloropropane	2.11	3.69	−16.7
Chlorobenzene	2.10	3.32	−6.3
1,1,1-Trichloroethane	2.02	2.65	10.2
Trichloroethene	2.00	2.60	10.9
Ethylbenzene	1.98	3.85	−29.4
Benzene	1.92	2.83	− 1.0
1,1,2-Trichloroethane	1.87	2.65	3.0
Tetrachloromethane	1.85	2.62	3.0
1,1,2,2-Tetrachloroethane	1.66	3.12	−27.0
Trichloromethane	1.65	2.15	9.0
1,2-Dibromethene	1.65	2.67	−10.7
1,2-Dibromoethane	1.56	2.72	−19.1
Dichloromethane	1.44	1.68	16.1
1,2-Dichloropropane	1.43	2.68	−28.1
1,3-Dichloropropane	1.42	1.90	5.2
1,2-Dichloroethane	1.28	2.18	−18.8

[a] % Error = [(1 − log $K_{OM, calc}$)/log K_{OM}] × 100.
[b] 1,1-Bis(4-chlorophenyl)-2,2,2-trichloroethylene.
[c] 2,2-Bis(4-chlorophenyl)-1,1-dichloroethylene.
After Okouchi, S. and Saegusa, H., 1989. *Bull. Chem. Soc. Jpn.*, **62**, 922–924.

Since the method does not distinguish between isomers, the estimated log K_{OM} of 3,4,5-trichlorophenol is also 2.91. The measured value is 3.56 (Sabljic, 1987). The estimate of log K_{OM} is in error by 18%, and the estimate of K_{OM} is in error by 78%.

10.3.2   Method of Bahnick and Doucette

Bahnick and Doucette (1988) describe a relationship between log K_{OC} and molecular connectivity indices. It is

$$\log K_{OC} = 0.64 + 0.53\,^1\chi + 2.09\Delta^1\chi^\nu \qquad (10.10)$$

n = 56, s = 0.34, r = 0.969, F(2,53) = 411

where K_{OC} = organic carbon–water partition coefficient, l/kg
$^1\chi$ = first-order molecular connectivity index of the molecule

$\Delta^1\chi^v = (^1\chi^v)_{np} - {}^1\chi^v$

${}^1\chi^v$ = first-order valence connectivity index of the molecule

$(^1\chi^v)_{np}$ = first-order valence connectivity index of the corresponding nonpolar hydrocarbon

n   = number of chemicals in the training set

s   = standard error of the estimate

r   = correlation coefficient

F   = F-ratio

The model applies to a diverse set of chemicals including haloalkanes, chlorobenzenes, polycyclic aromatic hydrocarbons, polychlorinated biphenyls, and pesticides with values of log K_{OC} in the range from 1 to 6.5. The training set is listed in Table 10.3. When applied to a test set of 40 chemicals that included 16 alkylbenzenes and haloalkanes and 24 hydrocarbons containing oxygen and nitrogen heteroatoms, the model produced a standard error of the estimate of 0.37.

As with the other connectivity models, the first-order molecular connectivity index, ${}^1\chi$, encodes the size of a molecule (Sabljic, 1984; 1987). The parameter $\Delta^1\chi^v$, on the other hand, encodes the contribution of electronegative atoms such as N and O in a molecule to nondispersive interactions between solute and soil.

Example

Estimate the value of K_{OC} for 2-chloroacetanilide using Equation 10.10.

1. Calculate the first-order connectivity index for 2-chloroacetanilide. The hydrogen-suppressed graph of the chemical is shown along with delta values in Figure 10.1.

$$
{}^1\chi = \Sigma_{bonds} \left(\delta_i\delta_j\right)^{-0.5}
$$

$$
= (4)\,(3 \cdot 2)^{-0.5} + (3)\,(2 \cdot 2)^{-0.5} + (3)\,(1 \cdot 3)^{-0.5} + (1)\,(3 \cdot 3)^{-0.5}
$$

$$
= 5.198
$$

2. Calculate the first-order valence connectivity index for 2-chloroacetanilide. The hydrogen-suppressed graph is shown along with valence delta values in Figure 10.1.

$$
{}^1\chi^v = \Sigma_{bonds} \left(\delta_i^v\delta_j^v\right)^{-0.5}
$$

$$
= (3)\,(3 \cdot 3)^{-0.5} + (3)\,(4 \cdot 4)^{-0.5} + (2)\,(3 \cdot 4)^{-0.5} + (1)\,(4 \cdot 0.78)^{-0.5}
$$

$$
+ (1)\,(4 \cdot 6)^{-0.5} + (1)\,(1 \cdot 4)^{-0.5}
$$

$$
= 3.597
$$

Table 10.3   Training Set for Equation 10.10

Chemical	$^1\chi$	$\Delta^1\chi^v$	log K_{OC} calc.[a]	log K_{OC} exp.[b]
3-Methylaniline	3.788	0.211	2.20	1.65
3-Chloro-4-methoxyaniline	4.736	0.660	1.77	1.93
Diphenylamine	6.449	0.207	3.62	2.78
Anisol	3.392	0.448	1.50	1.54
Acetophenone	4.305	0.489	1.90	1.63
Nitrobenzene	4.305	0.359	2.17	1.94
4-Aminonitrobenzene	4.698	0.569	1.94	1.88
3,4-Dichloronitrobenzene	5.109	0.358	2.60	2.53
Diallate	6.896	0.728	2.77	3.28
2-Chloroacetanilide	5.198	0.713	1.90	1.58
3-Chloroacetanilide	5.182	0.713	1.89	1.86
4-Methoxyacetanilide	5.720	1.161	1.24	1.40
3-(4-Chlorophenyl)-1-methyl-methoxyurea	6.630	1.400	1.22	1.84
3-Phenyl-1-cyclopentylurea	7.343	0.913	2.62	1.93
Phenylurea	4.788	0.924	1.24	1.35
3-(3,4-Dichlorophenyl)-1-methylurea	6.130	1.000	1.79	2.46
n-Propyl-N-phenylcarbamate	6.326	1.050	1.79	2.06
Carbaryl	7.309	1.092	2.23	2.02
Bromacil	6.485	0.722	2.56	1.86
Trietazine	7.206	0.889	2.60	2.78
Cyanazine	7.465	1.071	2.35	2.30
Butralin	9.574	0.940	3.75	3.91
Fluchloralin	10.628	0.966	4.25	3.56
3-(3-Chlorophenyl)-1,1-dimethylurea	6.092	1.005	1.76	1.79
2,4-D	6.092	1.142	1.48	1.30
Picloram	5.947	1.108	1.47	1.23
Tetrachloroguaiacol	6.002	0.724	2.30	2.85
2,4,6-Trichlorophenol	4.609	0.227	2.50	2.52
1-Naphthol	5.377	0.276	2.89	2.64
Pentachlorophenol	5.464	0.227	2.95	2.95
1,2,7,8-Dibenzocarbazole	10.416	0.223	5.69	6.11
Acridine	6.933	0.130	4.04	4.11
7,12-Dimethylbenzanthracene	9.771	0.000	5.82	5.37
3-Methylcholanthrene	10.330	0.000	6.12	6.25
1,3,5-Trimethylbenzene	4.182	0.000	2.85	2.82
1,3,5-Trichlorobenzene	4.182	0.000	2.85	2.85
Trichloroethene	2.270	0.000	1.84	2.00
1,4-Dimethylbenzene	3.788	0.000	2.654	2.52
Naphthalene	4.966	0.000	3.27	3.11
Anthracene	6.933	0.000	4.31	4.42
Tetracene	8.899	0.000	5.36	5.81
Pyrene	7.933	0.000	4.84	4.92
Chlorobenzene	3.394	0.000	2.44	2.41
1,3-Dichlorobenzene	3.788	0.000	2.65	2.47
1,4-Dichlorobenzene	3.788	0.000	2.65	2.44
1,2,3,5-Tetrachlorobenzene	4.609	0.000	3.08	3.20
2,5,2′,5′-Tetrachlorobiphenyl	7.575	0.000	4.65	4.91
2,3,4,2′,5′-Pentachlorobiphenyl	8.003	0.000	4.88	4.54
2,3,4,2′,5′-Heptachlorobiphenyl	8.841	0.000	5.33	5.95

Table 10.3   (continued)   Training Set for Equation 10.10

Chemical	$^1\chi$	$\Delta^1\chi^v$	log K_{OC} calc.[a]	log K_{OC} exp.[b]
DDT	8.874	0.000	5.34	5.38
Dichloromethane	1.414	0.000	1.39	1.44
1,2-Dibromoethane	1.914	0.000	1.65	1.64
Pentachlorobenzene	5.037	0.000	3.31	3.50
Tetrachloromethane	2.000	0.000	1.70	1.85
1,1,1-Trichloroethane	2.000	0.000	1.70	2.26
Hexachlorobenzene	5.464	0.000	3.53	3.59

[a] The calculated values of log K_{OC} were obtained using Equation 10.10.
[b] See Bahnick and Doucette (1988) for sources of experimental log K_{OC} values.

Reprinted from Bahnick, D.A. and Doucette, W.J., 1988. With kind permission from Elsevier Science Ltd., The Boulevard, Langford Lane, Kidlington OX5 1GB, U.K.

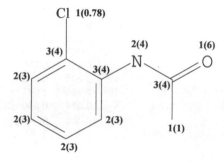

Figure 10.1   The hydrogen-suppressed graph of 2-chloroacetanilide. δ values are shown next to each atom. δ_V values are given in parentheses.

3. Calculate the first-order valence connectivity index and $\Delta^1\chi^v$ for the nonpolar analog of 2-chloroacetanilide. The hydrogen-suppressed structure of the analog is obtained by replacing the carbonyl oxygen with a methyl group. The valence delta values are shown in Figure 10.2.

$$\left(^1\chi^v\right)_{np} = \Sigma_{bonds} \left(\delta_i^v \delta_j^v\right)^{-0.5}$$

$$= (3)\,(3\cdot3)^{-0.5} + (2)\,(3\cdot4)^{-0.5} + (2)\,(1\cdot3)^{-0.5} + (1)\,(4\cdot0.78)^{-0.5} + (1)\,(4\cdot2)^{-0.5}$$

$$+ (1)\,(2\cdot3)^{-0.5} + (1)\,(4\cdot4)^{-0.5}$$

$$= 4.310$$

$$\Delta^1\chi^v = \left(^1\chi^v\right) - {}^1\chi^v$$

$$= 4.310 - 3.597$$

$$= 0.713$$

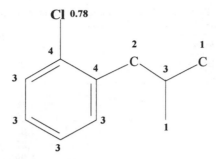

Figure 10.2 The hydrogen-suppressed graph of the nonpolar analog of 2-chloroacetanilide. δ^V values are shown next to each atom.

4. Estimate K_{OC} using Equation 10.10.

$$\log K_{OC} = 0.53^1\chi - 2.09\Delta^1\chi^\nu + 0.64$$

$$= (0.53)(5.198) - (2.09)(0.713) + 0.64$$

$$= 1.91$$

The measured value of log K_{OM} for 2-chloroacetanilide is 1.34 (Biggs, 1981). It was mentioned in Section 10.1 that $K_{OC} \approx 1.724 \cdot K_{OM}$ or log $K_{OC} \approx 0.24 + \log K_{OM}$. Using this relationship to convert K_{OM} values to K_{OC} values, the experimental value of log K_{OC} for 2-chloroacetanilide is 1.58 (Bahnick and Doucette, 1988). The estimate of the log K_{OC} value is in error by 21%. The estimate of the K_{OC} value is in error by 114%.

10.3.3 Method of Meylan et al.

Meylan et al. (1992) describe a structure-K_{OC} model that is an extension of a linear relationship between K_{OM} and the first-order molecular connectivity reported by Sabljic (1987). Sabljic's correlation is

$$\log K_{OM} - 0.53^1\chi + 0.54 \tag{10.11}$$

n = 72, s = 0.295, r = 0.976, F(1,70) = 1415

where K_{OM} = organic matter–water partition function, l/kg
  $^1\chi$ = first-order molecular connectivity
  n = number of chemicals in the training set
  s = standard error of the estimate
  r = correlation coefficient
  F = F-ratio

The model applies to chloroalkanes, chloroalkenes, chlorobenzenes, alkylbenzenes, halogenated phenols, polychlorinated biphenyls, polycyclic aromatic hydrocarbons

Table 10.4   Empirical Polarity Factor by Chemical Class for Polar and Ionic Chemicals

Chemical class	$K_{OM}^{calcd} - K_{OM}^{obsd}$ [a]	n [b]
Substituted benzenes and pyridines	1.00	6
Carbamates	1.05	5
Anilines	1.08	12
Nitrobenzenes	1.16	6
Phenylureas	1.88	15
Triazines	1.88	11
Acetanilides	1.97	17
Uracils	1.99	3
Alkyl-N-phenylcarbamates	2.01	10
3-Phenyl-methylureas	2.07	5
3-Phenyl-1-methyl-1-methoxyureas	2.13	4
Dinitrobenzenes	2.28	7
3-Phenyl-1,1'-dimethylureas	2.36	16
Organic acids	2.39	8
3-Phenyl-1-cycloalkylureas	2.76	4

[a] $K_{OM}^{calcd} - K_{OM}^{obsd}$ is the average difference between the estimated value of K_{OM} derived from Equation 10.11 and the measured value.
[b] n is the number of compounds within the chemical class of the training set.

After Sabljic, A., 1987. *Environ. Sci. Technol.*, **21**, 358–366.

(PAH), and heterocyclic and substituted PAH analogs with log K_{OM} values in the range from 1 to 6.5. A test set of 143 diverse chemicals was used for model validation. The test set included acids, phosphates, and pesticides. The model works well for nonpolar solutes, but the predictive error is high for polar chemicals. The error of the estimate appears to be roughly constant throughout the range of log K_{OM}, however.

The first-order molecular connectivity index of the molecule, $^1\chi$, encodes only the size of a molecule (Sabljic, 1984; 1987). Sabljic (1987) notes that $K_{OM}^{calcd} - K_{OM}^{obsd}$, the absolute difference between the estimated and the measured values of K_{OM}, is a function of the solute's chemical class. This is shown in Table 10.4. Also, the range of error of the K_{OM} estimate for each chemical class is comparatively small. The evidence suggests that Equation 10.11 does not account for the important polar interactions between solute and soil that control sorption.

In order to account for the influence of solute polarity on sorption, Sabljic proposes a hybrid model which incorporates an empirical "polarity" correction. The semiempirical model is

$$\log K_{OM} = 0.53\,^1\chi - 0.996P_f + 0.53 \tag{10.12}$$

n = 215, s = 0.276, r = 0.972

where K_{OM} = organic matter–water partition function, l/kg
   $^1\chi$ = first-order molecular connectivity
   P_f = chemical class polarity factor

$$= K_{OM}^{calcd} - K_{OM}^{obsd}$$

n   = number of chemicals in the training set

s   = standard error of the estimate

r   = correlation coefficient

Values of $P_f = K_{OM}^{calcd} - K_{OM}^{obsd}$ for various chemical classes are given in Table 10.4. The model applies to chloroalkanes, chloroalkenes, chlorobenzenes, alkylbenzenes, halogenated phenols, organic acids, organic phosphates, pesticides, polychlorinated biphenyls, polycyclic aromatic hydrocarbons, and their heterocyclic and substituted analogs with log K_{OM} values between 0.3 and 6.5. It provides reliable estimates of the K_{OM} values of all classes of chemicals except the organic phosphates. The model has not been validated with an independent test set of chemicals, however.

Sabljic's training set is not large, and many classes contain few representatives as Table 10.4 shows. On the other hand, the average deviation between measured and predicted values within each class of chemical is below 0.30, the model's standard error.

It is easy to estimate K_{OM} values of monofunctional chemicals using Sabljic's method, but it is not clear how to proceed when a chemical is multifunctional and falls into more than one class. This is a typical problem with correlation methods. Meylan et al. (1992) avoid this problem by using fragment contributions to estimate the value of soil–water distribution coefficients for polar chemicals. A simple model is described for estimating the K_{OC} values of nonpolar chemicals. It is

$$\log K_{OC} = 0.62 + 0.53\,{}^{1}\chi \qquad (10.13)$$

n = 64, s = 0.267, r = 0.978, F = 1371

where K_{OC} = organic carbon–water partition coefficient, l/kg

       ${}^{1}\chi$ = first-order molecular connectivity index of the molecule

       n   = number of chemicals in the training set

       s   = standard error of the estimate

       r   = correlation coefficient

       F   = F-ratio

The method applies to aromatic hydrocarbons, halogenated aromatic hydrocarbons, phenols, halogenated phenols, polycyclic aromatic hydrocarbons, and halogenated aliphatic hydrocarbons with log K_{OC} values between 1.75 and 6.03. The model was validated with a test set consisting of 41 nonpolar chemicals that included pesticides. The standard error of the estimate is 0.444 and the correlation coefficient is 0.930 for the test set. The test set is shown in Table 10.5.

A similar model was recently derived for hydrophobic chemicals by Sabljic et al. (1995). It is:

$$\log K_{OC} = 0.70 + 0.52\ {}^{1}\chi$$

n = 81, s = 0.264, r = 0.981, F(1,79) = 1993

Table 10.5   Test Set for Equation 10.13 — Chemicals with No
             Correction Factors

Chemical	log K_{OC} exp[a]	log K_{OC} calc[b]	$^1\chi$
biphenyl	3.27	3.80	5.966
2,2′,4-PCB	4.84	4.44	7.182
2,2′,5-PCB	4.57	4.44	7.182
benzo[a]pyrene	5.95	5.90	9.916
benz[a]anthracene	5.30	5.36	8.916
fluorene	3.85	4.05	6.449
fluoranthene	4.62	4.85	7.949
phenol	2.40	2.43	3.394
2-chlorophenol	2.60	2.65	3.805
3-chlorophenol	2.54	2.64	3.788
4-bromophenol	2.41	2.64	3.788
3,4-dichlorophenol	3.09	2.86	4.198
3,5-dimethylphenol	2.83	2.85	4.182
2,3,5-trimethylphenol	3.61	3.07	4.609
p-cresol	2.70	2.64	3.788
5-indanol	3.40	3.21	4.860
styrene	2.96	2.71	3.932
o-xylene	2.25	2.65	3.805
n-propylbenzene	2.87	2.98	4.432
aldrin	4.69	5.02	8.276
α-BHC (benzene hexachloride)	3.30	3.53	5.464
β-BHC (benzene hexachloride)	3.50	3.53	5.464
α-chlordane	4.77	4.94	8.114
mirex	6.00	5.67	9.500
1,3-dichloropropane	1.75	1.91	2.414
iodobenzene	3.10	2.43	3.394
2-aminoanthracene	4.50	4.52	7.327
1-methylnaphthalene	3.36	3.48	5.377
1-ethylnaphthalene	3.78	3.77	5.915
2-ethylnaphthalene	3.76	3.76	5.898
1-naphthylamine	3.51	3.48	5.377
6-aminochrysene	5.20	5.59	9.343
carbazole	3.40	4.05	6.449
7H-dibenzo[c,g]carbazole	6.03	6.16	10.416
13H-dibenzo[a,i]carbazole	6.02	6.16	10.416
quinoline	3.10	3.26	4.966
acridine	4.11	4.31	6.933
benzo[f]quinoline	4.64	4.32	6.949
benzo[b]thiophene	3.49	3.00	4.466

[a] See Meylan et al. (1992) for sources of experimental log K_{OC}
   values.
[b] The calculated values of log K_{OC} were obtained using
   Equation 10.13.

Reprinted with permission from Meylan, W., Howard, P.H., and
Boethling, R.S., 1992. *Environ. Sci. Technol.,* **26**, 1560–1567.
With permission of the American Chemical Society.

The model is reported to yield the most accurate estimates of any model that applies to nonpolar chemicals, in general. This is discussed further below.

Meylan et al. (1992) propose a hybrid method of estimating the K_{OC} values of polar chemicals. The method incorporates both a structure–K_{OC} relationship and fragment contributions. This is a refinement of Sabljic's approach. The model is

$$\log K_{OC} = 0.62 + 0.53 {}^1\chi + \sum_i n_i P_{f,i} \qquad (10.14)$$

n = 189, s = 0.230, r = 0.977

where K_{OC} = organic carbon–water partition coefficient, l/kg
  ${}^1\chi$ = first-order molecular connectivity index of the molecule
  n_i = number of groups i in molecular structure
  $P_{f,i}$ = polarity correction factor for group i
  n = number of chemicals in the training set
  s = standard error of the estimate
  r = correlation coefficient

Values of the polarity correction factor, $P_{f,i}$, are given in Table 10.6. The method applies to aromatic hydrocarbons, halogenated aromatic hydrocarbons, phenols, halogenated phenols, polycyclic aromatic hydrocarbons, halogenated aliphatic hydrocarbons, pesticides, and 26 additional classes of chemical that contain N, O, P, and S groups. The log K_{OC} values of the training set range from 0 and 6.5. The model was validated with a test set consisting of 164 diverse chemicals. A standard estimate error of 0.465 and a correlation coefficient of 0.895 were obtained. The validation set is listed in Table 10.7.

Lohninger (1994) describes a similar multiple linear regression model for estimating K_{OC} involving two topological descriptors (including the zeroth-order valence connectivity index, ${}^0\chi^v$) coupled with group contributions. The ${}^0\chi^v$ index is highly correlated with molecular volume as is the ${}^1\chi$ index. Smaller training and test sets were used, and the model appears to be no more accurate nor more generally applicable than the model of Meylan et al. In both cases, assembly of the training sets involved critical evaluation of measured K_{OC} values found in the literature, screening out estimated K_{OC} values, and converting K_d or K_{OM} values to K_{OC} values using the relationships given in Section 10.1.

The training and validation sets used by Meylan et al. (1992) and by Lohninger (1994) were carefully selected to eliminate estimated values and outliers. Lohninger's training and test sets are shown in Tables 10.8 and 10.9, respectively. Interestingly, there is little overlap with the test sets used by Meylan et al. given in Tables 10.5 and 10.7.

Table 10.6 Polar Fragments and Polarity Correction Factors, $P_{f,i}$, Used to Estimate Log K_{OC} Values

Fragment	$P_{f,i}$	Freq. in Training Set	Note
With N			
Azo	−1.028	2	
With N and C			
Nitrile/cyanide	−0.722	2	a
Nitrogen to noncyclic aliphatic carbon	−0.124	44	b
Nitrogen to cycloalkane	−0.822	3	
Nitrogen to nonfused aromatic ring	−0.777	69	a,b
Pyridine ring with no other fragments	−0.700	2	c
Aromatic ring with two nitrogen atoms	−0.965	2	
Triazine ring	−0.752	7	
With N and O			
Nitro	−0.632	18	
With N, C, and O			
Urea (N–CO–N)	−0.922	26	
Acetamide, aliphatic carbon (N–CO–C)	−0.811	12	
Uracil (–N–CO–N–CO–C=C–ring)	−1.806	3	
N–CO–O–N=	−1.920	2	
Carbamate (N–CO–O–phenyl)	−2.002	4	
N–phenyl carbamate (N–CO–O or N–CO–S)	−1.025	10	
With C and O			
Ether, aromatic	−0.643	11	d
Ether, aliphatic	−1.264	3	
Ketone	−1.248	4	
Ester	−1.309	4	
Aliphatic alcohol	−1.519	4	
Carboxylic acid	−1.751	6	a
Carbonyl	−1.200		e
With P and O			
Organophosphorous [P=O], aliphatic	−1.698	2	a,f
Organophosphorous [P=O], aromatic	−2.878	2	a
With P and S			
Organophosphorous [P=S]	−1.263	8	a
With C and S			
Thiocarbonyl	−1.100	2	
With S and O			
Sulfone	−0.995	4	

Table 10.6 (continued)   Polar Fragments and Polarity Correction Factors, $P_{f,i}$, Used to Estimate Log K_{OC} Values

[a] Counted only once per structure regardless of the number of occurrences.
[b] Any nitrogen attached by a double bond is not counted; carbonyl and thiocarbonyl are not counted as carbons.
[c] Counted only when no other listed fragment is present in structure.
[d] Either one or both carbons aromatic; if both carbons are aromatic, the structure cannot be cyclic.
[e] Not included in regression derivation; estimated from other carbonyl fragments; counted only when no other carbonyl-containing fragments are present.
[f] This is the only fragment counted, even if other fragments are present in the structure.

After Meylan, W., Howard, P.H., and Boethling, R.S., 1992. *Environ. Sci. Technol.,* **26**, 1560–1567.

Example

Estimate the K_{OC} value of 2-chloroacetanilide using the method of Meylan et al., Equation 10.14.

1. Calculate the first-order connectivity index for 2-chloroacetanilide. The hydrogen-suppressed structure of 2-chloroacetanilide is shown along with delta values in Figure 10.1.

$$^1\chi = \Sigma_{bonds} \left(\delta_i \delta_j\right)^{-0.5}$$

$$= (4)\,(3 \cdot 2)^{-0.5} + (3)\,(2 \cdot 2)^{-0.5} + (3)\,(1 \cdot 3)^{-0.5} + (1)\,(3 \cdot 3)^{-0.5}$$

$$= 5.198$$

2. Estimate $\Sigma_i n_i P_{f,i}$ using the polarity correction factors in Table 10.3. The acetanilide structure contains the following fragments.

Fragment	$P_{f,i}$
Nitrogen to nonfused aromatic ring	−0.777
Acetamide, aliphatic carbon (N–CO–C)	−0.811
TOTAL	−1.588

3. Estimate K_{OC} using Equation 10.14.

$$\log\,K_{OC} = 0.53\,^1\chi + 0.62 + \Sigma_i n_i P_{f,i}$$

$$= 0.53\,(5.198) + 0.62 + (-1.588)$$

$$= 1.79$$

The measured value of log K_{OM} for 2-chloroacetanilide is 1.34 (Briggs, 1981). In Section 10.1, it was noted that $K_{OC} \approx 1.724 \cdot K_{OM}$ or log $K_{OC} \approx 0.24 + \log K_{OM}$. Using

Table 10.7 Test Set for Equation 10.14 — Chemicals with Polarity Correction Factors

Chemical	Fragments[a]	log K_{oc} exp[b]	log K_{oc} calc[c]	$^1\chi$
acetanilides				
3-methylacetanilide	5,11	1.45	1.79	5.182
3-(trifluoromethyl)acetanilide	5,11	1.75	2.43	6.393
4-methoxyacetanilide	5,11,16	1.40	1.43	5.720
9-acetylanthracene	18	3.68	3.77	8.271
acids				
acetic acid	21	0.00	−0.21	1.732
anthracene-9-carboxylic acid	21	2.67	3.27	8.271
3,6-dichlorosalicylic acid	21	2.30	1.82	5.537
benzoic acid	21	1.50	1.16	4.305
4-hydroxybenzoic acid	21	1.43	1.37	4.715
4-methylbenzoic acid	21	1.77	1.37	4.715
4-nitrobenzoic acid	9,21	1.54	1.22	5.626
3,4-dinitrobenzoic acid	9,21	1.53	1.29	6.947
hexanoic acid	21	1.46	0.88	3.770
phenylacetic acid	21	1.45	1.42	4.788
phthalic acid	21	1.07	1.87	5.626
alcohols				
methanol	20	0.44	−0.36	1.000
ethanol	20	0.20	−0.14	1.414
1-propanol	20	0.48	0.12	1.914
1-butanol	20	0.50	0.39	2.414
1-pentanol	20	0.70	0.65	2.914
1-hexanol	20	1.01	0.92	3.414
1-heptanol	20	1.14	1.19	3.914
1-octanol	20	1.56	1.45	4.414
1-nonanol	20	1.89	1.72	4.914
1-decanol	20	2.59	1.98	5.414
sec-phenethyl alcohol	20	1.50	1.39	4.305
anilines				
4-chloroaniline	5	1.96	1.86	3.788
3,4-dichloroaniline	5	2.29	2.08	4.198
3,5-dichloroaniline	5	2.49	2.07	4.182
3,5-dinitroaniline	5,9	2.55	1.77	6.003
N.N-dimethylaniline	3,5	2.26	1.89	4.305
N-methylaniline	3,5	2.28	1.81	3.932
3-methyl-4-bromoaniline	5	2.26	2.08	4.198
2,3,4,5-tetrachloroaniline	5	3.03	2.52	5.037
2,3,4-trichloroaniline	5	2.60	2.31	4.626
3-(trifluoromethyl)aniline	5	2.36	2.50	4.999
aldicarb sulfone	3,13,27	0.50	0.88	6.205
asulam	5,15	2.48	2.52	6.954
azinphos methyl	1,3,22,25	2.28	1.84	9.093
benzamides				
benzamide	22	1.46	1.71	4.305
2-chlorobenzamide	22	1.51	1.93	4.715
2-nitrobenzamide	9,22	1.45	1.78	5.626
3-nitrobenzamide	9,22	1.95	1.78	5.626
4-nitrobenzamide	9,22	1.93	1.78	5.626
3,5-dinitrobenzamide	9,22	2.31	1.85	6.947

Table 10.7 (continued)   Test Set for Equation 10.14 — Chemicals with Polarity Correction Factors

Chemical	Fragments[a]	log K_{OC} exp[b]	log K_{OC} calc[c]	$^1\chi$
4-methylbenzamide	22	1.78	1.93	4.715
N-methylbenzamide	3,22	1.42	1.87	4.843
benzidine	5	3.46	3.44	6.754
BMPC	3,14	1.71	2.32	7.185
1-butylamine	3	1.88	1.78	2.414
butyranilide	5,11	1.71	2.13	5.826
butyl benzyl phthalate	19	4.23	3.97	11.220
captafol	11	3.32	3.44	8.343
captan	11	2.30	2.94	7.400
carbamates				
methyl N-(3-chlorophenyl)	5,15	2.15	1.86	5.720
methyl N-(3,4-dichlorophenyl)	5,15	2.74	2.08	6.130
carbophenothion	25	4.66	3.93	8.593
chlorfenvinphos	24	2.47	2.77	9.453
chlornitrofen	9,16	3.90	4.12	8.969
6-chloro-2,4-diaminomethylthiopyrimidine	5,7	2.08	1.62	5.147
chloroneb	16	3.10	2.36	5.685
3-chloro-4-bromonitrobenzene	9	2.60	2.72	5.109
chloroxuron	3,5,10,16	3.55	3.11	9.542
chlorpropham	5,15	2.80	2.32	6.675
crotoxyphos	19,24	2.00	1.70	9.898
cycloate	3,4,15	2.54	2.26	6.791
diamidaphos	3,24	1.51	0.79	6.200
diazinon	7,25	2.75	3.13	8.898
dicamba	5,21	1.50	1.46	6.075
3,3'-dichlorobenzidine	5	4.35	3.87	7.575
dieldrin	17	4.10	4.03	8.776
diethylacetamide	11	1.84	1.54	3.719
di-2-ethylhexyl phthalate	19	4.94	5.22	13.556
diflubenzuron	5,10,22	3.83	3.03	9.969
dimeton-S-methyl	23	1.49	1.95	5.682
dimethoate	3,11,25	1.20	1.39	5.576
2,6-dinitro-α,α,α-trifluoro-p-toluidine	5,9	2.56	2.65	7.642
2,6-dinitro-N-propyl-α,α,α-trifluoro-p-toluidine	3,5,9	3.61	3.34	9.180
diphenylamine	5	2.78	3.28	6.449
diphenyl ether	16	3.29	3.41	6.449
dipropetryn	3,5,8	3.07	3.10	8.007
disulfoton	25	3.22	2.91	6.682
DNOC (dinitro-o-cresol)	9	2.41	2.78	6.430
EPN	9,25	3.12	4.07	10.049
esters				
benzoic acid butyl ester	19	2.10	2.69	6.343
benzoic acid ethyl ester	19	2.30	2.16	5.343
benzoic acid methyl ester	19	2.10	1.89	4.943
benzoic acid phenyl ester	19	3.16	3.23	7.360
ethyl 4-hydroxybenzoate	19	2.21	2.36	5.736
ethyl 4-nitrobenzoate	9,19	2.48	2.22	6.647
ethyl 3,5-dinitrobenzoate	9,19	2.74	2.28	7.985
ethyl 4-methylbenzoate	19	2.59	2.36	5.753

Table 10.7 (continued)   Test Set for Equation 10.14 — Chemicals with Polarity Correction Factors

Chemical	Fragments[a]	log K_{OC} exp[b]	log K_{OC} calc[c]	$^1\chi$
ethyl phenylacetate	19	1.89	2.41	5.826
ethyl pentanoate	19	1.97	1.61	4.308
ethyl hexanoate	19	2.06	1.87	4.808
ethyl heptanoate	19	2.61	2.14	5.308
ethyl octanoate	19	3.02	2.40	5.808
ethoin	25	4.06	4.12	8.593
ipazine	3,5,8	2.91	2.74	7.562
maleic hydrazide	11	0.45	0.31	3.788
methabenzthiazuron	3,10	2.81	3.29	7.220
methidathion	3,15,17,25	1.53	0.96	7.542
methomyl	3,13	1.30	1.08	4.702
methoxychlor	16	4.90	4.63	9.952
methyl chloramben	5,19	2.71	1.76	6.058
metolachlor	3,5,11,17	2.46	2.46	9.061
molinate	3,15	1.92	0.46	5.843
neburon	3,5,10	3.40	2.95	8.041
4-nitrophenol	9	2.37	2.49	4.698
norflurazon	3,5,11	3.28	3.75	9.342
oxadiazon	5,15,16	3.51	3.54	10.091
permethrin	16,19	4.80	5.25	12.375
phenazine	7	3.37	3.34	6.933
3-phenyl-1-cyclohexylurea	4,5,10	2.07	2.27	7.843
3-phenyl-1,1-dimethylureas				
3-chlorophenyl	3,5,10	1.79	1.92	6.092
3,5-dimethylphenyl	3,5,10	1.73	2.12	6.486
3,5-dimethyl-4-bromophenyl	3,5,10	2.53	2.35	6.914
3-fluorophenyl	3,5,10	1.73	1.92	6.092
4-fluorophenyl	3,5,10	1.43	1.92	6.092
3-phenyl-1,1-dimethylureas				
3-methoxyphenyl	3,5,10,16	1.72	1.56	6.630
4-methoxyphenyl	3,5,10,16	1.40	1.56	6.630
4-methylphenyl	3,5,10	1.51	1.92	6.092
phenylureas				
3-chloro-4-methoxyphenylurea	5,10,16	2.00	1.54	6.130
3,4.dichlorophenylurea	5,10	2.49	1.90	5.592
3-methyl-4-bromophenylurea	5,10	2.37	1.90	5.592
3-methyl-4-fluorophenylurea	5,10	1.78	1.90	5.592
4-phenoxyphenylurea	5,10,16	2.56	2.66	8.237
3-(trifluoromethyl)phenylurea	5,10	1.96	2.32	6.393
phosalone	3,5,14,25	2.63	1.77	9.987
piperophos	23	3.44	3.63	10.021
pirimicarb	3,5,7,15	1.90	1.52	7.824
profluralin	3,5,9	3.93	4.26	11.146
prometon	3,5,8,16	2.60	2.20	7.507
pronamide	3,22	2.30	3.20	7.337
propachlor	3,5,11	2.42	2.45	6.664
propylene glycol methyl ether acetate	17,19	0.36	0.26	4.164
pyrazon	3,5,11	2.08	2.74	7.198
pyroxychlor	16	3.48	3.13	5.931
quintozene	9	4.30	3.38	6.375

Table 10.7 (continued)   Test Set for Equation 10.14 — Chemicals with Polarity Correction Factors

Chemical	Fragments[a]	log K_{OC} exp[b]	log K_{OC} calc[c]	$^1\chi$
secbumeton	3,5,8,16	2.78	2.29	7.689
silvex	16,21	1.75	1.91	6.914
sulfones				
SDI3207 (dinitroamino)	3,5,9,27	2.47	2.76	10.189
SD12030 (dinitroamino)	3,5,9,27	2.86	3.29	11.189
SD12346 (dinitroamino)	3,5,9,27	3.07	3.56	11.689
tebuthiuron	3,10	2.79	2.98	6.858
terbuthylazine	3,5,8	2.32	2.52	6.904
tetrachloroguaiacol	16	2.85	3.17	6.002
thiobencarb	3,15	3.27	3.43	7.668
triallate	3,15	3.35	3.22	7.269
trichlorfon	23	1.90	1.73	5.276
trichloroacetamide	11	0.99	1.38	2.943
4,5,6-trichloroguaiacol	16	2.80	2.94	5.575
3.5,6-trichloro-2-pyridinol	6	2.11	2.37	4.609
urea	10	0.50	0.62	1.732
veratrole	16	2.03	1.93	4.881

[a] Fragments listed in Table 10.6.
[b] See Meylan et al. (1992) for sources of experimental log K_{OC} values.
[c] The calculated values of log K_{OC} were obtained using Equation 10.17 with the polarity correction factors listed in Table 10.6.

Reprinted from Meylan, W., Howard, P.H., and Boethling, R.S., 1992. *Environ. Sci. Technol.*, **26**, 1560–1567. With the permission of the American Chemical Society.

this relationship to convert K_{OM} values to K_{OC} values, the experimental value of log K_{OC} for 2-chloroacetanilide is 1.58 (Bahnick and Doucette, 1988). The percentage error of the estimate of log K_{OC} using the method of Meylan et al. is 13%. The percentage error in the estimate of K_{OC} is 62%.

10.3.4   Quantitative K_{OC}–K_{OW} Relationships

Quantitative relationships between K_{OC} and K_{OW} are widely used directly or as part of the important computer programs CHEMEST (Lyman and Potts, 1987), DTEST in E4CHEM (Bruggemann and Munzer, 1988), and AUTOCHEM in the GEMS Graphical Exposure Modeling System and PCCHEM in the PCGEMS system (GEMS, 1986).

In Section 10.2, we saw that K_{OC} and K_{OW} are related by $K_{OC} = (\gamma_O/\gamma_{OC})\,K_{OW}$. Since n-octanol and soil-sediment organic matter are both nonpolar and have roughly similar carbon/oxygen ratios, we expect that $\gamma_O/\gamma_{OC} \approx 1$ and that the ratio is not sensitive to solute and sorbate concentrations. Karickhoff (1981) first showed that organic chemicals are sorbed primarily on organic matter in sediments and that K_{OC} for chemicals sorbed to sediments is related to K_{OW}. Correlations between K_{OC} and K_{OW} have since been reported by many workers. Some of these are given in Table 10.10. Sabljic et al. (1995) reported class-specific K_{OC}–K_{OW} models for all hydrophobic classes of chemicals. The K_{OC}–K_{OW} relationship most commonly used is (Lyman, 1990)

Table 10.8   Training Data Set: Aqueous Solubility and log K_{OC} of 120 Pesticides[a]

Name	CAS-Nr	Solubility (mg/l)	log K_{OC}
1 -Naphthaleneacetamide	86-86-2	100	2.00
2,4-D acid	94-75-7	890	1.85
2,4-DB butoxyethyl ester	32357-46-3	8	2.70
Alachlor	15972-60-8	240	2.23
Aldoxycarb (aldicarb sulfon)	1646-88-4	10000	1.00
Amitraz	33089-61-1	1	3.00
Ancymidol	12771-68-5	650	2.08
Asulam	3337-71-1	4000	1.60
Azinphos-methyl	86-50-0	29	3.00
Benefin (benfluralin)	1861-40-1	0.1	4.25
Bensulide	741-58-2	5.6	3.00
Bifenthrin	82657-04-3	0.1	5.38
Bromoxynil octanoate ester	1689-99-2	0.08	4.00
Captan	133-06-2	5.1	2.30
Carbofuran	1563-66-2	351	1.34
Chloramben	133-90-4	700	1.56
Chlorobenzilate	510-15-6	13	3.30
Chloropicrin	76-06-2	2270	1.79
Chloroxuron	1982-47-4	2.5	3.75
Chlorpyrifos	2921-88-2	0.4	4.37
Cycloate	1134-23-2	95	2.63
Cypermethrin	52315-07-8	0.004	5.00
DBCP	96-12-8	1000	1.85
DCPA (chlorthal-dimethyl)	1861-32-1	0.5	3.70
Diazinon	333-41-5	60	2.36
Dichlorprop (214-DP) butoxyethyl ester	120-36-5	50	3.00
Dicofol	115-32-2	0.8	3.70
Diethathyl-ethyl	58727-55-8	105	3.15
Dinoseb	88-85-7	100	2.70
Dipropetryn	4147-51-7	16	3.31
Diuron	330-54-1	42	2.68
EPTC	759-94-4	344	2.30
Ethalfluralin	69481-52-3	0.3	3.60
Ethofumesate	26225-79-6	50	2.53
Etridiazole	2593-15-9	50	3.00
Fenamiphos	22224-92-6	400	2.00
Fenoxaprop-ethyl	66441-23-4	0.8	3.98
Fenthion	55-38-9	4.2	3.18
Fluazifop-p-butyl	79241-46-6	2	3.76
Flumetralin	62924-70-3	0.1	4.00
Fluridone	59756-60-4	10	3.00
Fomesafen	72178-02-0	50	1.78
Methabenzthiazuron	18691-97-9	59	2.81
Hexythiazox	78587-05-0	0.5	3.79
Imazapyr acid	81334-34-1	11000	2.00
Isazofos	42509-80-8	69	2.00
Isopropalin	33820-53-0	0.1	4.00
Lindane	58-89-9	7	3.04

Table 10.8 (continued)   Training Data Set: Aqueous Solubility and log K_{oc} of 120 Pesticides[a]

Name	CAS-Nr	Solubility (mg/l)	log K_{oc}
Malathion	121-75-5	130	3.26
Piperophos	24151-93-7	25	3.44
Mecoprop (MCCP)	7085-19-0	620	1.30
Metaldehyd	9002-91-9	230	2.38
Methazole	20354-26-1	1.5	3.48
Methiocarb	2032-65-7	30	2.32
Methoxychlor	72-43-5	0.1	4.90
Methyl isothiocyanate	556-61-6	7600	0.78
Metolachlor	51218-45-2	530	2.30
Metsulfuron methyl	74223-64-6	9500	1.54
Molinate	2212-67-1	970	2.28
NAA ethyl ester	2122-70-5	105	2.48
Naled	300-76-5	2000	2.26
Neburon	555-37-3	4.8	3.40
Norflurazon	27314-13-2	28	3.52
Oxadiazon	19666-30-9	0.7	3.51
Oxycarboxin	5259-88-1	1000	1.98
Oxyfluorfen	42874-03-3	0.1	5.00
Parathion (ethyl parathion)	56-38-2	24	3.70
Pebulate	1114-71-2	60	2.63
Permethrin	52645-53-1	0.006	5.00
Phorate	298-02-2	22	2.82
Phosmet	732-11-6	20	2.91
Piperalin	3478-94-2	20	3.70
Prochloraz	67747-09-5	34	2.70
Prometon	1610-18-0	720	2.77
Pronamide (propyzamide)	23950-58-5	15	2.54
Propanil	709-98-8	200	2.17
Propazine	139-40-2	8.6	2.44
Propiconazole	60207-90-1	110	2.81
Pyrazon (chloridazon)	1698-60-8	400	2.08
Sethoxydim	74051-80-2	4390	2.00
Simazine	122-34-9	6.2	2.37
Sulprofos	35400-43-2	0.31	4.08
Temephos	3383-96-8	0.001	5.00
Terbufos	13071-79-9	5	2.70
Thiabendazole	148-79-8	50	3.40
Thiobencarb	28249-77-6	28	2.95
Thiophanate-methyl	23564-05-8	3.5	3.26
Tralomethrin	66841-25-6	0.001	5.00
Triallate	2303-17-5	4	3.38
Trichlorfon	52-68-6	120000	1.00
Trifluralin	1582-09-8	0.3	4.37
Trimethacarb	2686-99-9	58	2.60
Chlor-thiamide	1918-13-4	950	2.26
2,4,5-T acid	93-76-5	150	1.72
Triclopyr, acid	55335-06-3	440	1.67
Trietazine	1912-26-1	20	3.01

Table 10.8 (continued)   Training Data Set: Aqueous Solubility and log K_{OC} of 120 Pesticides[a]

Name	CAS-Nr	Solubility (mg/l)	log K_{OC}
Butralin	33629-47-9 1		4.15
Profluralin	26399-36-0 .... 0.1		4.16
Chlorfenvinphos	470-90-6 145		2.47
Fenamiphos	22224-92-6 400		2.00
Benomyl	17804-35-2 4		3.28
Cyanazine	21725-46-2 170		2.54
Fenoprop (Silvex)	93-72-1 140		1.75
Diphenamid	957-51-7 260		2.32
Flucythrinate	70124-77-5 .... 0.06		5.00
Maleic hydrazide	123-33-1 6000		0.45
Monocrotophos	6923-22-4 1000000		0.00
Phenmedipharm	13684-63-4 .... 4.7		3.38
Quizalofop-ethyl	76578-14-8 .... 0.31		2.71
Tribufos	78-48-8 .... 2.3		3.70
Crotoxyphos	7700-17-6 1000		2.47
Ametryn	834-12-8 185		2.83
Chloroneb	2675-77-6 8		3.22
Dicrotofos	141-66-2 1000000		1.88
Fenoxycarb	72490-01-8 6		3.00
Linuron	330-55-2 75		2.60
Dieldrin	60-57-1 .... 0.18		4.10
Phosphamidon	13171-21-6 1000000		0.85
Tebuthiuron	34014-18-1 2500		1.90
Diallate	2303-16-4 14		3.52
Sulfometuron-methyl	74222-97-2 70		1.89

[a] See Lohninger (1994) for sources of experimental values.

$$\log K_{OC} = 0.544 \ \log K_{OW} + 1.377 \qquad (10.15)$$

n = 45, r^2 = 0.74

where K_{OC} = organic carbon–water partition coefficient, l/kg
   K_{OW} = octanol–water partition coefficient
   n   = number of chemicals in training set
   r   = correlation coefficient

Equation 10.15 applies to a wide variety of chemicals including aliphatic and aromatic halocarbons, haloalkanes, haloalkenes, substituted benzenes, nitrogen heterocylic compounds, halobenzenes, halogenated bipheyls, halogenated diphenyl oxides, carboxylic acids and esters, dinitroanalines, ureas, uracils, symmetric triazines, phosphorus-containing insecticides, carbamates, thiocarbamates, and carbamoyl

Table 10.9 Test Data Set: Aqueous Solubility and log K_{OC} of 81 Pesticides[a]

Name	CAS-Nr	Solubility (mg/l)	log K_{OC}
1,3-Dichloropropene	542-75-6	2250	1.51
2,4,5-T acid	93-76-5	150	1.72
Acephate	30560-19-1	818000	0.30
Aldicarb	116-06-3	6000	1.48
Amitrole (Aminotriazole)	61-82-5	360000	2.00
Anilazine	101-05-3	8	3.00
Atrazine	1912-24-9	33	2.40
Bendiocarb	22781-23-3	40	2.76
Bifenox	42576-02-3	0.398	4.00
Bromacil acid	314-40-9	700	2.09
Butylate	2008-41-5	46	2.60
Carbaryl	63-25-2	120	2.71
Carboxin	5234-68-4	195	2.41
Chlorimuron ethyl	90982-32-4	1200	2.04
Chlorothalonil	1897-45-6	0.6	3.14
Chlorpropham (CIPC)	101-21-3	89	2.60
Cyfluthrin	68359-37-5	0.002	5.00
Cyromazine	66215-27-8	136000	2.30
DCNA (Dicloran)	99-30-9	7	3.00
Desmedipham	13684-56-5	8	3.18
Dichlobenil	1194-65-6	21.2	2.60
Diclofop-methyl	51338-27-3	3	4.20
Diflubenzuron	35367-38-5	0.08	4.06
Dimethoate	60-51-5	39800	1.00
Disulfoton	298-04-4	25	3.49
Endosulfan	115-29-7	0.32	4.09
Esfenvalerate	66230-04-4	0.002	3.72
Ethion	563-12-2	1.1	4.43
Ethoprop (Ethoprophos)	13194-48-4	750	1.85
Quintozene	82-68-8	0.1	4.30
Fenarimol	60168-88-9	14	2.78
Fenvalerate	51630-58-1	0.002	3.72
Fluometuron	2164-17-2	110	2000
Fluvalinate	69409-94-5	0.005	6.00
Fonofos	944-22-9	16.9	2.94
Hexazinone	51235-04-2	33000	1.73
Hydramethylnon (Amdro)	67485-29-4	0.006	5.86
Iprodione	36734-19-7	13.9	2.85
Isofenphos	25311-71-1	24	2.78
Cyhalothrin	91465-08-6	0.003	5.26
Pirimicarb	23103-98-2	2700	1.90
Metalaxyl	57837-19-1	8400	1.70
Methamidophos	10265-92-6	1000000	0.70
Methidathion	950-37-8	220	2.60
Methomyl	16752-77-5	58000	1.86
Methyl bromide	74-83-91	13400	0.34
Methyl parathion	298-00-0	60	3.71
Metribuzin	21087-64-9	1220	2.21
Mevinphos	7786-34-7	600000	1.64
Napropamide	15299-99-7	74	2.85

Table 10.9   Test Data Set: Aqueous Solubility and log K_{OC} of 81 Pesticides[a]

Name	CAS-Nr	Solubility (mg/l)	log K_{OC}
Nitrapyrin	1929-82-4	40	2.76
Oryzalin	19044-88-3	2.5	2.78
Oxamyl	23135-22-0	282000	1.40
Oxydemeton-methyl	301-12.2	1000000	1.00
Oxythioquinox (Quinomethionate)	2439-01-2	0.1	3.36
PCNB	82-68-8	0.44	3.70
Pendimethalin	40487-42-1	0.275	3.70
Phosalone	2310–17-0	3.0	3.26
Pirimiphos-methyl	29232-93-7	9	3.00
Profenofos	41198-08-7	28	3.30
Prometryn	7287-19-6	33	3.15
Propachlor	1918-16-7	613	2.62
Propargite	2312-35-8	0.5	3.60
Propham (IPC)	122-42-9	250	2.30
Propoxur	114-26-1	1800	1.48
Siduron	1982-49-6	18	2.62
Terbacil	5902-51-2	710	1.95
Terbutryn	886-50-0	22	3.30
Thidiazuron	51707-55-2	20	2.04
Thiodicarb	59669-26-0	19.1	2.54
Thiram	137-26-8	30	2.83
Triadimefon	43121-43-3	71.5	2.48
Tridiphane	58138-08-2	1.8	3.75
Triforine	26644-46-2	30	2.30
Vernolate	1929-77-7	108	2.41
Chlorpyrifos-methyl	5598-13-0	4	3.76
Carbophenothion	786-19-6	1	4.90
Secbumeton	26259-45-0	600	2.78
Fluchloralin	33245-39-5	0.5	3.80
Dinitramin	29091-05-2	1.1	3.84
Dimethipin	55290-64-7	3000	0.48

[a] See Lohninger (1994) for sources of experimental values.

Reprinted from *Chemosphere*, **29**, Loninger, H. Estimation of soil partition coefficints of pesticides, 1623–1626, copyright 1994. With kind permission from Elsevier Science Ltd., The Boulevard, Langford Lane, Kidlington OX5 1GB, U.K.

oximes with log K_{OW} values in the range from –3 to 6.6. The training set, given in Table 10.11, consists primarily of pesticides.

Muller and Kordel (1996) compared quantitative K_{OW}–K_{OC} relationships with a molecular connectivity–K_{OC} relationship (Sabljic et al., 1995) and with the group contribution methods of Meylan et al. (1992) and Lohninger (1994). The K_{OW} values were either taken from the Medchem data base or were estimated with CLOGP (Leo and Weininger, 1989). Only the K_{OW}–K_{OC} method proved to be "suitably" accurate when used to estimate the K_{OC} values of 66 diverse chemicals. Sabljic et al. (1995) also report that class-specific quantitative K_{OW}–K_{OC} relationships are superior to

Table 10.10 Some Relationships Between Sorption Coefficients and Octanol–Water Partition Coefficients

$\log K_{OC} = A + B \log K_{OW}$

Chemical Class	A	B	n	s	r^2	Note
Hydrophobic $1 < \log K_{OW} < 7.5$	0.10	0.81	81	0.451	0.887	a
Nonhydrophobic $-2.0 < \log K_{OW} < 8.0$	1.02	0.52	390	0.557	0.631	b
Acetanilides $0.9 < \log K_{OW} < 5.0$	1.12	0.40	21	0.339	0.491	c
Alcohols $-1.0 < \log K_{OW} < 5.0$	0.50	0.39	13	0.397	0.747	d
Amides $-1.0 < \log K_{OW} < 4.0$	1.25	0.33	28	0.491	0.440	e
Anilines $1.0 < \log K_{OW} < 5.0$	0.85	0.62	20	0.341	0.808	f
Carbamates $-1.0 < \log K_{OW} < 5.0$	1.14	0.365	43	0.408	0.568	g
Dinitroanilines $0.5 < \log K_{OW} < 5.5$	1.92	0.38	20	0.242	0.817	h
Esters $1.0 < \log K_{OW} < 8.0$	1.05	0.49	25	0.463	0.753	i
Nitrobenzenes $1.0 < \log K_{OW} < 4.5$	0.55	0.77	10	0.583	0.666	j
Organic acids $-0.5 < \log K_{OW} < 4.0$	0.32	0.60	23	0.336	0.736	
Phenols, benzonitriles $0.5 < \log K_{OW} < 5.5$	1.08	0.57	24	0.373	0.737	k
Phenylureas $0.5 < \log K_{OW} < 4.2$	1.05	0.49	52	0.335	0.616	l
Phosphates $0 < \log K_{OW} < 6.5$	1.17	0.49	41	0.452	0.726	
Triazines $1.5 < \log K_{OW} < 4.0$	1.50	0.30	16	0.379	0.273	m
Triazoles $-1.0 < \log K_{OW} < 5.0$	1.405	0.47	15	0.482	0.631	n

a Equation 10.11 is more reliable when $\log K_{OW}$ values are below 4.
b Overestimates K_{OC} values of n-alkyl alcohols and organic acids. Underestimates values of amino-polycyclic aromatic hydrocarbons, aliphatic amines, and alkyl ureas.
c With CH_3O, Cl, Br, NO_2, CF_3, and CH_3 substituents.
d With alkyl, phenalkyl, and OH substituents.
e Acetamides (with F, Cl, Br, CH_3O, and alkyl substituents) and benzamides (with NO_2 and N-methyl substituents).
f With Cl, Br, CF_3, CH_3, N-methyl, and N,N-dimethyl substituents.
g With alkyl, alkenyl, Cl, Br, N-methyl, CH_3O substituents.
h With CF_3, alkyl-SO_2, NH_2SO_2, CH_3, and t-butyl substituents.
i Phthalates (with alkyl, phenyl, Cl), benzoates (with alkyl, phenyl, NO_2, OH, Cl, NH_2), phenylacetates (with alkyl, phenalkyl), hexanoates, heptanoates, and octanoates with alkyl substituents.
j With Cl, Br, and NH_2 substituents.
k Phenols (with Cl, Br, NO_2, CH_3, CH_3O, OH) and benzoniriles with Cl.
l With CH_3, CH_3O, F, Cl, Br, cyclic-alkyl, CF_3, and PhO substituents.
m With Cl, CH_3O, CH_3S, NH_2, and N-alkyl substituents.
n With alkyl, CH_3O, F, Cl, CF_3, and NH_2 substituents.

From Sabljic, A., Gusten, H., Verhaar, H., and Hermans, J., 1995. *Chemosphere*, **31**, 4489–4514.

Table 10.11 Training Set for Equation 10.15

Chemical	Chemical
Anthracene	2,4,5-T
Benzene	Trichlopyr
9-Methylanthracene	Trifluralin
2-Methylnaphthalene	Diuron
Naphthalene	Fenuron
Phenanthrene	Fluometuron
Pyrene	Linuron
Tetracene	Monolinuron
2,2',4,5,5'-Pentachlorobiphenyl, Aroclor-1254	Monuron
2,2',4,4',5,5'-Hexachlorobiphenyl	Urea
DDT	Atrazine
Methoxychlor	Cyanazine
Chlorpyrifos	Ipazine
Chlorpyrifos-methyl	Propazine
Leptophos	Simazine
Methyl parathion	Trietazine
Parathion	2-Methoxy-3,5,6-trichloropyridine
Carbaryl	Nitrapyrin
Methomyl	3,5,6-Trichloro-2-pyridinol
6-Chloropicolinic acid	Dinoseb
2,4-D Acid	Arochlor
Picloram	Propachlor
Hexachlorobenzene	

From Lyman, W.J. 1990.

other models when used with the Medchem data base. They caution that K_{OW}–K_{OC} correlations that apply to nonpolar chemicals, in general, are not reliable because published K_{OW} values of nonpolar chemicals are unreliable, especially for log K_{OW} values below 4. Of course, specific relationships are accurate only when used within their range of applicability. Generally applicable relationships such as Equation 10.15 are likely to produce large estimation errors, as the low correlation coefficients show.

Example

Estimate the value of K_{OC} for pentachlorophenol using Equation 10.15.

1. The value of log K_{OC} is estimated using Equation 10.15. For pentachlorophenol, log $K_{OW} = 5.04$ (Xie et al., 1984).

$$\log K_{OC} = 0.544 \ \log K_{OW} + 1.377$$

$$= 0.544 \ (5.04) + 1.377$$

$$= 4.12$$

Kenaga and Goring (1980) report log K_{OC} of pentachlorophenol as 2.95. The estimate of log K_{OC} is in error by 40%. The estimate of K_{OC} is in error by 139.

In general, estimates of K_{OC} values based solely on K_{OW} values are not highly reliable. For instance, a K_{OC}–K_{OW} correlation of Karickhoff et al. (1979) (not given in Table 10.11) was evaluated with a test set of 16 volatile and semivolatile organic solvents by Walton et al. (1992). The relationship was found to underpredict sorption of chemicals with log K_{OW} values below 1 and overpredict sorption of chemicals with log K_{OW} values above 3.

Sorption is sensitive to soil characteristics to some degree. This is responsible, in part, for the poor agreement between measured K_p values reported by different laboratories and for the large predictive errors of the estimation methods described so far. Bintein and Devillers (1994) report a K_p–K_{OW} correlation that applies to acids and bases and specifically accounts for soil pH and organic content. The correlation is

$$\log K_P = 0.25 + 0.93\log K_{OW} + 1.09f_{OC} + 0.32CFa - 0.55CFb' \quad (10.16)$$

n = 229, s = 0.433, r = 0.966, F = 786.07

where K_p = chemical's sorption coefficient
K_{OW} = octanol–water partition coefficient
f_{OC} = fraction of organic carbon in soil
CFa = correction factor for acids
= $-\log[1 + 10^{pH-pKa}]$
CFb' = correction factor for bases
= $-\log[1 + 10^{pKa-(pH-2)}]$
pKa = $-\log$ Ka
Ka = chemical's acid dissociation constant
n = number of chemicals in training set
s = standard error of the estimate
r = correlation coefficient
F = F-ratio

The CFa correction factor in Equation 10.16 applies to organic acids only. It corrects for the pH dependence of the distribution of acid between neutral and anionic species. The CFb' correction factor applies to organic bases only. It accounts for the pH dependence of the distribution of base between neutral and cationic species. If the chemical is not an acid or base, both terms are set equal to zero. The model applies to alkyl- and chlorobenzenes, polycyclic aromatic hydrocarbons, haloalkanes, polychlorobiphenyls, organic acids, and organic bases with log K_p values in the range of –1.5 to 5. All the training set data are derived from linear Freundlich isotherms, and $K_p = K_d$ in all cases. The training set is shown in Table 10.12. The model applies to soils with organic carbon content greater than 0.1%. The predictive accuracy of the model was confirmed with a test set of 87 diverse chemicals. The test set is listed in Table 10.13.

Table 10.12 Training Set for Equation 10.16[a]

Chemical	n[b]	log K_{OW}	pKa	%OC	pH
Acetophenone	12	1.59		0.15–2.38[c]	
Acridine	12	3.40	5.68	0.48–2.38	4.54–7.83[c]
Acrylonitrile	2	0.12		0.66–1.49	
2-Aminoanthracene	13	4.13	4.10	0.15–2.38	4.54–8.32
6-Aminochrysene	13	4.98	3.70	0.15–2.38	4.54–8.32
Anisole	1	2.11		1.10	
Anthracene–9-carboxylic acid	12	4.30	3.65	0.15–2.38	4.54–8.32
Benzamide	1	0.64		1.25	
Benzene	2	2.13		0.66–1.49	
	1	2.13		1.10	
n-Butylbenzene	1	4.13		0.15	
Carbon tetrachloride	2	2.64		0.66–1.49	
Chlorobenzene	2	2.84		0.66–1.49	
	1	2.84		1.10	
	1	2.84		0.15	
2-Chlorobiphenyl	1	4.51		1.10	
Chloroform	2	1.97		0.66–1.49	
3-Chlorophenol	1	2.50	8.85	2.80	8.00
1,2:5,6-Dibenzanthracene	14	6.50		0.11–2.38	
1,2-Dichlorobenzene	2	3.38		0.66–1.49	
	1	3.38		1.10	
	1	3.38		0.93	
1,3-Dichlorobenzene	1	3.38		1.10	
1,4-Dichlorobenzene	1	3.39		1.10	
	1	3.39		0.15	
2,2'-Dichlorobiphenyl	1	4.80		1.10	
2,4'-Dichlorobiphenyl	1	5.10		1.10	
1,2-Dichloroethane	1	1.45		0.93	
3,4-Dichlorophenol	2	3.44	7.39	2.80	6.00–8.00
7,12-Dimethylbenzanthracene	13	5.98		0.15–2.38	
1,4-Dimethylbenzene	2	3.15		0.66–1.49	
	1	3.15		0.15	
2,4-Dinitro-o-cresol	12	2.85	4.46	0.15–3.04	4.27–8.29
Ethylbenzene	1	3.15		1.10	
Furan	2	1.34		0.66–1.49	
Hexachlorobenzene	1	5.73		0.66	
Hexanoic acid	I	1.90	4.85	4.85	2.80
Methoxychlor	13	5.08		0.13–3.29	
3-Methylcholanthrene	12	6.42		0.11–2.38	
Nitrobenzene	2	1.87		0.66–1.49[c]	
Pentachlorophenol	1	5.04	4.92	3.20	4.70
Pyrene	14	5.09		0.11–2.38	
	12	5.09		0.13–3.29	
Silvex	12	3.41	3.07	0.15–3.04	4.27–8.29[c]
2,3,7,8-TCDD	1	6.42		0.66	
1,2,3,4-Tetrachlorobenzene	1	4.64		0.15	
1,2,4,5-Tetrachlorobenzene	1	4.72		0.15	
1,1,2,2-Tetrachloroethane	1	2.39		0.93	
Tetrachloroethylene	1	2.60		0.15	
2,3,4,6-Tetrachlorophenol	2	4.42	5.38	1.70–3.20	3.40–4.70

Table 10.12 (continued) Training Set for Equation 10.16[a]

Chemical	n[b]	log K_{OW}	pKa	%OC	pH
Tetrahydrofuran	2	0.46		0.66–1.49	
1,2,4,5-Tetramethylbenzene	1	4.05		0.15	
Toluene	2	2.71		0.66–1.49	
	1	2.71		0.15	
1,2,3-Trichlorobenzene	1	4.14		0.15	
	1	4.14		4.70	
1,2,4-Trichlorobenzene	1	4.02		1.10	
	1	4.02		0.15	
2,4,4'-Trichlorobiphenyl	1	5.62		1.10	
1,1,1-Trichloroethane	1	2.47		0.93	
2,4,5-Trichlorophenol	1	3.72	7.43	2.80	8.00
1,2,3-Trichloropropane	2	2.01		0.66–1.49	
1,2,3-Trimethylbenzene	1	3.60		0.15	
1,3,5-Trimethylbenzene	1	3.60		0.15	

[a] See Bintein and Devillers (1994) for sources of experimental values.
[b] Number of K_p values
[c] Range

Reprinted from *Chemosphere*, **28**, Bintein, S. and Devillers, J., QSAR for organic chemical sorption in soils and sediments, 1173–1174, copyright 1994. With kind permission from Elsevier Science Ltd., The Boulevard, Langford Lane, Kidlington OX5 1GB, U.K.

Example

Estimate the value of K_{OC} for pentachlorophenol using Equation 10.16. Assume that the soil has a pH of 3.7 and an organic carbon content of 2.70%.

1. Pentachlorophenol is acidic and its pKa = 4.74 (Lagas, 1988). Calculate the value of CFa. CFb' = 0.

$$CFa = -\log\left[1 + 10^{pH-pKa}\right]$$
$$= -\log\left[1 + 10^{3.7-4.74}\right]$$
$$= -\log\left[1 + 0.09\right]$$
$$= -0.04$$

2. The value of log K_{OC} is estimated using Equation 10.16. For pentachlorophenol, log K_{OW} = 5.04 (Xie et al., 1984). For the soil, pH = 3.7 and f_{OC} = 0.0270.

$$\log K_p = 0.25 + 0.93 \ \log K_{OW} + 1.09 \ \log f_{OC} + 0.32 \ CFa - 0.55 \ CFb'$$
$$= 0.25 + 0.93 \ (5.04) + 1.09 \ \log (0.0270) + 0.32 \ (-0.04)$$
$$= 3.21$$

Table 10.13 Test Set for Equation 10.16[a]

Chemical	n[b]	log K_{OW}	pKa	N	%OC	pH
Acetanilide	2	1.16		0.82–0.84[c]	1.58–4.85[c]	
Acridine	1	3.40	5.68	1.02	0.58	7.64
Alachlor	1	2.64		1.01	1.17	
Aldicarb	1	1.08		d	2.05	
Aldrin	1	5.66		d	2.05	
Ametryne	34	2.58	4.00	d	0.35–20.88	4.50–9.00
Aniline	1	0.90	4.60	0.83	0.84	5.40
Atrazine	4	2.33	1.68	1	0.40–7.60	4.40–8.00
	1	2.33	1.68	0.92	0.43	6.05
	25	2.33	1.68	d	0.35–2.84	4.30–7.10
Benzamide	1	0.64		0.88	4.85	
Benzoic acid (B.a.)	1	1.97	4.20	0.90	4.85	2.80
B.a. ethyl ester	3	2.64		0.81–0.88	1.25–4.85	
B.a. methyl ester	3	2.20		0.81–0.85	1.25–4.85	
B.a. phenyl ester	2	3.59		0.91–0.93	1.25–1.58	
4-Bromophenol	1	2.59	9.34	d	1.46	6.70
Captafol	1	3.83		d	2.05	
Captan	1	2.54		d	2.05	
Carbendazim	15	1.52	4.48	d	0.63–3.47	4.14–7.54
Carbaryl	1	2.32		d	2.05	
Carbofuran	2	1.63		0.90–0.94	1.60–2.50	
Chlorfenvinphos	1	3.10		d	2.05	
Chlorobenzene	1	2.84		0.87	4.70	
2-Chlorophenol	1	2.17	8.49	0.80	2.96	5.70
3-Chlorophenol	5	2.50	8.85	0.80	1.70–3.20	3.40–6.00
	3(34)	2.50	8.85	0.85–0.96	2.15–9.05	3.60–5.90
	1	2.50	8.85	0.83	2.96	5.70
Chlorsulfuron	1	2.20	3.60	0.87	1.42	4.60
	23	2.20	3.60	d	0.95–20.36	5.20–7.90
3-Cresol	1	1.98	10.0	0.87	2.96	5.70
2,4-D	3	2.81	2.64	0.89–1.06	1.25–4.85	2.80–7.10
	19	2.81	2.64	0.82–1.16	0.28–2.73	6.30–8.30
Diazinon	1	3.11		d	2.05	
Dibenzothiophene	11	4.38		0.88–1.10	0.15–2.38	
2.4-Dichloroaniline	3	2.78	2.05	0.85–1.11	3.54–9.05	3.60–5.90
3,5-Dichloroaniline	1	2.69	2.37	0.91	0.84	5.40
1,4-Dichlorobenzene	1	3.39		0.96	2.15	
2,4-Dichlorophenol	3	3.21	7.68	d	0.20–3.70	4.20–7.40
3,4-Dichlorophenol	5	3.44	7.39	0.80–0.90	0.90–3.20	3.40–7.50
	4	3.44	7.39	0.81–0.94	2.15–9.05	3.60–5.90
Dieldrin	1	4.60		I$	2.05	
Dimethoate	1	0.79		d	2.05	
N,N-Dimethylaniline	3	2.31	5.15	0.87–0.91	1.25–4.85	2.80–7.10
Diuron	34	2.60		d	0.35–20.88	
Fenamiphos	1	3.18		d	2.05	
Folpet	1	3.63		d	2.05	
Hexachlorobenzene	1	5.73		d	1.53	
Hexanoic acid	2	1.90	4.85	1.01	1.25–1.58	6.70–7.10
3-Methoxyphenol	1	1.58	9.65	0.89	2.96	5.70
N-Methylaniline	1	1.66	4.85	0.89	4.85	2.80
4-Methylaniline	1	1.39	5.08	0.96	4.85	2.80
4-Methylbenzoic acid	1	2.34	4.36	0.82	4.85	2.80

Table 10.13 (continued)   Test Set for Equation 10.16[a]

Chemical	n[b]	log K_{OW}	pKa	N	%OC	pH
Metsulfuron-methyl	23	2.20	3.30	[d]	0.95–20.36	5.20–7.90
4-Nitroaniline	1	1.39	1.01	0.81	1.25	6.70
4-Nitrobenzoic acid	1	1.89	3.44	1.14	4.85	2.80
2-Nitrophenol	1	1.79	7.22	0.89	2.96	5.70
Parathion	1	3.76		I$	2.05	
	6	3.76		1.02–1.11	0.44–14.28	
2,2',5,5'-PCB	2	6.10		[d]	0.16–1.87	
	5	6.10		[d]	0.16–1.87	
3,3',4,4'-PCB	5	6.10		[d]	0.16–1.87	
2,2',3,4,5'-PCB	2	6.50		[d]	0.16–1.87	
2,2',3,5',6-PCB	2	6.40		[d]	0.16–1.87	
2,2',3,4,5-PCB	2	6.60		[d]	0.16–1.87	
2,2',4,5,5'-PCB	2	6.40		[d]	0.16–1.87	
	5	6.40		[d]	0.16–1.87	
2,3,3',4',6-PCB	2	6.53		[d]	0.16–1.87	
2,3',4,4',5-PCB	2	6.40		[d]	0.16–1.87	
2,2',3,4,4',5-PCB	2	7.00		[d]	0.16–1.87	
2,2',3,4,5,5'-PCB	2	7.59		[d]	0.16–1.87	
2,2',3,4',5',6-PCB	2	6.80		[d]	0.16–1.87	
2,2',4,4',5,5'-PCB	2	6.90		[d]	0.16–1.87	
	5	6.90		[d]	0.16–1.87	
Pentachlorobenzene	2	4.94		[d]	0.16–1.87	
Pentachlorophenol	6	5.04	4.92	0.80–0.90	0.90–29.80	3.40–7.50
	4	5.04	4.92	0.82–0.92	2.15–9.05	3.60–5.90
	3	5.04	4.92	[d]	0.20–3.70	4.20–7.40
Phenol	1	1.46	9.99	[d]	1.46	6.70
Phenylacetic acid	2	1.41	4.31	0.95–1.12	2.80–6.70	1.25–4.85
Phorate	1	3.83		[d]	2.05	
Prometone	2	1.94	4.28	0.92–0.99	0.33–0.43	6.05–6.30
	25	1.94	4.28	[d]	0.35–2.84	4.30–7.10
Prometryne	25	2.99	4.05	[d]	0.35–2.84	4.30–7.10
Propazine	25	3.09	1.85	[d]	0.35–2.84	4.30–7.10
Pyridine	1	0.65	5.23	1.04	0.58	7.64
Quinoline	1	2.03	4.92	0.87	0.58	7.64
	2	2.03	4.92	0.84–0.91	0.35–0.58	7.46–7.64
Simazine	1	2.27	1.65	[d]	2.05	6.10
	25	2.27	1.65	[d]	0.35–2.84	4.30–7.10
2,4,5-T	3	3.36	2.90	0.84–1.14	1.25–4.85	2.80–7.10
	2	3.36	2.90	0.81–0.85	0.46–1.74	5.90–7.70
1,2,3,4-TCB[e]	2	4.64		0.91–1.13	2.15–4.70	
	2	4.64		[d]	0.16–1.87	
1,2,3,5-TCB[e]	2	4.46		[d]	0.16–1.87	
Tetrachloroguaiacol	3	4.45	5.97	[d]	0.20–3.70	4.20–7.40
2,3,4,5-TCP[f]	2	4.82	6.96	0.81–0.82	2.15–3.54	4.80–5.60
2,3,4,6-TCP[f]	5	4.42	5.38	0.80–0.90	0.90–29.80	4.60–7.50
	3	4.42	5.38	[d]	0.20–3.70	4.20–7.40
3,4,5-Trichloroaniline	6	3.49	1.78	1	0.12–6.34	4.50–5.10
1,2,3-Trichlorobenzene	3	4.14		0.96–1.01	12.15–9.05	
	2	4.14		[d]	0.16–1.87	
1,2,4-Trichlorobenzene	2	4.02		[d]	0.16–1.87	
1,3,5-Trichlorobenzene	2	4.19		[d]	0.16–1.87	
4,5,6-Trichloroguaiacol	3	3.74	7.40	[d]	0.20–3.70	4.20–7.40

Table 10.13 (continued)   Test Set for Equation 10.16[a]

Chemical	n[b]	log K_{OW}	pKa	N	%OC	pH
2,4,5-Trichlorophenol	8	3.72	7.43	0.80–0.90	0.90–29.80	3.40–7.50
	4	3.72	7.43	0.81–0.98	2.15–9.05	3.60–5.90
2,4,6-Trichlorophenol	3	3.75	7.42	[d]	0.20–3.70	4.20–7.40
Trifluralin	4	5.07		1	0.40–7.60	

[a] See Bintein and Devillers (1994) for sources of experimental values.
[b] Number of K_P values.
[c] Range
[d] Single concentration.
[e] Tetrachlorobenzene
[f] Tetrachlorophenol

Reprinted from *Chemosphere*, **28**, Bintein, S. and Devillers, J., QSAR for organic chemical sorption in soils and sediments, 1173–1174, copyright 1994. With kind permission from Elsevier Science Ltd., The Boulevard, Langford Lane, Kidlington OX5 1GB, U.K.

Lagas (1988) reports the log K_p of pentachlorophenol as 2.30 on soil with a pH of 3.7 and a f_{OC} of 0.0270. The estimate of log K_{OC} is in error by 40%. However, the log K_{OC} value of pentachlorophenol does increase rapidly from a value of about 2.8 at pH of 7 to a value of about 4.5 at pH 3 (Lee et al., 1990), and the measured value may not be accurate.

All K_{OC}–K_{OW} regression models are based on the assumption that only the K_{OC} values are subject to significant measurement error. However, K_{OW} values are hard to measure accurately, especially when log K_{OW} is greater than 5. Table 9.7 in Chapter 9 shows that reported values of log K_{OW} for individual chemicals vary by as much as 3.3 log units. Such errors in the log K_{OW} values of the training sets used to derive the K_{OC}–K_{OW} correlations will lead to significant uncertainty in the slope of the regression models (Sabljic, 1991).

10.3.5   Quantitative K_{OC}–Aqueous Solubility Relationships

We saw above that the K_{OC} values of hydrophobic chemicals are roughly linearly correlated with aqueous solubility. A few such correlations have been reported, and many K_{OC} values reported in the literature are estimated in this way. The correlations are reviewed by Karickhoff (1984), Sabljic (1989), and Lyman (1990). Several are given in Table 10.14. The K_{OC}–C_W^S relationship widely used as part of the computer program CHEMEST (Lyman and Potts, 1987) is (Lyman, 1990)

$$\log K_{OC} = 3.64 - 0.55 \log C_W^S \qquad (10.17)$$

n = 106, r^2 = 0.71

where K_{OC} = organic carbon–water partition coefficient, L/kg
$\quad\quad\ C_W^S$ = aqueous solubility, mg/L
$\quad\quad\ $ n   = number of chemicals in training set
$\quad\quad\ $ r   = correlation coefficient

Table 10.14 Some Relationships Between Sorption Coefficients and Aqueous Solubility

log K_{OC} = A + B log C_W^S

Chemical Class	A	B	n	r^2	Ref.
Aromatic, PCBs 2.13 < log K_{OC} < 5.62, −1.64 < log S_{WS} < −5.98	0.001	−0.729	12	0.996	Chiou et al., 1983
Amino- and carboxy-substituted PNA 1 < log K_{OC} < 7, −3.5 < log C_W^S < 4 C_W^S in µg/ml	4.273	−0.686	22	0.933	Means et al., 1982
Chlorinated hydrocarbons 1.5 < log K_{OC} < 5.6, −2.7 < log C_W^S < 5 C_W^S in µmol/l	4.277	−0.557	15	0.99	Chiou et al., 1979
Polychlorobiphenyls 4.5 < log K_h < 5.8, 0.5 < log C_W^S < 1.75 K_h in mg C/l, C_W^S in µg/l	6.186	−0.973	16	0.95	Lara and Ernst, 1989

Equation 10.17 applies to a wide variety of chemicals, including aliphatic and aromatic halocarbons, haloalkanes, haloalkenes, substituted benzenes, nitrogen heterocylic compounds, halobenzenes, halogenated biphenyls, halogenated diphenyl oxides, carboxylic acids and esters, dinitroanalines, ureas, uracils, symmetric triazines, phosphorus-containing insecticides, carbamates, thiocarbamates, and carbamoyl oximes with log C_W^S values in the range from −3.3 to 6. The training set, shown in Table 10.11, consists primarily of pesticides. As the low correlation coefficient shows, Equation 10.17 is not highly reliable and is likely to produce large estimation errors.

Karickhoff (1981, 1984) reports a correlation between K_{OC} and aqueous solubility of solid and liquid organic chemicals. The model includes a melting-point correction for solid chemicals. The origin of the correction is discussed in Chapter 7. The model is

$$\log K_{OC} = -0.93 + -0.01(T_m - 25) - 0.83 \log X^S \qquad (10.18)$$

n = 47, r^2 = 0.93

where K_{OC} = organic carbon–water partition coefficient, L/kg
X^S = aqueous solubility, solute mole fraction
T_m = melting point, °C
n = number of chemicals in training set
r = correlation coefficient

The relationship applies to hydrocarbons, chlorocarbons, chloro-s-triazines, carbamates, organophosphates, and phenyl ureas with log X_S values between −2.8 and −10.5. The training set is given in Table 10.15. In using Equation 10.18, liquid chemicals are assigned a melting point of 25°C.

Equations 10.17 and 10.18 and the K_{OC}–aqueous solubility relationships given in Table 10.14 are not reliable as the low correlation coefficients show. The models

Table 10.15   Training Set for Equation 10.18

Chemical	$T_m(°C)$ exp[a]	$-\log X_S$ exp[a]	$\log K_{OW}$ exp[a]	$\log K_{OC}$ exp[a]	$\log K_{OC}$ calc[b]
Hydrocarbons and Chlorinated Hydrocarbons					
3-methyl cholanthrene	179	9.66	6.42	6.09	6.02
dibenz[a,h]anthracene	269	9.79	6.50	6.22	5.30
7,12-dimethylbenz[a]anthracene	123	8.77	5.98	5.35	5.74
tetracene	335	10.43	5.90	5.81	5.25
9-methylanthracene	82	7.61	5.07	4.71	5.06
pyrene	156	7.92	5.18	4.83	4.64
phenanthrene	101	6.89	4.57	4.08	4.22
anthracene	216	8.12	4.54	4.20	4.25
naphthalene	80	5.35	3.36	2.94	3.00
benzene	25c	3.39	2.11	1.78	1.72
1,2-dichloroethane	25c	2.82	1.45	1.51	1.19
1,1,2,2-tetrachloroethane	25c	3.50	2.39	1.90	1.80
		3.46			
1,1,1-trichloroethane	25c	3.74	2.47	2.25	2.04
tetrachlorothylene	25c	4.28	2.53	2.56	2.54
		4.66			
γ BHC (lindane)	113	6.34	3.72	3.30	3.60
α BHC	160	6.90	3.81	3.30	3.66
β BHC	309	7.91	3.80	3.30	3.17
1,2-dichlorobenzene	25c	4.74	3.39	2.54	2.96
pp' DDT	109	9.79	6.19	5.38	6.81
methoxychlor	89	8.20	5.08	4.90	5.54
22',44',66' PCB	114	10.35	6.34	6.08	7.28
22',44',55' PCB	103	9.30	6.72	5.62	6.42
Chloro-s-triazines					
atrazine	172	5.56	2.33	2.17	2.31
			2.71	2.21	
				2.33	
propazine	213	6.17	2.94	2.20	2.49
				2.19	
				2.56	
simazine	226	6.35	2.16	2.13	2.53
				2.14	
				2.33	
trietazine	103	4.89	3.35	2.74	2.36
ipazine	87	5.53	3.94	3.22	3.10
				2.91	
cyanazine	167	4.89	2.24	2.30	1.75
				2.26	
Carbamates					
carbaryl	142	5.45	2.81	2.36	2.50
carbofuran	151	4.47	2.07	1.46	1.51
chlorpropham	41	5.13	3.06	2.77	3.17
		5.04			3.08
Organophosphates					
malathion	25c	5.10	2.89	3.25	3.29
parathion	25c	5.81	3.81	3.68	3.95
				4.03	
methylparathion	36	5.41	3.32	3.71	3.47
				3.99	

Table 10.15 (continued)   Training Set for Equation 10.18

Chemical	$T_m(^\circ C)$ exp[a]	$-\log X_s$ exp[a]	$\log K_{OW}$ exp[a]	$\log K_{OC}$ exp[a]	$\log K_{OC}$ calc[b]
chlorpyrifos	42	6.99	3.31	4.13	4.87
			4.82		
			5.11		
Phenyl Ureas					
diuron	155	5.49	1.97	2.60	2.41
			2.81	2.58	
fenuron	128	3.37	1.00	1.43	0.72
		3.50		1.63	0.84
linuron	93	5.27	2.19	2.91	2.80
				2.94	
monolinuron	76	4.31	1.60	2.30	2.08
				2.45	
monuron	170	4.68	1.46	2.00	1.52
			2.12	2.26	
fluometuron	164	5.16	1.34	2.24	2.02
Miscellaneous Compounds					
13Hdibenzo[a,i]carbazole	220	9.15	6.40	6.02	5.16
2,2' biquinoline	194	7.14	4.31	4.02	3.56
dibenzothiophene	98	6.84	4.38	4.05	4.20
acetophenone	25[c]	3.09	1.59	1.54	1.40
terbacil	176	4.23	1.89	1.71	1.05
				1.61	
bromacil	158	4.25	2.02	1.86	1.34

[a] See Karickhoff (1981) for sources of experimental data.
[b] The calculated values of log K_{OC} are obtained using Equation 10.18.
[c] Chemical is liquid at 25°C.

Reprinted from *Chemosphere*, **10**, Karickhoff, S.W., Semi-empirical estimation of sorption of hydrophobia, 840–842, copyright 1981. With kind permission from Elsevier Science Ltd., The Boulevard, Langford Lane, Kidlington OX5 1GB, U.K.

are likely to produce large estimation errors. As with the K_{OC}–K_{OW} correlations discussed above, the models are based on the assumption that only values of the dependent variable, K_{OC}, are subject to significant measurement error. As with K_{OW} values, C_W^S values are hard to measure reliably, especially when C_W^S is less than 1 mg/L. Table 7.10 in Chapter 7 shows that reported values of log C_W^S for individual chemicals vary by as much as 3.0 log units. Such errors in the values of C_W^S used in the training sets will lead to significant uncertainty in the slope of the regression models (Sabljic, 1991).

Example

Estimate the value of K_{OC} for pyrene using Equation 10.18.

1. Estimate the mole fraction solubility of pyrene. Two values of C_W^S are reported. They are 6.6×10^{-7} and 1.6×10^{-7} mol/L (Sabljic, 1987). We will use the average value, 4.1×10^{-7} mol/L. The solute concentration is very low, and we will assume that the solution density equals the density of water, $\rho_W = 1$ kg/L. Also, at such low

solute levels, the mole fraction of solute equals the mole ratio of solute to water. The molar mass of water, MW_w, is 55.56 mol/kg.

$$X^S \approx n_{pyrene}/n_{water}$$

$$\approx C_W^S/(\rho_w \cdot MW_w)$$

$$\approx 4.1 \times 10^{-7}/[55.56]$$

$$\approx 7.4 \times 10^{-9}$$

2. Estimate the value of K_{OC} with Equation 10.18. The melting point of pyrene, T_m, is 156°C (Lide, 1994).

$$\log K_{OC} = -0.93 - 0.01 (T_m - 25) - 0.83 \; \log X^S$$

$$= -0.93 - 0.01 (156 - 25) - 0.83 \; \log (7.4 \times 10^{-9}$$

$$= 4.51$$

The reported value of log K_{OC} for pyrene is 4.9 (Means et al., 1982). The estimate is in error by 8%. Using the K_{OC}–X_S correlation for aromatic and polynuclear aromatic hydrocarbons given in Table 10.14, an estimate of 6.83 is obtained. Karickhoff (1981) reports a K_{OC}–X_S correlation for condensed ring aromatics. It is

$$\log K_{OC} = -1.405 - 0.00953 (T_m - 25) - 0.921 \; \log X^S$$

This model yields an estimate of 3.23 for the mole fraction solubility of pyrene.

As shown in Section 7.2, aqueous solubility and the octanol–water partition coefficient are closely correlated. Therefore, the K_{OC}–C_W^S correlations are related to the K_{OC}–K_{OW} correlations described in Section 10.3.4. Since a large data base of accurate K_{OW} values is available, the K_{OC}–K_{OW} correlations are most often used in K_{OC} estimation methods.

10.4 SENSITIVITY TO ENVIRONMENTAL FACTORS AND METHOD ERROR

Lohninger (1994) critically reviewed the accuracy of published K_{OC} values. The standard deviation of published values was determined for 70 pesticides. The standard deviations of log K_{OC} ranged from 0.09 to 1.22 log units. The average standard deviation was 0.44 log units. Schwarzenbach et al. (1993) present evidence that the K_{OM} value of a chemical can vary by a factor of ±2 from soil to soil. Published K_{OC} values appear to be accurate only to within an order of magnitude, and the estimates of models derived with such data will be no more accurate.

Sorption is sensitive to soil type and age. The existence of competitive sorption processes and nonlinear sorption isotherms suggests that sorption involves, at least in part, site-specific interactions with surfaces (Pignatello, 1991; Brusseau, 1992; McGinley et al., 1993). Evidence has been presented that the relative importance of partitioning and adsorption may depend on the age of the soil organic matter (Young and Weber, 1995). Partitioning occurs primarily on amorphous humic materials, while adsorption occurs primarily on condensed microcrystalline structures formed by digenesis.

Soils and sediments have a limited sorption capacity. In deriving sorption models, it is assumed that chemical concentrations are well below the medium's sorption capacity. None of the models described here apply to heavily contaminated soils and sediments. Also, nonlinear sorption is observed at relatively low concentrations on air-dry soils (Shonnard et al., 1993).

Systematic studies of the temperature dependence of sorption have not been performed. Soil–water sorption coefficients are reported for $25 \pm 5°C$. Since sorption on soil organic matter is assumed to be much like partitioning between n-octanol and water, the temperature dependence of sorption should be similar to the temperature dependence of K_{OW}. We expect K_{OM} (and K_{OC}) to decrease with increasing temperature.

Sorption of ionic and ionizable organic chemicals on natural sorbents has been discussed by Schellenberg et al. (1984) and by Westall (1987). The colloidal surfaces of natural soils are negatively charged and preferentially sorb cationic solutes (Bintein and Devillers, 1994). On the other hand, no affinity for anionic solutes is observed. The degree of dissociation of organic bases to form cationic species is dependent on pH and ionic strength. Sorption should increase as the degree of dissociation of organic bases decreases.

Sorption of organic acids also depends on pH and ionic strength. For instance, the sorption of pentachlorophenol increases with increasing ionic strength and decreasing pH (Lee et al., 1990).

The presence of colloids in water enhances the transport of organic chemicals in soils and sediments. Quantitative models of colloid transport have not been reported as yet. Also, the presence of micelle-forming surfactants was shown to enhance the transport of pesticides in soils (Iglesias-Jimenez et al., 1996).

Sabljic et al. (1989) report a relationship between the association of polychlorinated biphenyl (PCB) congeners with dissolved marine humic substances and the first-order molecular connectivity index. The model is

$$\log K_h = -19.44 + 4.83\,^1\chi - 0.22 \left(^1\chi\right)^2 - 0.16 N_{o-Cl} \qquad (10.19)$$

n = 26, s = 0.052, r = 0.995, F (3,22) = 717, EV = 98.9%

where K_h = humic substance association coefficient
       $^1\chi$ = first-order simple connectivity index
       N_{o-Cl} = number of ortho-chlorine substituents

The model applies to PCB congeners with five to nine chlorine substituents. While $^1\chi$ encodes molecular size, the parameter N_{o-Cl} is correlated with the degree of nonplanarity of a PCB molecule. The level of cross-correlation between the two topological parameters, $^1\chi$ and N_{o-Cl}, is low ($r^2 = 0.26$).

K_{OM} values should be sensitive to the salting-out effect as are K_{OW} values. Not much experimental evidence is available, however. Karickhoff et al. (1979) found that the K_{OM} value of pyrene in 0.34 M NaCl solution was increased by 15% over the K_{OM} value of pyrene in distilled water. In view of this result, typical levels of dissolved salts may increase K_{OM} values by only a few percent.

Cosolvents enhance the aqueous solubility of organic chemicals and reduce K_{OM} values. An inverse correlation is observed between log K_{OM} and the fraction of organic cosolvent (Nkedi-Kizza, 1985; Lee et al., 1990):

$$\ln\left(\frac{K_{pM}}{K_{pW}}\right) = -A \cdot f_C \tag{10.20}$$

where K_{Pm} = sorption coefficient from mixed solvent, l/kg

   K_{pW} = sorption coefficient from water, l/kg

   A   = constant

   f_C = volume fraction of cosolvent

The cosolvent effect is observed for sorption of various chemicals, including pentachlorophenol, naphthalene, and diuron from methanol–water mixtures by soils and organo-clays ((Lee et al., 1990; Nzengung et al., 1996). For example, the K_{OM} value of anthracene in a 10% methanol solution is reduced by about 40% over the K_{OM} value of anthracene in distilled water.

REFERENCES

Abdul, S.L., Gibson, T.L., and Rai, D.N., 1987. Statistical correlations for predicting the partition coefficient for nonpolar organic contaminants between aquifer organic carbon and water. *Hazard. Waste Hazard. Mater.*, **4**, 211–222.

Angemar, Y., Rebhun, M., and Horowitz, M., 1984. Adsorption, phytotoxicity, and leaching of bromacil in some Israeli soils. *J. Environ. Qual.*, **13**, 321–327.

Bahnick, D.A. and Doucette, W.J., 1988. Use of molecular connectivity indices to estimate soil sorption coefficients for organic chemicals. *Chemosphere*, **17**, 1703–1715.

Bintein, S. and Devillers, J., 1994. QSAR for organic chemical sorption in soils and sediments. *Chemosphere*, **28**, 1171–1188.

Briggs, G.G., 1981. Theoretical and experimental relationships between soil adsorption, octanol-water partition coefficients, water solubilities, bioconcentration factors, and the parachor. *J. Agric. Food Chem.*, **29**, 1050–1059.

Bruggemann, R. and Munzer, B., 1988. Physico-chemical data estimation for environmental chemicals. In *Physical Property Prediction in Organic Chemistry*. Proceedings of the Beilstein Workshop 16-20th May, 1988, Schloss Korb, Italy. Jochum, C., Hicks, M.G., and Sunkel, J., Eds. Springer-Verlag, Berlin.

Brusseau, M.L., 1992. Factors influencing the transport and fate of contaminants in the subsurface. *J. Hazardous Mater.*, **32**, 137–143.

Calvet, R., 1989. Adsorption of organic chemicals in soils. *Environ. Health Perspect.*, **83**, 145–177.

Chiou, C.T., 1989. Chapter 1. In *Reactions and Movement of Organic Chemicals in Soil.* Sawhney, B.L. and Brown, K., Eds. Soil Science Society of America, Madison, WI.

Chiou, C.T., Peters, L.J., and Freed, V.H., 1979. A physical concept of soil–water equilibria for nonionic organic compounds. *Science*, **206**, 831–832.

Chiou, C.T., Porter, P.E., and Schmedding, D.W., 1983. Partition equilibria of nonionic organic compounds between soil organic matter and water. *Environ. Sci. Technol.*, **17**, 227–231.

Dean, J.A., 1996. *Lange's Handbook of Chemistry*, 14th Edition. McGraw-Hill, New York.

Drefahl, A. and Reinhard, M., 1993. Similarity-based search and evaluation of environmentally relevant properties for organic compounds in combination with the group contribution approach. *J. Chem. Inf. Comput. Sci.,* **33**, 886–895.

GEMS, 1986. Graphical Exposure Modeling System (GEMS) User's Guide, Vol. 5. Estimation Appendix E. AUTOCHEM, Task 3-2 under U.S. EPA Contract 68-02-3970, U.S. Environmental Protection Agency, Washington, D.C.

Grathwohl, P., 1990. Influence of organic matter from soils and sediments from various origins on the sorption of some chlorinated aliphatic hydrocarbons; implications on K_{OC} correlations. *Environ. Sci. Technol.,* **24**, 1687–1693.

Hamaker, J.W. and Thompson, J.M., 1972. Adsorption. In *Organic Chemicals in the Soil Environment.* Vol. 1. Goring, C.A.J. and Hamaker, J.W., Eds. Marcel Dekker, New York.

Hodson, J. and Williams, N.A., 1988. The estimation of the adsorption coefficient (Koc) for soils by high-performance liquid chromatography. *Chemosphere*, **17**, 67–77.

Iglesias-Jimenez, E., Sanchez-Martin, M.J., and Sanchez-Camazano, M., 1996. Pesticide adsorption in a soil–water system in the presence of surfactants. *Chemosphere*, **32**, 1771–1782.

Karickhoff, S.W., 1979. Sorption of hydrophobic pollutants on natural sediments. *Water Res.*, **13**, 241–248.

Karickhoff, S.W., 1981. Semiempirical estimation of sorption of hydrophobic pollutants on natural sediments and soils. *Chemosphere*, **10**, 833–849.

Karickhoff, S.W., 1984. Organic pollutant sorption in aquatic systems. *J. Hydraul. Eng.*, **110**, 707–735.

Karickhoff, S.W., Brown, D.S., and Scott, T.A., 1979. Sorption of hydrophobic pollutants on natural sediments and soils. *Water Res.*, **13**, 241–248.

Kenaga, E.E. and Goring, C.A.I., 1980. Relationship between water solubility, soil sorption, octanol-water partitioning, and concentrations of chemicals in biota. In *Aquatic Toxicology.* Proceedings of the Third Annual Symposium on Aquatic Toxicology, ASTM Special Technical Publication 707. Eaton, J.G., Parrish, P.R., and Hendricks, A.C., Eds., 79–129.

Kozak, J., 1983. Adsorption of prometryne and metolachlor by selected soil organic matter fractions. *Soil Sci.*, **136**, 94–101.

Lagas, P., 1988. Sorption of chlorophenols in the soil. *Chemosphere*, **14**, 205–216.

Lambert, S.M., 1967. Functional relationship between sorption in soil and chemical structure. *J. Agric. Food Chem.*, **15**, 572–576.

Lambert, S.M., 1968. Omega (Ω), a useful index of soil sorption equilibria. *J. Agric. Food Chem.*, **16**, 340–343.

Lara, R. and Ernst, W., 1989. Interaction between polychlorinated biphenyls and marine humic substances. Determination of association coefficients. *Chemosphere*, **19**, 1655–1664.

Lee, L.S., Rao, P.S.C., NKedi-Kizza, P., and Delfino, J.J., 1990. Influence of solvent and sorbent characteristics on distribution of pentachlorophenol in octanol–water and soil–water systems. *Environ. Sci. Technol.*, **24**, 654–661.

Leo, A.J. and Weininger, D., 1989. *CLOGP, Version 3.54 — User Reference Manual,* Medicinal Chemistry Project. Pomona College, Claremont, CA.

Lide, D.R., 1994. *CRC Handbook of Chemistry and Physics.* 74th edition. CRC Press, Boca Raton, FL.

Lohninger, H., 1994. Estimation of soil partition coefficients of pesticides from their chemical structure. *Chemosphere,* **29,** 1611–1626.

Lyman, W. J. and Potts, R.G., 1987. *CHEMEST: User's Guide — A Program for Chemical Property Estimation,* Version 2.1, A. D. Little, Cambridge, MA; Boethling, R.S., Campbell, S.E., Lynch, D.G., and LaVeck, G.D., 1988. Validation of CHEMEST, an On-Line System for the Estimation of Chemical Properties. *Ecotoxicol. Environ. Safety,* **15,** 21–30.; Lynch, D.G., Tirado, N.F., Boethling, R.S., Huse, G.R., and Thom, G.C., 1991. Performance of on-line chemical property estimation methods with TSCA premanufacture notice chemicals. *Sci. Total Environ.,* **109/110,** 643–648.

Lyman, W.J., 1990. Adsorption coefficient for soils and sediments. Chapter 4. In *Handbook of Chemical Property Estimation Methods. Environmental Behavior of Organic Compounds.* Lyman, W.J., Reehl, W.F., and Rosenblatt, D.H., Eds. American Chemical Society, Washington, D.C.

McGinley, P.M., Katz, L.E., and Weber, W.J., Jr., 1993. A distributed reactivity model for sorption by soils and sediments. II. Multicomponent systems and competitive effects. *Environ. Sci. Technol.,* **27,** 1524–1531.

Means, J.C., Wood, S.G., Hassett, J.J., and Banwart, W.L., 1982. Sorption of amino- and carboxy-substituted polynuclear aromatic hydrocarbons by sediments and soils. *Environ. Sci. Technol.,* **16,** 93–98.

Meylan, W., Howard, P.H., and Boethling, R.S., 1992. Molecular topology/fragment contribution method of predicting soil sorption coefficients. *Environ. Sci. Technol.,* **26,** 1560–1567.

Muller, M. and Kordel, W., 1996. Comparison of screening methods for the estimation of adsorption coefficients on soil. *Chemosphere,* **32,** 2493–2504.

Nkedi-Kizza, P., Rao, P.S.C., and Hornsby, A.G., 1985. Influence of organic cosolvents on sorption of hydrophobic organic chemicals by soils. *Environ. Sci. Technol.,* **19,** 975–979.

Nzengung, V.A., Voudrias, E.A., Nkedi-Kizza, P., Wampler, J.M., and Weaver, C.E., 1996. Organic cosolvent effects on sorption of hydrophobic organic chemicals by organoclays. *Environ. Sci. Technol.,* **30,** 89–96.

Okouchi, S. and Saegusa, H., 1989. Prediction of soil sorption coefficients of hydrophobic organic pollutants by absorbability index. *Bull. Chem. Soc. Jpn.,* **62,** 922–924.

Paya-Perez, A.B., Raiz, M., and Larsen, B.R., 1991. Soil sorption of 20 PCB congeners and six chlorobenzenes. *Ecotoxicol. Environ. Saf.,* **21,** 1–17.

Paya-Perez, A.B., Cortes, A., Sala, M.N., and Larsen, B., 1992. Organic matter fractions controlling the sorption of atrazine in sandy soils. *Chemosphere,* **25,** 887–898.

Pignatello, J.J., 1991. Chapter 16. In *Organic Substances and Sediments in Water.* Baker, R.A., Ed. Lewis Publishers, Chelsea, MI.

Rutherford, D.W., Chiou, C.T., and Kile, D.E., 1992. Influence of soil organic matter composition on the partition of organic compounds. *Environ. Sci. Technol.,* **26,** 336–340.

Sabljic, A., 1984. Predictions of the nature and strength of soil sorption of organic pollutants by molecular topology. *J. Agric. Food Chem.,* **32,** 243–246.

Sabljic, A., 1987. On the prediction of soil sorption coefficients of organic pollutants from molecular structure: application of molecular topology model. *Environ. Sci. Technol.,* **21,** 358–366.

Sabljic, A., 1989. *Environ. Health Perspect.,* **83,** 179.

Sabljic, A., 1991. Chemical topology and ecotoxicology. *Sci. Total Environ.*, **109/110**, 197–220.

Sabljic, A., Lara, R., and Ernst, W., 1989. Modeling association of highly chlorinated biphenyls with marine humic substances. *Chemosphere*, **19**, 1665–1676.

Sabljic, A., Gusten, H., Verhaar, H., and Hermans, J., 1995. QSAR modelling of soil sorption. Improvements and systematics of log K_{OC} vs. log K_{OW} correlations. *Chemosphere*, **31**, 4489–4514.

Schellenberg, K., Leuenberger, C., and Schwarzenbach, R.P., 1984. Sorption of chlorinated phenols by natural sediments and aquifer materials. *Environ. Sci. Technol.*, 18, 652–657.

Schwarzenbach, R.P. and Westall, J., 1981. Transport of nonpolar compounds from surface water to groundwater. Laboratory sorption studies. *Environ. Sci. Technol.*, **11**, 1360–1367.

Schwarzenbach, R.P., Gschwend, P.M., and Imboden, D.M., 1993. *Environmental Organic Chemistry*. John Wiley & Sons, New York.

Shonnard, D.R., Bell, R.L., and Jackman, A.P., 1993. Effects of nonlinear sorption on the diffusion of benzene and dichloromethane from two air-dry soils. *Environ. Sci. Technol.*, **27**, 457–466.

Suffett, I.H. and McCarthy, P., 1989. *Aquatic Humic Substances. Influence on Fate and Treatment of Pollutants*. Advances in Chemistry Series 219. American Chemical Society, Washington, D.C.

Tanford, C., 1980. *The Hydrophobic Effect*. John Wiley & Sons, New York.

Travis, C.C. and Etnier, E.L., 1981. A survey of sorption relationships for reactive solutes in soil. *J. Environ. Qual.*, **10**, 8–17.

Valsaraj, K.T., 1995. *Elements of Environmental Engineering*. Lewis Publishers, Boca Raton, FL.

von Open, B., Kordel, W., and Klein, W., 1991. Sorption of nonpolar and polar compounds to soils: process, measurement, and experience with the applicability of the modified OECD-guideline 106. *Chemosphere,* **22**, 285–304.

Vowles, P.D. and Mantoura, R.F.C., 1987. Sediment-water partition coefficients and HPLC retention factors of aromatic hydrocarbons. *Chemosphere*, **16**, 109–116.

Walton, B.T., Hendricks, M.S., Anderson, T.A., Griest, W.H., Merriweather, R., Beauchamp, J.J., and Francis, C.W., 1992. Soil sorption of volatile and semivolatile organic compounds in a mixture. *J. Environ. Qual.*, **21**, 552–558.

Westall, J.C., 1987. Adsorption mechanisms in aquatic surface chemistry. Chapter 1. In *Aquatic Surface Chemistry. Chemical Processes at the Particle–Water Interface*. Stumm, W., Ed. John Wiley & Sons, New York.

Xie, T.M., Hulthe, B., and Folestad, S., 1984. Determination of partition coefficients of chlorinated phenols, guiacols, and catechols by shake-flask GC and HPLC. *Chemosphere*, **13**, 445–459.

Young, T.M. and Weber, W.J., Jr., 1995. A distributed reactivity model for sorption by soils and sediments. III. Effects of diagenic processes on sorption energetics. *Environ. Sci. Technol.*, **29**, 92–97.

CHAPTER 11

Bioconcentration Factor
and Related Parameters

11.1 INTRODUCTION

This chapter describes methods of estimating bioconcentration and bioaccumulation factors in fish, animals, and plants. **Bioconcentration** is *the accumulation of chemicals from environmental media by nondietary routes.* The **bioconcentration factor**, BCF_{ORGN} (kg medium/kg organism or m^3 medium/kg organism), is *the ratio of the steady-state concentration of chemical in an organism to the concentration of chemical in the organism's environment*:

$$BCF_{ORGN} = \frac{C_{ORGN}}{C_M} \qquad (11.1)$$

where C_{ORGN} (kg of chemical/kg of organism) is the concentration of chemical in the organism and C_M (kg of chemical/kg of medium or kg of chemical/m^3 of medium) is the concentration of chemical in the environmental medium; water or air.

The literature prior to 1980 focused on bioconcentration in fish; it has been reviewed by Bysshe (1990) and Briggs (1981).

Bioconcentration in fish occurs primarily during intake of water through the gills. Ingestion of contaminants is believed to be relatively unimportant. Typically, BCF in fish is reported in units of g water/g of whole fish or mL water/g of whole fish. Bioconcentration factors of chemicals in fish range from 1 to 10^6 g/g and are assumed to be independent of fish species, exposure level, and time after steady state is reached.

Plant bioconcentration factors relate the concentration of a chemical in the plant to the concentration of the chemical in soil solution or air. The **root concentration factor** for root crops, BCF_R, is *the ratio of the concentration of chemical in plant root (g of chemical/g wet root) to the chemical concentration in soil solution (g chemical/mL soil solution)*. The **soil-to-plant concentration factor** or **plant uptake factor**, B_v, is often a more useful measure of bioconcentration in plants. It is *the ratio of the concentration of chemical in the plant (g of chemical/kg dry plant) to the concentration of chemical in agricultural soil (g of chemical/kg soil).*

Bioaccumulation is *the accumulation of chemicals via all exposure routes: ingestion, respiration, and direct contact.* Bioaccumulation may produce toxic levels of xenobiotics in fish and animals. Hydrophobic chemicals, in particular, may be found at high concentrations in the fat tissue of predators. The **bioaccumulation factor**, BAF (kg medium/kg organism or m^3 medium/kg organism), is defined exactly like the bioconcentration factor in Equation 11.1. It is *the steady-state ratio of the concentration of chemical in an organism (kg of chemical/kg of organism) to the concentration of chemical in the organism's environment (kg of chemical/kg of medium or kg of chemical/m³ of medium).* If ingestion is the major route of exposure to a chemical, it is useful to relate steady-state levels of chemical in the animal to the rate of dietary intake. The **transfer factor**, TF (day/kg animal product), is *the ratio of the chemical concentration in an animal product (meat or milk) to the mass of chemical consumed per day.*

Sources of data, mainly on bioconcentration in fish, include the CIS AQUIRE data base (AQUIRE, 1996), the Hazardous Substance Data Base (HSDB, 1996) found on the National Library of Medicine's Toxicology Data Network (TOXNET), the CHEMFATE data base (CHEMFATE, 1986), and the CIS ENVIROFATE data base (ENVIROFATE, 1985). Some information is also available on topics such as microbial bioconcentration in aquatic systems (Baughman and Paris, 1981) and uptake of chemicals from air, rain, and soil by plants (Schonherr and Riederer, 1989; Paterson et al., 1990; Hope, 1995). Octanol–air partition coefficients of hydrophobic chemicals, K_{OA}, have been reported for chlorobenzenes, polychlorinated biphenyls, and DDT (Harner and Mackay, 1995).

In the following section, the theoretical basis of the methods of estimating bioconcentration factors is discussed. Then estimation methods for bioconcentration and bioaccumulation factors are described. The presentation is divided into three main subtopics: chemical uptake in fish, chemical uptake in animals, and chemical uptake in plants. In each subsection, method applicability and accuracy are discussed, as is the sensitivity of the various bioconcentration and bioaccumulation factors to environmental parameters.

11.2 A BIOCONCENTRATION MODEL

Imagine that an organism consists of several phases, each with a different composition. At equilibrium, the fugacity of a chemical that is passively acquired by partitioning between the organism and its environment must be the same in each phase of the organism as it is in the environmental medium. For example, the fugacity of a chemical in the organism's lipid phase, f_{LIP} (Pa), equals the fugacity of the chemical in the environmental medium in contact with the organism, f_M (Pa); $f_{LIP} = f_M$ at equilibrium (Mackay, 1982).

For the lipid phase, $f_{LIP} = \gamma_{LIP} x_{LIP} f_R$, where γ_{LIP} is the activity coefficient of the chemical in the lipid phase, x_{LIP} is the mole fraction of the chemical in the lipid, and f_R (Pa) is a reference fugacity. For the water phase, $f_W = \gamma_W x_W f_R$, where γ_W is the activity coefficient of chemical in water, x_W is the mole fraction of chemical in water, and f_R (Pa) is the reference fugacity. Equating the two fugacity terms, $f_{LIP} = f_W$ and $\gamma_{LIP} x_{LIP} f_R = \gamma_W x_W f_R$. Rearranging this expression:

$$\frac{x_{LIP}}{x_W} = \frac{\gamma_W}{\gamma_L} \tag{11.2}$$

Typical environmental concentrations of xenobiotics are low enough that $x_W \approx n_C/n_W = C_W V_{W,m}$, where n_C/n_W is the mole ratio of chemical to water in the aqueous phase and $V_{W,m}$ (L/mol) is the molar volume of water, and $x_{LIP} \approx n_C/n_{LIP} = C_{LIP} V_{LIP,m}$, where n_C/n_{LIP} is the mole ratio of chemical to lipid in the lipid phase and $V_{LIP,m}$ (L/mol) is the molar volume of lipid. Making these substitutions in Equation 11.2,

$$BCF_{LIP} = \frac{C_{LIP}}{C_W} = \frac{\gamma_W V_{W,m}}{\gamma_{LIP} V_{LIP,m}} \tag{11.3}$$

where BCF_{LIP} is the bioconcentration factor of a hydrophobic chemical in lipid.

The octanol–water system is widely believed to be a good surrogate for the fish lipid–water system (Mackay, 1982). Octanol has roughly the same atomic C/H/O ratio as lipid, and it has been suggested that chemicals behave similarly in octanol and in lipid. Most hydrophobic organic chemicals form nearly ideal solutions with n-octanol. Measured values of chemical activity coefficients in octanol, γ_O, are found to lie within the narrow range from 1 to about 10, depending on chemical structure (Schwarzenbach et al., 1993). It seems probable that most hydrophobic organic chemicals form nearly ideal solutions with lipid, also, and $\gamma_{LIP} \approx 1$. On the other hand, aqueous solutions of hydrophobic organic chemicals are not ideal. The value of γ_W ranges from 10^1 to over 10^{11}, depending on molecular structure. Therefore, variation in BCF_{LIP} value between chemicals is almost entirely due to variation in the value of γ_W with chemical structure: $BCF_{LIP} \approx (V_{W,m}/V_{LIP,m}) \cdot \gamma_W$. Since the discussion in Chapter 7 shows that γ_W is inversely correlated with aqueous solubility, BCF_{LIP} must also be inversely correlated with aqueous solubility.

The octanol–water partition coefficient, K_{OW}, is defined as the concentration ratio of a chemical in the octanol phase to that in the water phase of an octanol–water system at equilibrium. In Section 9.2, reasoning identical to that used to model lipid–water partitioning is used to derive Equation 9.3, $K_{OW} = \gamma_W \cdot V_{W,m}/\gamma_O \cdot V_{O,m}$, where γ_O is the activity coefficient of chemical in octanol and $V_{O,m}$ is the molar volume of octanol. Combining Equations 9.3 and 11.4 and rearranging, we find that $BCF_{LIP} = (\gamma_O V_{O,m}/\gamma_{LIP} V_{LIP,m}) K_{OW}$.

If hydrophobic organic chemicals do behave ideally in both n-octanol and in lipid, $\gamma_O/\gamma_{LIP} \approx$ constant. In addition, octanol–water partition coefficients are always reported for dilute solutions at 25°C, and the molar volume of octanol is nearly identical for every measurement: $V_{O,m} \approx$ constant. Bioconcentration factors are reported for dilute solutions at between 15 and 25°C, and the molar volume of lipid is approximately the same for every measurement: $V_{LIP,m} \approx$ constant. We conclude that $\gamma_O V_{O,m}/\gamma_{LIP} V_{LIP,m} \approx$ constant, and $BCF_{LIP} \approx$ constant$\cdot K_{OW}$.

Hamelink et al. (1971) suggest that hydrophobic chemicals accumulate in fish primarily by partitioning between water and fish lipid. This view is supported by evidence that the magnitude of bioconcentration is linked to the lipid content of fish. Although bioconcentration factors are often found to be only weakly correlated with the lipid content in fish, showing that other controls are also important (Barron, 1990), we assume for the sake of simplicity that bioconcentration of hydrophobic chemicals is determined in large measure by the lipid content of living organisms, $BCF_{ORGN} \approx L \cdot BCF_{LIP}$ or

$$\log BCF_{ORGN} \approx \log L + \log BCF_{LIP} \qquad (11.4)$$

where L is the mass fraction of lipid in the organism. Since BCF_{LIP} is correlated with K_{OW}, we conclude that

$$\log BCF_{ORGN} = A + B \log K_{OW} \qquad (11.5)$$

where A and B are constants. Comparing Equations 11.4 and 11.5, we see that the value of the constant A depends upon the lipid content of the organism, L, as well as on the other major determinants of bioconcentration. When the chemical is found primarily in the organism's lipids, $B \approx 1$. If a significant amount of the chemical is found in other tissue, the value of B may differ from 1. Equation 11.5 has been used to model bioconcentration in fish (Mackay 1982), mammals (Kenaga, 1980), and plants (Paterson et al., 1990).

We are assuming that the thermodynamics of lipid–water partitioning are similar to the thermodynamics of octanol–water partitioning. This is not strictly true (Opperhuizen et al., 1988). The correlation between BCF and K_{OW} may depend upon chemical class and temperature and may not always be linear.

When $\log K_{OW} \approx 6$, the log BCF values of chlorobenzenes and polychlorinated biphenyls reach a maximum value of about 5.5 in fish (Tulp and Hutzinger, 1978; Konemann and van Leeuwen, 1980). Above $\log K_{OW} \approx 6$, BCF declines as K_{OW} increases. A similar observation has been reported for the bioaccumulation of chlorinated organic chemicals in animals (McLachlan, 1993). Banerjee and Baughman (1991) suggest that this is mainly because chemical solubility in lipid reaches a maximum when $BCF \approx 6$. As a result, Equation 11.5 overestimates BCF values when $\log K_{OW}$ is greater than 6. Models describing the correlation between log BCF and log K_{OW} that extend beyond this point are either parabolic or bilinear (Bintein et al., 1993).

11.3 METHODS OF ESTIMATING FISH
BIOCONCENTRATION FACTORS

11.3.1 BCF–K_{OW} Correlations

Several researchers have reviewed the correlation between the bioconcentration factors of nonpolar chemicals in fish and octanol–water partition coefficients

Table 11.1   Some Relationships Between Fish Bioconcentration Factors and Octanol–Water Partition Coefficients

log BCF = A + B log K_{OW}

Model No./Chemical Class/Test Species	A	B	n	r	Ref.
Pesticides, PCBs 0 > logBCF > 6, 1 > logK_{OW} > 7/fathead minnows	−0.70	0.95	55	0.95	Veith et al., 1979
Mixed 0 > log BCF > 5, 1 > logK_{OW} > 7/fathead minnow, bluegill sunfish	−0.23	0.76	84	0.95?	Veith et al., 1980
Halogenated chemicals 2.4 > logBCF > 4.3, 3.4 > logK_{OW} > 5.5/rainbow trout	−0.68	0.94	18	0.95	Oliver, 1984
Mixed 0 > log BCF > 5, 1 > logK_{OW} > 7/several freshwater fish	−0.52	0.80	107	0.904	Isnard and Lambert, 1988

(Kenaga and Goring, 1980; Briggs, 1981; Isnard and Lambert, 1988; Bysshe, 1990; Devillers et al., 1996). Neely et al. (1974) first modeled the relationship between BCF and K_{OW} using the muscle tissue of rainbow trout. Since then, many correlations have been reported between whole-fish bioconcentration factors and K_{OW}. Some widely used models are shown in Table 11.1. Despite the variety of training-set chemicals and target animals, the correlations all have a slope of about one, i.e., $B \approx 1$ in Equation 11.5 as anticipated.

Mackay (1982) describes a correlation between whole-fish bioconcentration factors, BCF_f, and K_{OW} that uses only one empirical constant, $BCF_f = 0.048 \cdot K_{OW}$ or

$$\log BCF_f = -1.32 + \log K_{OW} \qquad (11.6)$$

n = 50?, r^2 = 0.95

where BCF_f = fish bioconcentration factor, g/g
   K_{OW} = octanol–water partition coefficient
   n   = number of chemicals in the training set
   r   = correlation coefficient

Equation 11.6 applies to a diverse set of chemicals including nitrogen- and halo-substituted aromatics and aliphatics with K_{OW} values in the range from 10 to 10^6. Pesticides are heavily represented in the training set. Most of the BCF_f values were measured with fathead minnows (Veith et al., 1979).

The discussion in Section 11.2 shows that if chemical uptake in fish is determined entirely by lipid–water partitioning, $BCF_f \approx L \cdot K_{OW}$. Equation 11.6 implies that a fish with an average lipid content of about 4.8% is a good model animal for bioconcentration relationships. Of course, the lipid content of fish varies with species, size, diet, and age of the animal. Also, there is significant chemical uptake by nonlipid tissue in fish.

Devillers et al. (1996) compared several of the models given in Table 11.1 with Mackay's model, Equation 11.6, and with several nonlinear models. For log K_{OW} < 6, all models give equivalent results. With critically evaluated data from a test set of 227 diverse chemicals, all of the models produce a root-mean-square (RMS) error of 0.6 log units. The test set, along with log K_{OW} and log BCF_f values, is shown in Tables 11.2 and 11.3.

Example

Estimate the value of the bioconcentration factor of pentachlorobenzene in fish using Mackay's model, Equation 11.6.

1. For pentachlorobenzene, log K_{OW} = 5.00 (Mackay et al., 1992). The BCF_f value is

$$\log \ BCF_f = -1.32 + \log \ K_{OW}$$

$$= -1.32 + 5.00$$

$$= 3.68$$

Reported BCF_f values of pentachlorobenzene range from 2.94 to 4.23 (Devillers et al., 1996). The estimated value of BCF_f lies at the middle of the range of reported values.

As noted in Section 11.2, log BCF_f values of highly hydrophobic chemicals are not linearly correlated with log K_{OW}. When log K_{OW} > 6.5, chemical uptake decreases (Konemann and van Leeuwen, 1980) and BCF values decrease with increasing K_{OW}. The linear BCF_f–K_{OW} correlations significantly overestimate the BCF_f value in this K_{OW} range. Various nonlinear models have been proposed that give better results. (Konemann and van Leeuwen, 1980; Connell and Hawker, 1988; Nendza, 1991; Bintein et al., 1993).

Equation 11.7 proposed by Bintein et al. (1993) was compared with Mackay's model, Equation 11.6, with several of the correlations given in Table 11.1 and with two nonlinear bioconcentration models (Devillers et al., 1996). With a test set of 227 diverse chemicals for which log K_{OW} < 6, the model produces an RMS error of 0.6 log units as do the other models. With a test set of 75 diverse chemicals having log K_{OW} > 6, it produces an RMS error of 0.86 log units, the lowest of all models tested.

$$\log BCF_f = 0.91 \ \log K_{OW} - 1.975 \log\left(6.8 \times 10^{-7} K_{OW} + 1\right) - 0.786 \qquad (11.7)$$

n = 154, s = 0.347, r = 0.95

where BCF_f = bioconcentration factor, g/g
       K_{OW} = octanol–water partition coefficient
       n   = number of chemicals in training set
       s   = standard deviation
       r   = correlation coefficient

Table 11.2 Octanol–Water Partition Coefficients and Bioconcentration Factors Obtained Under Flow-Through Conditions

Chemical	log K_{OW}[a]	log BCF[a]	Species
Acenaphthene	3.92	2.59	*Lepomis macrochirus*
Acridine	3.45	2.10	*Pimephales promelas*
Acrolein	0.90	2.54	*Lepomis macrochirus*
Benzo[a]pyrene	5.97	3.51	*Lepomis macrochirus*
Bis(2-chloroethyl)ether	1.12	1.04	*Lepomis macrochirus*
BPMC	2.78	1.41	*Pseudorasbora parva*
Bromacil	2.11	0.51	*Pimephales promelas*
5-Bromoindole	3.00	1.15	*Pimephales promelas*
4-Bromophenol	2.59	1.56	*Brachydanio rerio*
Bromophos	4.88	4.60[bc]	*Poecilia reticulata*
	4.88	4.65[b]	*Poecilia reticulata*
2-t-Butoxy ethanol	0.39	−0.22	*Cyprinus carpio*
Butyl benzyl phthalate	4.05	2.89	*Lepomis macrochirus*
t-Butyl isopropyl ether	2.14	0.76	*Cyprinus carpio*
t-Butyl methyl ether	1.24	0.18	*Cyprinus carpio*
4-t-Butyl phenol	3.31	1.86	*Brachydanio rerio*
Carbaryl	2.31	0.95	*Pseudorasbora parva*
Carbon tetrachloride	2.73	1.48	*Lepomis macrochirus*
Chlordane	6.00	4.58	*Pimephales promelas*
α-Chlordane	6.00	4.45	*Oncorhynchus mykiss*
γ-Chlordane	6.00	4.20	*Oncorhynchus mykiss*
Chlornitrofen	4.70	3.04	*Pseudorasbora parva*
2-Chloroaniline	1.93	0.30	*Cyprinus carpio*
	1.93	0.57	*Cyprinus carpio*
3-Chloroaniline	1.91	−0.10	*Cyprinus carpio*
	1.91	0.34	*Cyprinus carpio*
4-Chloroaniline	1.88	−0.10	*Cyprinus carpio*
	1.88	0.23	*Cyprinus carpio*
p-Chlorobiphenyl	4.61	3.54[c]	*Oryzias latipes*
Chloroform	1.90	0.78	*Lepomis macrochirus*
2-Chloronitrobenzene	2.52	2.10	*Oncorhynchus mykiss*
3-Chloronitrobenzene	2.50	1.89	*Oncorhynchus mykiss*
4-Chloronitrobenzene	2.39	2.00	*Oncorhynchus mykiss*
2-Chlorophenol	2.15	2.33	*Lepomis macrochirus*
3-Chlorophenol	2.50	1.25	*Brachydanio rerio*
Chlorothion	3.63	2.46[bc]	*Poecilia reticulata*
	3.63	2.61[h]	*Poecilia reticulata*
Cresyldiphenyl phosphate	4.51	2.00	*Alburnus alburnus*
	4.51	2.34	*Alburnus alburnus*
4-Cyanophenol	1.60	0.91	*Brachydanio rerio*
Cyanophos	2.71	2.49[bc]	*Poecilia reticulata*
	2.71	2.62[b]	*Poecilia reticulata*
Cypermethrin	6.05	2.89	*Oncorhynchus mykiss*
	6.05	2.92	*Oncorhynchus mykiss*
2,7-DCDD[d]	6.38	2.56	*Poecilia reticulata*
p,p'-DDE	5.83	4.71	*Pimephales promelas*
	5.83	4.91	*Oncorhynchus mykiss*
o,p'-DDT	6.00	4.57	*Pimephales promelas*
p,p'-DDT	6.00	4.47	*Pinuophalespromelas*
	6.00	4.81	*Oncorhynchus mykiss*
	6.00	4.86	*Oncorhynchus mykiss*

Table 11.2 (continued)   Octanol–Water Partition Coefficients and Bioconcentration Factors Obtained Under Flow-Through Conditions

Chemical	log K_{ow}[a]	log BCF[a]	Species
	6.00	4.95	Oncorhynchus mykiss
	6.00	4.97	Oncorhynchus mykiss
	6.00	4.99	Oncorhynchus mykiss
Decachlorobiphenyl	8.27	3.92[c]	Poecilia reticulata
	8.27	4.00[c]	Poecilia reticulata
Deltamethrin	6.20	2.62	Oncorhynchus mykiss
	6.20	2.70	Oncorhynchus mykiss
Diazinon	3.81	1.24	Poecilia reticulata
	3.81	1.34	Oryzias latipes
	3.81	1.41	Misgurnus anguillicaudatus
	3.81	1.45	Oryzias latipes
	3.81	1.56	Cyprinus auratus
	3.81	1.80	Oncorhynchus mykiss
	3.81	1.81	Cyprinus carpio
	3.81	2.08	Cyprinus carpio
	3.81	2.18	Pseudorasbora parva
	3.81	2.18	Pseudorasbora parva
1,3-Dibromobenzene	3.75	2.82	Oncorhynchus mykiss
1,4-Dibromobenzene	3.79	1.96[c]	Poecilia reticulata
4,4′-Dibromobiphenyl	5.72	4.24[c]	Poecilia reticulata
2,6-Dibromo-4-cyanophenol	2.61	1.67	Brachydanio rerio
Dicapthon	3.58	2.95[b]	Poecilia reticulata
	3.58	3.09[bc]	Poecilia reticulata
1,2-Dichlorobenzene	3.43	1.95	Lepomis macrochirus
	3.43	2.43	Oncorhynchus mykiss
	3.43	2.75	Oncorhynchus mykiss
1,3-Dichlorobenzene	3.53	1.82	Lepomis macrochirus
	3.53	2.62	Oncorhynchus mykiss
	3.53	2.87	Oncorhynchus mykiss
1,4-Dichlorobenzene	3.44	1.78	Lepomis macrochirus
	3.44	2.47	Jordanella floridae
	3.44	2.57	Oncorhynchus mykiss
	3.44	2.71	Oncorhynchus mykiss
	3.44	2.86	Oncorhynchus mykiss
	3.44	2.95	Oncorhynchus mykiss
2,5-Dichlorobiphenyl	5.16	4.00	Oncorhynchus mykiss
	5.16	4.14[c]	Carassius auratus
	5.16	4.24[c]	Poecilia reticulata
	5.16	4.53	Oncorhynchus mykiss
3.5-Dichlorobiphenyl	5.37	3.79	Oncorhynchus mykiss
	5.37	3.82	Oncorhynchus mykiss
1,2-Dichloroethane	1.45	0.30	Lepomis macrochirus
4,5-Dichloroguaiacol	3.18	2.03	Oncorhynchus mykiss
1,4-Dichloronaphthalene	4.66	3.75	Oncorhynchus mykiss
2,3-Dichloronitrobenzene	3.05	2.16	Oncorhynchus mykiss
2,4-Dichloronitrobenzene	3.05	2.07	Oncorhynchus mykiss
2,5-Dichloronitrobenzene	3.03	2.05	Oncorhynchus mykiss
3,4-Dichloronitrobenzene	3.04	2.07	Oncorhynchus mykiss
3,5-Dichloronitrobenzene	3.09	2.23	Oncorhynchus mykiss
Dichlorvos	1.47	−0.10	Gnathopogon aerulescens

Table 11.2 (continued)   Octanol–Water Partition Coefficients and Bioconcentration Factors Obtained Under Flow-Through Conditions

Chemical	log K_{ow}^a	log BCF[a]	Species
Dieldrin	5.40	3.65	*Pseudorosbora parva*
Di(2-ethylhexyl)phthalate	7.45	2.19	*Pimephales promelas*
	7.45	2.95	*Pimephales promelas*
Diethyl phthalate	2.47	2.07	*Lepomis macrochirus*
2,4-Dimethylphenol	2.42	2.18	*Lepomis macrochirus*
Dimethyl phthalate	1.56	1.76	*Lepomis macrochirus*
Diphenylamine	3.42	1.48	*Pimephales promelas*
Diuron	2.68	2.16	*Pimephales promelas*
	2.68	2.20	*Pimephales promelas*
EPN	3.85	3.37	*Pseudorasbora parva*
Fenitrothion	3.47	1.65	*Oryzias latipes*
	3.47	1.68	*Oryzias latipes*
	3.47	3.36[b]	*Poecilia reticulata*
	3.47	2.39	*Pseudorasbora parva*
	3.47	2.48	*Oryzias latipes*
	3.47	3.54[bc]	*Poecilia reticulata*
	3.47	2.74	*Oryzias latipes*
	3.47	2.75	*Oryzias latipes*
Fenthion	4.17	1.96	*Oryzias latipes*
	4.17	2.02	*Oryzias latipes*
	4.17	2.68	*Gnathopogon aerulescens*
	4.17	4.17[bc]	*Poecilia reticulata*
	4.17	4.22[b]	*Poecilia reticulata*
Fenthion-S2145	3.74	3.65[bc]	*Poecilia reticulata*
	3.74	3.84[b]	*Poecilia reticulata*
Fenvalerate	6.20	2.61	*Oncorhynchus mykiss*
	6.20	2.96	*Oncorhynchus mykiss*
1,2,3,4,7,8-HCDD[e]	7.79	3.29	*Oncorhynchus mykiss*
	7.79	3.36	*Oncorhynchus mykiss*
	7.79	3.63	*Pimephales promelas*
	7.79	4.13[c]	*Poecilia reticulata*
1,2,3,4,6,7,8-HCDD[e]	8.20	2.71	*Pimephales promelas*
	8.20	2.97	*Oncorhynchus mykiss*
	8.20	3.15	*Oncorhynchus mykiss*
	8.20	3.16	*Oncorhynchus mykiss*
	8.20	3.75[c]	*Poecilia reticulata*
1,2,3,4,6,7,8-HCDF[f]	7.92	3.62[c]	*Poecilia reticulata*
Heptachlor	5.44	3.98	*Pimephales promelas*
Heptachlor epoxide	5.40	4.16	*Pimephales promelas*
Heptachloronorbornene	5.28	4.05	*Pimephales promelas*
Hexabromobenzene	6.07	3.04	*Oncorhynchus mykiss*
Hexabromobiphenyl[g]	6.39	4.26	*Pimephales promelas*
2,2′,4,4′,6,6′-Hexabromobiphenyl	7.20	4.66[c]	*Poecilia reticulata*
Hexabromocyclododecane	5.81	4.26	*Pimephales promelas*
Hexachlorobenzene	5.73	3.74	*Oncorhynchus mykiss*
	5.73	4.16[c]	*Oryzias latipes*
	5.73	4.21	*Pimephales promelas*
	5.73	4.30	*Oncorhynchus mykiss*
	5.73	4.34	*Lepomis cyanellus*
	5.73	4.43	*Poecilia reticulata*

Table 11.2 (continued) Octanol–Water Partition Coefficients and Bioconcentration Factors Obtained Under Flow-Through Conditions

Chemical	log K_{OW}[a]	log BCF[a]	Species
2,2′,4,4′,5,5′-Hexachlorobiphenyl	6.90	4.84	*Oncorhynchus mykiss*
	6.90	5.30[c]	*Poecilia reticulata*
	6.90	5.65	*Poecilia reticulata*
1,1,2,3,4,4-Hexachloro-1,3-butadiene	4.78	3.76	*Oncorhynchus mykiss*
	4.78	4.23	*Oncorhynchus mykiss*
α-Hexachlorocyclohexane	3.80	2.33	*Oncorhynchus mykiss*
	3.80	3.04	*Brachydanio rerio*
	3.80	3.04	*Brachydanio rerio*
	3.80	3.20	*Oncorhynchus mykiss*
	3.80	3.38	*Oncorhynchus mykiss*
β-Hexachlorocyclohexane	3.78	3.16	*Brachydanio rerio*
	3.78	3.18	*Brachydanio rerio*
δ-Hexachlorocyclohexane	4.14	2.45	*Oncorhynchas mykiss*
	4.14	3.21	*Brachydanio rerio*
	4.14	3.25	*Brachydanio rerio*
γ-Hexachlorocyclohexane	3.72	2.16	*Oncorhynchus mykiss*
	3.72	2.26	*Pimephales promelas*
	3.72	2.93	*Brachydanio rerio*
	3.72	2.96	*Brachydanio rerio*
	3.72	3.08	*Oncorhynchus mykiss*
	3.72	3.10	*Pseudorasbora parva*
	3.72	3.30	*Oncorhynchus mykiss*
	3.72	3.32	*Oncorhynchas mykiss*
Hexachloroethane	3.93	2.14	*Lepomis macrochirus*
	3.93	2.71	*Oncorhynchus mykiss*
	3.93	3.08	*Oncorhynchus mykiss*
Hexachloronorbornadiene	5.28	3.81	*Pimephales promelas*
IBP	3.21	0.60	*Pseudorasbora parva*
Iodophenphos	5.16	4.30[bc]	*Poecilia reticulata*
	5.16	4.68[b]	*Poecilia reticulata*
Isophorone	1.67	0.85	*Lepomis macrockirus*
Leptophos	5.88	3.78	*Pseudorasbora parva*
Methidation	2.42	1.26	*Gnathopogon aerulescens*
Methoxychlor	5.08	3.92	*Pimephales promelas*
2-Methyl-4,6-dinitrophenol	2.13	0.16	*Brachydanio rerio*
Methylisocyanothion	3.58	3.39[b]	*Poecilia reticulata*
	3.58	3.53[bc]	*Poecilia reticulata*
Methylparathion	2.94	2.98[b]	*Poecilla reticulata*
	2.94	3.04[bc]	*Poecilia reticulata*
2-Methylphenol	1.95	1.03	*Brachydanio rerio*
Mirex	6.89	4.26	*Pimephales promelas*
	6.89	4.31[c]	*Poecilia reticulata*
Molinate	3.21	1.41	*Pseudorasbora parva*
Naphthalene	3.36	2.49	*Lepomis macrochirus*
	3.36	2.51	*Lepomis macrochirus*
3-Nitrophenol	2.00	1.40	*Brachydanio rerio*
2,2′,3,3′,4,4′,5,5′-Octachlorobiphenyl	7.67	4.33[c]	*Poecilia reticulata*
Octachlorodibenzo-p-dioxin	7.59	1.93	*Oncorhynchus mykiss*
	7.59	2.59	*Oncorhynchus mykiss*
	7.59	2.67	*Oncorhynchus mykiss*
	7.59	2.85	*Poecilia reticulata*

Table 11.2 (continued)   Octanol–Water Partition Coefficients and Bioconcentration
Factors Obtained Under Flow-Through Conditions

Chemical	log K_{ow}[a]	log BCF[a]	Species
	7.59	3.35	*Pimephales promelas*
	7.59	3.36[c]	*Poecilia reticulata*
Octachlorodibenzofuran	7.97	2.77	*Poecilia reticulata*
	7.97	3.10[c]	*Poecilia reticulata*
Octachloronaphthalene	6.42	2.52	*Oncorhynchus mykiss*
Octachlorostyrene	6.29	4.52	*Pimephales promelas*
1,2,3,4,7-PCDD [h]	7.44	2.91	*Oncorhynchus mykiss*
	7.44	3.16	*Pimephates promelas*
1,2,3,7,8-PCDD [h]	6.64	4.50[c]	*Poecilia reticulata*
2,3,4,7,8-PCDF [i]	6.92	4.36[c]	*Poecilia reticulata*
Pentachloroaniline	5.08	2.56	*Poecilia reticulata*
Pentachlorobenzene	5.18	2.94[c]	*Poecilia reticulata*
	5.18	3.53	*Lepomis macrochirus*
	5.18	4.11	*Oncorhynchus mykiss*
	5.18	4.30	*Oncorhynchus mykiss*
	5.18	4.36	*Poecilia reticulata*
Pentachloroethane	2.89	1.83	*Lepomis macrochirus*
Pentachloronitrobenzene	4.77	2.23	*Oncorhynchus mykiss*
	4.77	2.38	*Pseudorasbora parva*
	4.77	2.41	*Oncorhynchus mykiss*
	4.77	2.77	*Oncorhynchus mykiss*
Pentachlorophenol	5.01	2.33	*Jordanella floridae*
	5.01	2.58[c]	*Oryzias latipes*
	5.01	2.89	*Pimephales promelas*
	5.01	2.99	*Brachydanio rerio*
	5.01	3.23	*Oryzias latipes*
Permethrin	6.50	3.23	*Pimephales promelas*
	6.50	3.29	*Oncorhynchus mykiss*
	6.50	3.39	*Oncorhynchus mykiss*
	6.50	3.49	*Pimephales promelas*
	6.50	3.52	*Pimephales promelas*
Phenol	1.46	1.24	*Brachydanio rerio*
Phenthoate	3.69	1.56	*Pseudorasbora parva*
2-Phenyldodecane	8.19	2.65	*Oncorhynchus mykiss*
N-Phenyl-2-naphthylamine	4.38	2.17	*Pimephales promelas*
Phosmet	2.78	1.56	*Gnathopogon aerulescens*
Ronnel	4.81	4.55[bc]	*Poecilia reticulata*
	4.81	4.64[b]	*Poecilia reticulata*
Salithion	2.67	1.88	*Gnathopogon aerulescens*
Simetryne	2.54	0.31	*Gnathopogon aerulescens*
SV 5	3.00	2.95[b]	*Poecilia reticulata*
	3.00	3.22[bc]	*Poecilia reticulata*
1,2,3,4-TCDD [i]	6.20	2.90	*Poecilia reticulata*
1,3,6,8-TCDD [i]	6.29	3.20	*Oncorhynchus mykiss*
	6.29	3.32	*Oncorhynchus mykiss*
	6.29	3.32	*Oncorhynchus mykiss*
	6.29	3.39	*Oncorhynchus mykiss*
	6.29	3.57	*Oncorhynchus mykiss*
	6.29	3.76	*Pimephales promelas*
2,3,7,8-TCDD [i]	6.42	4.01[c]	*Poecilia reticulata*
	6.42	4.56	*Oncorhynchus mykiss*

Table 11.2 (continued)   Octanol–Water Partition Coefficients and Bioconcentration Factors Obtained Under Flow-Through Conditions

Chemical	log K_{OW}[a]	log BCF[a]	Species
	6.42	4.57	*Oncorhynchus mykiss*
	6.42	4.59	*Oncorhynchus mykiss*
	6.42	4.93	*Oncorhynchus mykiss*
2,3,7,8-TCDF [k]	6.53	3.31[c]	*Poecilia reticulata*
	6.53	3.42	*Oncorhynchus mykiss*
	6.53	3.65	*Oncorhynchus mykiss*
1,2,4,5-Tetrabromobenzene	5.13	3.57	*Oncorhynchus mykiss*
	5.13	3.81	*Oncorhynchus mykiss*
2,2',5,5'-Tetrabromobiphenyl	6.50	4.97[c]	*Poecilia reticulata*
2,3,4,5-Tetrachloroanillne	4.57	2.10	*Poecilia reticulata*
2,3,5,6-Tetrachloroaniline	4.46	2.46	*Poecilia reticulata*
1,2,3,4-Tetrachlorobenzene	4.64	3.72	*Oncorhynchus mykiss*
	4.64	3.80	*Oncorhynchus mykiss*
	4.64	3.82	*Poecilia reticulata*
	4.64	3.91	*Oncorhynchus mykiss*
	4.64	4.08	*Oncorhynchus mykiss*
1,2,3,5-Tetrachlorobenzene	4.66	4.40[b]	*Poecilia reticulata*
	4.66	4.61[b]	*Poecilia reticulata*
	4.66	3.64	*Poecilia reticulata*
	4.66	4.81[bc]	*Poecilia reticulata*
	4.66	4.89[bc]	*Poecilia reticulata*
1,2,4,5-Tetrachlorobenzene	4.60	3.61	*Jordanella floridae*
	4.60	3.72	*Oncorhynchus mykiss*
	4.60	4.11	*Oncorhynchus mykiss*
Tetrachlorobenzyltoluene	7.80	3.36	*Brachydanio rerio*
2,2',3,3'-Tetrachlorobiphenyl	6.18	4.69	*Oncorhynchus mykiss*
2,2',5,5'-Tetrachlorobiphenyl	6.09	4.63	*Poecilia reticulata*
	6.09	4.69[c]	*Carassius auratus*
	6.09	4.84[c]	*Poecilia reticulata*
	6.09	4.90[c]	*Poecilia reticulata*
	6.09	5.30	*Oncorhynchus mykiss*
2,3',4',5-Tetrachlorobiphenyl	6.23	4.62[c]	*Carassius auratus*
1,1,2,2-Tetrachloroethane	2.39	0.90	*Lepomis macrochirus*
Tetrachloroethylene	2.53	1.69	*Lepomis macrochirus*
Tetrachloroguaiacol	4.45	2.26	*Oncorhynchus mykiss*
1,2,3,4-Tetrachloronaphthalene	5.75	3.71	*Oncorhynchus mykiss*
2,3,4,5-Tetrachloronitrobenzene	3.93	1.87	*Oncorhynchus mykiss*
	3.93	1.90	*Oncorhynchus mykiss*
2,3,5,6-Tetrachloronitrobenzene	4.38	3.13	*Oncorhynchus mykiss*
	4.38	3.20	*Oncorhynchus mykiss*
	4.38	3.34	*Oncorhynchus mykiss*
2,3,5,6-Tetrachlorophenol	4.39	2.15	*Jordanellafloridae*
Thiobencarb	3.40	1.82	*Gnathopogon aerulescens*
	3.40	2.23	*Pseudorasbora parva*
Toluene diamine	3.16	1.96	*Pimephales promelas*
2,4,6-Tribromoanisole	4.48	2.94	*Pimephales promelas*
1,3,5-Tribromobenzene	4.51	3.23[c]	*Poecilia reticulata*
	4.51	3.70	*Oncorhynchus mykiss*
	4.51	3.97	*Oncorhynchus mykiss*
	4.51	4.08	*Oncorhynchus mykiss*
2,4,6-Tribromobiphenyl	6.03	3.88[c]	*Poecilia reticulata*

Table 11.2 (continued)   Octanol–Water Partition Coefficients and Bioconcentration Factors Obtained Under Flow-Through Conditions

Chemical	log K_{OW}[a]	log BCF[a]	Species
2,4,6-Tribromophenol	4.23	2.71	*Brachydanio rerio*
2,3,4-Trichloroaniline	3.68	2.00	*Poecilia reticulata*
2,4,5-Trichloroaniline	3.69	2.33	*Poecilia reticulata*
2,4,6-Trichloroaniline	3.69	2.33	*Poecilia reticulata*
3,4,5-Trichloroaniline	3.32	2.36	*Poecilia reticulata*
1,2,3-Trichlorobenzene	4.14	2.90	*Poecilia reticulata*
	4.14	3.08	*Oncorhynchus mykiss*
	4.14	3.28	*Poecilia reticulata*
	4.14	3.41	*Oncorhynchus mykiss*
1,2,4-Trichlorobenzene	4.05	2.95	*Oncorhynchus mykiss*
	4.05	3.11	*Oncorhynchus mykiss*
	4.05	3.31	*Jordanella floridae*
	4.05	3.32	*Pimephales promelas*
	4.05	3.36	*Lepomis cyanellus*
	4.05	3.36	*Oncorhynchus mykiss*
	4.05	3.51	*Oncorhynchus mykiss*
	4.05	3.57	*Oncorhynchus mykiss*
1,3,5-Trichlorobenzene	4.19	3.26	*Oncorhynchus mykiss*
	4.19	3.48	*Poecilia reticulata*
	4.19	3.61	*Oncorhynchus mykiss*
2,2′,5-Trichlorobiphenyl	5.60	4.30[c]	*Carassius auratus*
	5.60	4.91	*Oncorhynchus mykiss*
2,4,5-Trichlorobiphenyl	5.90	3.78[c]	*Poecilia reticulata*
	5.90	4.15	*Poecilia reticulata*
2,4′,5-Trichlorobiphenyl	5.79	4.63[c]	*Carassius auratus*
1,1,1-Trichloroethane	2.47	0.95	*Lepomis macrochirus*
1,1,2-Trichloroethylene	2.42	1.23	*Lepomis macrochirus*
3,4,5-Trichloroguaiacol	4.11	2.41	*Oncorhynchus mykiss*
4,5,6-Tricigoroguaiacol	3.74	1.97	*Oncorhynchus mykiss*
2,3,4-Trichloronitrobenzene	3.61	2.20	*Oncorhynchus mykiss*
2,4,5-Trichloronitrobenzene	3.40	1.84	*Oncorhynchus mykiss*
2,4,6-Trichloronitrobenzene	3.69	2.88	*Oncorhynchus mykiss*
2,4,6-Trichlorophenol	3.75	1.94	*Jordanella floridae*
	3.75	2.37[c]	*Oryzias latipes*
Tricresyl phosphate	5.11	2.22	*Pimephales promelas*
	5.11	2.90	*Alburnus alburnus*
Trifluralin	5.34	3.50	*Pseudoraobora parva*
Triphenyl phosphate	4.59	2.60	*Alburnus alburnus*

[a] See Devillers et al. (1996) for sources of experimental values.
[b] Log BCF expressed on a lipid content basis.
[c] Log BCF estimated from rate constants.
[d] Dichlorodibenzo-p-dioxin
[e] Hexa- and hepta-chlorodibenzo-p-dioxin
[f] Heptachlorodibenzofuran
[g] Isomer not specified
[h] Pentachlorodibenzo-p-dioxin
[i] Pentachlorodibenzofuran
[j] Tetrachlorodibenzo-p-dioxin
[k] Tetrachlorodibenzofuran

Reprinted from *Chemosphere*, **33**, Devillers, J., Bintein, S., and Domine, D., Comparison of BCF models based on log P, 1047–1065, copyright 1996. With kind permission from Elsevier Science Ltd., The Boulevard, Langford Lane, Kidlington OX5 1GB, U.K.

Table 11.3   Octanol–Water Partition Coefficients and Bioconcentration Factors
Obtained Under Static and Semi-Static Conditions

Chemical	log K_{OW}[a]	log BCF[a]	Species
Acridine	3.45	3.11	*Poecilia reticulata*
Aniline	0.90	0.41	*Brachydanio rerio*
Anthracene	4.45	2.95	*Lepomis macrochirus*
	4.45	3.86	*Poecilia reticulata*
Atrazine	2.61	0.78	*Brachydanio rerio*
Benzo(b)furan	2.86	2.56	*Poecilia reticulata*
Benzo[b]naphtho(2,3-d)thiophene	5.07	4.17	*Poecilia reticulata*
Benzo[a]pyrene	5.97	3.69	*Lepomis macrochirus*
Benzo[b]thiophene	3.26	2.53	*Poecilia reticulata*
2,2′-Bithiophene	3.75	3.55	*Poecilia reticutata*
p-sec-Butylphenol	3.08	1.57	*Salmo salar*
t-Butylphenyldiphenyl phosphate	5.12	2.89	*Pimephales promelas*
	5.12	3.04	*Oncorhynchus mykiss*
Carbazole	3.84	2.70	*Poecilia reticulata*
2-Chloroaniline	1.93	1.18	*Brachydanio rerio*
3-Chloroaniline	1.91	1.06	*Brachydanio rerio*
4-Chloroaniline	1.88	0.91	*Brachydanio rerio*
2-Chloronaphthalene	4.14	3.63	*Poecilia reticulata*
2-Chloronitrobenzene	2.52	2.29[b]	*Poecilia reticulata*
3-Chloronitrobenzene	2.50	2.42[b]	*Poecilia reticulata*
4-Chloronitrobenzene	2.39	2.46[b]	*Poecilia reticulata*
2-Chloro-6-nitrotoluene	3.09	3.09[b]	*Poecilia reticulata*
4-Chloro-2-nitrotoluene	3.05	3.02[b]	*Poecilia reticulata*
2,8-DCDD[d]	5.60	2.83	*Carassius auratus*
Dibenzo(1,4)dioxan	4.19	3.85	*Poecilia reticulata*
Dibenzofuran	4.21	3.54	*Poecilia reticulata*
Dibenzothiophene	4.49	3.82	*Poecilia reticulata*
1,2-Dibromobenzene	3.64	2.70	*Pimephales promelas*
	3.64	3.50	*Poecilia reticulata*
1,4-Dibromobenzene	3.79	2.70	*Poecilia reticulata*
	3.79	3.40	*Pimephales promelas*
2,4-Dichloroaniline	2.91	1.98	*Brachydanio rerio*
3,4-Dichloroaniline	2.78	1.48	*Brachydanio rerio*
1,2-Dichlorobenzene	3.43	2.40	*Poecilia reticulata*
	3.43	2.70	*Pimephales promelas*
1,3-Dichlorobenzene	3.53	3.78[b]	*Poecilia reticulata*
1,4-Dichlorobenzene	3.44	1.70	*Poecilia reticulata*
	3.44	2.40	*Pimephales promelas*
1,4-Dichloronaphthalene	4.66	3.36	*Poecilia reticulata*
1,8-Dichloronaphthalene	4.19	3.79	*Poecilia reticulata*
2,3-Dichloronaphthalene	4.51	4.04	*Poecilia reticulata*
2,3-Dichloronitrobenzene	3.05	3.01[b]	*Poecilia reticulata*
2.4-Dichloronitrobenzene	3.05	3.02[b]	*Poecilia reticulata*
2,5-Dichloronitrobenzene	3.03	2.92[b]	*Poecilia reticulata*
3,5-Dichloronitrobenzene	3.09	3.01[b]	*Poecilia reticulata*
2,3-Dimethylnitrobenzene	2.83	2.86[b]	*Poecilia reticulata*
3,4-Dimethylnitrobenzene	2.91	2.84[b]	*Poecilia reticulata*
1,2-Dinitrobenzene	1.69	1.02[b]	*Poecilia reticulata*
1,3-Dinitrobenzene	1.49	1.87[b]	*Poecilia reticulata*
2,4-Dinitrotoluene	1.98	2.31[b]	*Poecilia reticulata*
2,6-Dinitrotoluene	2.10	2.44[b]	*Poecilia reticulata*

Table 11.3 (continued)   Octanol–Water Partition Coefficients and Bioconcentration Factors Obtained Under Static and Semi-Static Conditions

Chemical	log K_{ow}[a]	log BCF[a]	Species
p-Dodecylphenol	7.91	3.78	Salmo salar
Fluorene	4.18	3.35	Poecilia reticulata
Hexachlorobenzene	5.73	5.62[b]	Poecilia reticulata
γ-Hexachlorocyclohexane	3.72	2.34	Brachydanio rerio
	3.72	2.41	Salmo salar
	3.72	2.84	Salmo salar
2-Nitroaniline	1.85	0.91	Brachydanio rerio
3-Nitroaniline	1.37	0.92	Brachydanio rerio
4-Nitroaniline	1.39	0.64	Brachydanio rerio
Nitrobenzene	1.85	1.47[b]	Poecilia reticulata
2-Nitrotoluene	2.36	2.28[b]	Poecilia reticulata
3-Nitrotoluene	2.42	2.21[b]	Poecilia reticulata
4-Nitrotoluene	2.37	2.37[b]	Poecilia reticulata
p-Nonylphenol	5.76	2.45	Salmo salar
Pentachloroaniline	5.08	3.78	Poecilia reticulata
Pentachlorobenzene	5.18	3.45	Carassius auratus
	5.18	5.18[h]	Poecilia reticulata
	5.18	4.23	Poecilia reticulata
Pyrene	5.18	3.68	Poecilia reticulata
2,3,7,8-TCDD[e]	6.42	3.90	Pimephales promelas
2,3,4,5-Tetrachloroaniline	4.57	3.28	Poecilia reticulata
2,3,5,6-Tetrachloroaniline	4.46	3.59	Poecilia reticulata
1,2,3,4-Tetrachlorobenzene	4.64	4.74[b]	Poecilia reticulata
3,3',4,4'-Tetrachlorobiphenyl	6.63	4.59	Poecilia reticulata
3,3',4,4'-Tetrachlorodiphenylether	5.78	4.51	Poecilia reticulata
Tetrachloroguaiacol	4.45	3.15	Salmo salar
Thianthrene	4.47	3.56	Poecilia reticulata
2,3,4-Trichloroaniline	3.68	2.61	Poecilia reticulata
2,4,5-Trichloroaniline	3.69	2.88	Poecilia reticulata
2,4,6-Trichloroaniline	3.69	3.13	Poecilia reticulata
3,4,5-Trichloroaniline	3.32	3.04	Poecilia reticulata
1,3,5-Trichlorobenzene	4.19	4.35[b]	Poecilia reticulata
2,4,5-Trichlorobiphenyl	5.90	4.26	Poecilia reticulata
2,4,5-Trichlorodiphenylether	5.44	4.18	Poecilia reticulata
1,3,7-Trichloronaphthalene	5.35	4.43	Poecilia reticulata
2,4,6-Trichlorophenol	3.75	2.84	Salmo salar
Tri-m-cresyl phosphate	5.11	2.78	Pimephales promelas
	5.11	2.89	Oncorhynchus mykiss
Tri-p-cresyl phosphate	5.11	2.97	Pimephales promelas
	5.11	3.15	Oncorhynchus mykiss
Triphenyl phosphate	4.59	2.75	Pimephales promelas
	4.59	2.76	Oncorhynchus mykiss
Xanthene	4.23	3.62	Poecilia reticulata

[a] See Devillers et al. (1996) for sources of experimental values.
[b] Log BCF expressed on a lipid content basis.
[c] Log BCF estimated from rate constants.
[d] Dichlorodibenzo-p-dioxin.
[e] Tetrachlorodibenzo-p-dioxin.

Reprinted from *Chemosphere,* **33**, Devillers, J., Bintein, S., and Domine, D., Comparison of BCF models based on log P, 1047–1065, copyright 1996. With kind permission from Elsevier Science Ltd., The Boulevard, Langford Lane, Kidlington OX5 1GB, U.K.

The BCF–K_{OW} regression models are based on the assumption that the K_{OW} values in the training set have no significant measurement error associated with them. However, K_{OW} values are difficult to measure, especially when log K_{OW} is greater than 5. Table 9.10 in Chapter 9 shows that recently reported values of log K_{OW} of individual chemicals vary by as much as 3.3 log units. This leads to significant uncertainty in the slope of the regression model.

In Section 11.2, we saw that BCF_f is correlated with γ_W, and in Section 7.2, we found that γ_W was inversely correlated with aqueous solubility. Therefore, the bio-concentration factor is inversely correlated with aqueous solubility. Several BCF_f–aqueous solubility correlations have been reported (Isnard and Lambert, 1988). The correlation between BCF_f and aqueous solubility can be derived by combining Equation 11.6 or 11.7 with any suitable K_{OW}–aqueous solubility correlation presented in Chapter 7.

Aqueous solubilities below 1.0 mg/L are difficult to measure. Wide discrepancies are found between solubility values reported in this range by different laboratories. This is partly responsible for some discrepancy between the various BCF_f–solubility correlations that have been reported and for the comparatively high uncertainty in the slopes of the regression models.

11.3.2 Structure-BCF Correlations

Bioconcentration is determined by molecular structure. The nature and arrangement of the fragments of a molecule determine the molecule's size and polarity and the way the chemical partitions between lipid and water. Structure–BCF_f correlations have been reported by several workers (Koch, 1983; Govers et al., 1984). Sabljic (1987) describes a model to estimate bioconcentration factors in fish. The model is

$$\log BCF_f = -2.131 + 2.124 \, ^2\chi^v - 0.160 \left(^2\chi^v \right)^2 \tag{11.8}$$

$n = 84$, $s = 0.345$, $r = 0.966$, $F(2,81) = 599$

where BCF_f = bioconcentration factor, g/g
      $^2\chi^v$ = second-order valence connectivity index
      n = number of chemicals in the training set
      s = standard deviation
      r = correlation coefficient
      F = F-ratio

The model applies to chlorinated benzenes, polychlorinated biphenyls, chlorinated diphenyl oxides, halocarbons, alkyl and alkenyl benzenes, polycyclic aromatic hydrocarbons, substituted phenols, anilines, phthalates, and aromatic nitro compounds with log BCF_f in the range 0 to 5.

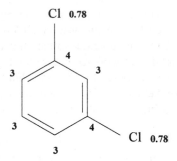

Figure 11.1 Hydrogen-suppressed structure of 1,4-dichlorobenzene. δ^v values are shown next to each atom.

Example

Estimate the bioconcentration factor of 1,4-dichlorobenzene in fish using Equation 11.8.

1. Calculate the first-order connectivity index for dichlorobenzene. The hydrogen-suppressed structure is shown along with valence delta values in Figure 11.1.

$$^2\chi^v = \Sigma_{bonds} \left(\delta_i^v \delta_j^v \delta_k^v\right)^{-0.5}$$

$$= (1)(3 \cdot 3 \cdot 3)^{-0.5} + (4)(3 \cdot 3 \cdot 4)^{-0.5} + (4)(4 \cdot 3 \cdot 0.78)^{-0.5}$$

$$= 2.167$$

2. The BCF_f value is estimated using Equation 11.8.

$$\log BCF_f = -2.131 + 2.124 \ ^2\chi^v - 0.160 \left(^2\chi^v\right)^2$$

$$= -2.131 + (2.124)(2.167) - (0.160)(2.167)$$

$$= 2.12$$

Reported log BCF_f values of 1,4-dichlorobenzene range from 1.78 to 2.95 (Devillers et al., 1996). The estimated value of log BCF_f lies at the midpoint of this range.

11.3.3 Sensitivity to Environmental Parameters and Method Error

While BCF_f measurements are made with standard procedures (Veith et al., 1979), all important parameters may not be controlled, and discrepancies between BCF_f values reported by different laboratories may be as great as three orders of magnitude (Bysshe, 1990). This is evident in the data reported in Tables 11.2 and 11.3. The standard error of log BCF_f estimates made with the methods described in this chapter is about 0.5 log unit, comparable to the uncertainty in the training data. Furthermore, lipid–water partitioning of chemicals with log K_{OW} greater than 5 does not reach steady state quickly (Oliver, 1984), and reported BCF_f values of highly hydrophobic chemicals may be in error. The models are used primarily for screening purposes, and order-of-magnitude estimates of BCF_f values are satisfactory for this purpose.

The models described in this chapter are best applied to nonpolar chemicals. Active transport and other processes may control the bioconcentration of polar chemicals causing BCF_f values to differ significantly from model estimates.

Bioconcentration factors vary with ambient temperature. The temperature dependence appears to differ from species to species and from chemical to chemical (Davies and Dobbs, 1984). Like K_{OW}, BCF_f depends upon ionic strength and pH.

For some compounds, chemical conversion competes efficiently with uptake. In this case, BCF_f values may vary with chemical exposure levels. Oliver (1984) reports that BCF_f values of halogenated chemicals in trout depend on exposure level in the ng/L range. The value of BCF_f roughly doubles on increasing the exposure level by a factor of 20.

Because the lipid content of the fish's diet determines in part the lipid content of the animal, bioconcentration is sensitive to the lipid content of the fish's diet. Lipid content also varies with species, size, and age of the animal. When lipid-normalized, bioconcentration factors appear not to be greatly dependent on these factors (Esser and Moser, 1982).

Chemical uptake is enhanced in predatory fish and mollusks that are exposed to elevated levels of xenobiotics in their diet. For chemicals with log K_{OW} values in the range from 5 to 7, contaminated food is a significant source of chemical residue in top predator fish, and contaminant biomagnification, the increase of xenobiotic concentrations in organisms with trophic level, results in higher residue levels than predicted with the BCF_f models described in Section 11.3 (Oliver, 1984; Thomann, 1989). Below log $K_{OW} = 6$, diet appears not to be a significant source of exposure. Above log $K_{OW} = 7$, model results show that bioaccumulation varies with assimilation efficiency, bioconcentration in the food source (phytoplankton), and top predator growth rate. Even so, food-chain accumulation remains a significant source of chemical residue in fish.

Oliver and Niimi (1988) found elevated levels of polychlorinated biphenyl residues in salmonids (coho salmon, lake trout, brown trout, and rainbow trout) in Lake Ontario that could not be attributed to bioconcentration alone. They suggested that contaminated food is a major source of chemical uptake by these animals. A correlation between bioaccumulation factor and K_{OW} was suggested. It is

$$\log BAF_f = 1.07 \log K_{OW} - 0.21 \qquad (11.9)$$

$n = 18, r = 0.93$

where BAF_f = fish bioaccumulation factor, g/g

$\qquad$ = C_{org}/C_W

C_{org} = concentration in organism, g chemical/g fish

C_W = concentration in water, g chemical/g water

K_{OW} = octanol–water partition coefficient

The relationship applies to polychlorinated biphenyls and other chlorinated organics with log K_{OW} in the range from 4 to 7.

Comparison of Equation 11.9 with 11.6 shows that residue levels in salmonids are about one order of magnitude higher than residue levels in minnows and sunfish. Some of the difference is attributed to the high lipid content of salmonids (about 11%). An increase in residue level of a factor of 5, about half the difference, is attributed to dietary exposure.

Some fish species metabolize xenobiotics to form water-soluble chemicals. In this case, BCF_f values might be overestimated by the methods described in Section 11.3.

The estimation methods underestimate BCF values in marine organisms slightly due to the salting-out effect which reduces the solubility of neutral organic chemicals. (See Chapter 9.) The effect is not large (Knezovich, 1994). Evidence has been presented to show that freshwater models can be used to obtain reliable order-of-magnitude estimates of bioaccumulation in marine species (Zaroogian et al., 1985).

The BCF_f models described in Section 11.3 do not apply to very small organisms, such as plankton, which have large area/volume ratios and may have higher apparent partition coefficients because of surface sorption. Models proposed for aquatic organisms such as plankton are shown in Table 11.4. The correlations shown for *Daphnia magna* and *Mytilius edulis* in Table 11.4 are similar to those reported earlier for *D. pulex* and for a marine mollusk by Hawker and Connell (1986) and to Equation 11.6 for fish.

Some of the differences between the models described in Table 11.4 and the models given in Table 11.1 may be due to the differences in protein content of the organism as well as to surface adsorption. Earthworms contain about ten times more protein than lipid. For muscle and protein, log $BCF_{prot} = 0.1 + 0.56$ log K_{OW} (Neely et al., 1974).

Pesticides are believed to partition passively between water and microbial cells. Baughman and Paris (1981) report a correlation between the BCF_{bact} values of chemicals in bacteria and K_{OW}; log $BCF_{bact} = -0.361 + 0.907$ log K_{OW}, n = 14 pesticides, r = 0.98, 1.8 > log BCF_{bact} > 5.0, 3 > log K_{OW} > 7. This model is similar to the correlations given in Table 11.1 for fish.

11.4 METHODS OF ESTIMATING BIOACCUMULATION FACTORS IN CATTLE AND DAIRY PRODUCTS

Mammals are exposed to xenobiotics primarily through ingestion of food and soil. Nonpolar organic chemicals are hydrophobic and are found concentrated in

Table 11.4 Some Relationships Between Bioconcentration (BCF) and Bioaccumulation (BAF) Factors and Octanol–Water Partition Coefficients for Species Other Than Fish

Log BCF = A + B Log K_{OW} Chemical Class/Test Species	A	B	n	r	Ref.
Mixed 0 > log BCF > 5, 1 > logK_{OW} > 7/water flea (*Daphnia magna*)	−1.10	0.85	52	0.96	Geyer et al., 1991
Mixed 0 > log BCF > 4, 2 > logK_{OW} > 6/mussel (*Mytilus edulis*)	−0.808	0.858	16	0.96	Geyer et al., 1991
PCBs, chlorinated hydrocarbons 3 > log BAF > 7, 4 > log K_{OW} > 7/plankton	0.33	0.68	26	0.91	Oliver and Niimi, 1988
PCBs, chlorinated hydrocarbons 3 > log BAF > 7, 4 > log K_{OW} > 7/mysids (*Mysis relicta*)	0.53	0.77	25	0.88	Oliver and Niimi, 1988
PCBs, chlorinated hydrocarbons 3 > log BAF > 7, 4 > log K_{OW} > 7/amphipods (*Pontoporeia affinis*)	0.61	0.61	27	0.89	Oliver and Niimi, 1988
PCBs, chlorinated hydrocarbons 3 > log BAF > 7, 4 > log K_{OW} > 7/oligochaetes (*Tubifex tubifex* and *Limnodrilus hoffmeisteri*)	0.44	0.73	26	0.82	Oliver and Niimi, 1988

lipids in animal fat and milk. The concentration of a chemical in meat and milk products is estimated using transfer factors:

$$C_{prd} = TF_{prd}\left(C_{feed}Q_{feed} + C_W Q_W + C_S Q_S\right)$$ (11.10)

where C_{prd} = chemical concentration in food product, mg/kg
TF_{prd} = product (meat or milk) transfer factor, day/kg
C_{feed} = chemical concentration in cattle feed, mg/kg
Q_{feed} = cattle's consumption rate of feed, kg/day
C_W = chemical concentration in drinking water, mg/l
Q_W = cattle's consumption rate of water, l/day
C_S = chemical concentration in soil, mg/kg
Q_S = cattle's consumption rate of soil, kg/day

The transfer factor is

$$TF_{prd} = \frac{F \cdot BCF_{fat}}{Q_f}$$ (11.11)

where TF_{prd} = product (meat or milk) transfer factor, day/kg
BCF_{fat} = bioconcentration factor in beef fat, kg/kg
F = mass fraction of fat in the food product (beef or milk)
Q_f = animal's consumption rate of feed, kg/day

Kenaga (1980) and Garten and Trabalka (1983) conducted early studies of the correlation between bioaccumulation factors and K_{OW} in mammals. Poor correlations were obtained, in part, because the available K_{OW} data were poor. Garten and Trabalka conclude that the models cannot be used reliably for screening.

More recently, Travis and Arms (1988) described correlations of BCF_{fat} with K_{OW} and aqueous solubility for a variety of animals. They suggested the following correlation for meat products:

$$\log TF_{beef} = -7.6 + \log K_{OW} \qquad (11.12)$$

$n = 36$, $r^2 = 0.81$

where TF_{beef} = feed-to-meat transfer factor, day/kg
      K_{OW} = octanol–water partition coefficient
      n   = number of chemicals in the training set
      r^2 = correlation coefficient

The training set consists of pesticides with $\log K_{OW}$ values in the range from 0 to 7. The correlation is based on the assumptions that beef is 25% fat and that the dry feed ingestion rate of beef cattle is 8 kg/day.

Travis and Arms propose the following correlation for milk:

$$\log TF_{milk} = -8.1 + \log K_{OW} \qquad (11.13)$$

$n = 28$, $r^2 = 0.74$

where TF_{milk} = feed-to-milk transfer factor, day/kg
      K_{OW} = octanol–water partition coefficient
      n   = number of chemicals in the training set
      r^2 = correlation coefficient

The training set consists of pesticides with $\log K_{OW}$ in the range from 0 to 7. The correlation is based on the assumptions that milk is 3.68% fat and that the dry feed ingestion rate of lactating cows is 16 kg/day.

The uncertainty in estimating TF_{beef} and TF_{milk} using Equations 11.11 and 11.12 is large. McKone and Ryan (1989) report that the 95% prediction interval is 2 log units (two orders of magnitude in the value of the transfer factor) for both models.

Kenaga (1980) reports that bioaccumulation in beef fat is highly correlated with bioconcentration in fish lipid. As with bioconcentration in fish, bioaccumulation in animals may reach a maximum when $\log K_{OW}$ is about 6. McLachlan (1993) reports that bioaccumulation of chlorinated organic chemicals in lactating cows is correlated with K_{OW} up to $\log K_{OW} = 6.5$. At higher K_{OW} values, bioaccumulation is inversely correlated with K_{OW}.

The bioaccumulation models for chemicals in mammals are order-of-magnitude estimates at best. The ingestion of xenobiotics by animals depends upon diet, season, and climate and is not uniform throughout the year. Also, the relative importance of food ingestion and soil ingestion as a source of exposure varies with animal species and feeding method.

11.5 METHODS OF ESTIMATING BIOACCUMULATION FACTORS IN PLANTS AND VEGETABLES

The subject of modeling chemical uptake by plants was reviewed by Paterson et al. (1990). Important routes of bioaccumulation in plants are uptake of chemical through the root system and wet or dry deposition from the atmosphere (Mogghissi et al., 1980; Hope, 1995). In addition, the direct uptake by foliage of chemical vapor in air is important (Simonich and Hites, 1994).

The chemical concentration in vegetation due to uptake from soil is given by:

$$C_{veg} = B_v \cdot F_{dw} \cdot C_{as}$$ (11.14)

where C_{veg} = chemical concentration in vegetation, mg/kg wet plant

B_v = soil-to-plant bioconcentration factor, kg dry root-zone soil/kg dry plant

F_{dw} = dry-material weight fraction of plant, g/g

C_{as} = concentration of chemical in root-zone soil, mg/kg

Typically, F_{dw} = 0.15 g/g.

A correlation between B_v, the ratio of chemical concentration in aboveground plant parts to the chemical concentration in soil, and the soil–water partition coefficient is described by Baes (1982). The correlation is

$$\log B_v = 1.54 - 1.18 \log K_d$$ (11.15)

n = 21, r² = 0.88

where B_v = soil-to-plant bioconcentration factor, kg dry root-zone soil/kg dry aboveground plant parts

K_d = soil–water partition coefficient, l/kg

n = number of chemicals in the training set

r^2 = correlation coefficient

The training set consists of diverse chemicals with K_d values in the range from 1 to 10^6.

A correlation between B_v and K_{OW} was proposed by Travis and Arms (1988). It is

$$\log B_v = 1.588 - 0.578 \log K_{OW}$$ (11.16)

n = 29, r² = 0.73

where B_v = soil-to-plant bioconcentration factor, kg dry root-zone soil/kg dry
                aboveground plant parts
       K_{OW} = octanol–water partition coefficient
       n   = number of chemicals in the training set
       r^2 = correlation coefficient

The training set consists of diverse pesticides with log K_{OW} values in the range from 1 to 9. The correlation is not highly reliable, as the low correlation coefficient shows. Model estimates of B_v may agree with measured values within no more than a factor of 100. Topp et al. (1986) and McKone (1993) report similar relationships for uptake of chemicals from root-zone soil and soil solution.

The chemical concentration in root crops due to uptake from soil solution is determined by the crop's root concentration factor, BCF_r [(g chemical/g wet root)/(g chemical/mL soil solution)], the ratio of the root concentration of chemical to the concentration of chemical in soil solution. $C_{root} = C_{SW} \cdot BCF_r$. Briggs et al. (1982, 1983) proposed the following nonlinear correlation for plants with nonwoody stems:

$$\log\left(BCF_r - 0.82\right) = -1.55 + 0.77 \log K_{OW} \qquad (11.17)$$

n = 35, r^2 = 0.98

where BCF_r = root concentration factor, g wet root/mL soil solution
       K_{OW} = octanol–water partition coefficient
       n   = number of chemicals in the training set
       r^2 = correlation coefficient

The training set consists of diverse chemicals with log K_{OW} values in the range from –2 to 5. The data were collected in a laboratory study of uptake of nonionic chemicals from culture solutions and may not accurately reflect the uptake of chemicals from soils.

Sorption to soil particles reduces the availability of chemicals for uptake by plant roots, and the plant–soil bioconcentration factor is inversely proportional to the organic-carbon–water partition coefficient K_{OC} (Topp et al., 1986). The soil-to-plant concentration factor for root crops, B_v (g soil/g wet root), is calculated on a wet-root basis, and $F_{dw} = 1$ in Equation 11.14. Bioaccumulation of chemicals by root crops from contaminated soil is given by

$$B_v = \frac{BCF_r}{K_d} \qquad (11.18)$$

where B_v = soil-to-plant bioconcentration factor, kg soil/kg dry plant
       BCF_r = root concentration factor, g wet root/mL soil
       K_d = soil–water partition coefficient, L/kg

Gobas et al. (1991) studied the bioconcentration of aromatic hydrocarbons in a submerged aquatic macrophyte, *Myriophyllum spicatum*. Uptake kinetics were studied in a continuous-flow chamber. Concentrations in both water and plants were measured in units of μg/L. Gobas et al. report the following correlation between the plant-water bioconcentration factor and K_{OW}:

$$\log BCF_{PW} = -2.24 + 0.98 \log K_{OW} \tag{11.19}$$

$n = 9$, $r^2 = 0.97$

where BCF_{PW} = plant-water bioconcentration factor, L of plant matter/L of water
      K_{OW} = octanol–water partition coefficient
      n   = number of chemicals in training set
      r   = correlation coefficient

The training set consisted of aromatic hydrocarbons with log K_{OW} values in the range from 4 to 8.26. Unlike the models developed for fish and animals, the linear relationship between log BCF_{PW} and log K_{OW} does not break down when log $K_{OW} > 5$.

Uptake of atmospheric pollutants by wet or dry deposition on plant surfaces is described by the cuticle water partition coefficient, BCF_{CW}. Schonherr and Riederer (1989) report a correlation between K_{CW} and K_{OW}. It is

$$\log BCF_{CW} = 0.057 + 0.970 \log K_{OW} \tag{11.20}$$

$n = 13$, $r = 0.987$, $s = 0.339$, $F(1,11) = 383$

where BCF_{CW} = cuticle-water bioconcentration factor, L of plant matter/L of water
      K_{OW} = octanol–water partition coefficient
      n   = number of chemicals in the training set
      s   = standard deviation
      r   = correlation coefficient
      F   = F ratio

The relationship applies to aromatic hydrocarbons, halo-substituted aromatic hydrocarbons, nitrophenols, chlorophenols, herbicides, and phthalates with log K_{OW} in the range from −2 to 8. Schonherr and Riederer also report a correlation between BCF_{CW} and aqueous solubility.

Sabljic et al. (1990) report a structure/cuticle–water partition coefficient relationship. It is

$$\log BCF_{CW} = -0.37 + 1.31\,^3\chi^v - 1.49 N_{aliph-OH} \tag{11.21}$$

$n = 47$, $s = 0.250$, $r = 0.992$, $F(2,44) = 1418$

where BCF_{CW} = cuticle–water partition coefficient
      $^3\chi^v$ = third-order valence connectivity
      $N_{aliph-OH}$ = number of aliphatic –OH groups
      n   = number of chemicals in the training set

s   = standard deviation
r   = correlation coefficient
F   = F ratio

The model applies to aliphatic and aromatic alcohols, phenols, and acids with log BCF_{CW} in the range from -2 to 8. The topological parameter $^3\chi^v$ is correlated with molecular size. The cross-correlation between the indicator variable $N_{aliph-OH}$ and $^3\chi^v$ is low ($r^2 = 0.03$)

Several workers studied the partitioning of volatile chemicals between foliage and air. Bacci et al. (1990) studied the bioconcentration of pesticide vapor in azalea leaf. They propose a relationship between the leaf–air bioconcentration factor, the air–water partition coefficient, and the octanol–water partition coefficient. The relationship is

$$\log\left(BCF_{LA} \cdot K_{AW}\right) = -1.95 + 1.14 \log K_{OW} \tag{11.22}$$

n = 9, r = 0.96

where BCF_{LA} = leaf–air bioconcentration factor
       K_{AW} = air–water partition coefficient
       K_{OW} = octanol–water partition coefficient
       n   = number of chemicals in training set
       r   = correlation coefficient

The training set consists of diverse pesticides with K_{AW} in the range from 3.44×10^{-1} to 3.57×10^{-5} and with log K_{OW} from 1.2 to 6.9.

The leaf–air bioconcentration factor is the ratio of chemical concentration in foliage, C_{lf} (g/m³), to the chemical concentration in air, C_a (g/m³), and $BCF_{LA} = C_{lf}/C_a = (C_{lf}/C_w)/(C_a/C_w) = (L \cdot K_{OW})/K_{AW}$, where L is the lipid mass fraction of the foliage (Connell and Hawker, 1988; Travis and Hattemer-Frey, 1988). The model, with $L = 0.022$, appears to accurately predict the levels of hexachlorobenzenes, hexachlorocyclohexanes, and DDTs sorbed on plant foliage in 300 plant samples collected at 26 locations worldwide (Calamari et al.,1991).

Foliage-to-air partitioning of hydrophobic chemicals on azalea leaves was also studied by Paterson et al. (1991). Introducing the octanol–air partition coefficient, $K_{OA} = K_{OW}/K_{AW}$, they report:

$$BCF_{LA} = 0.19 + 0.7K_{WA} + 0.05K_{OA} \tag{11.23}$$

n = 14

where BCF_{LA} = vegetation bioconcentration factor, [(g/m³ of wet leaf)/(g/m³ of air)]
       K_{AW} = air–water partition coefficient
       K_{OA} = octanol–air partition coefficient
                 = K_{OW}/K_{AW}
       K_{OW} = octanol–water partition coefficient
       n   = number of chemicals in training set

The training set consists of pesticides with BCF_{LA} values in the range 1×10^5 to 1×10^6, Henry's law constant, H, in the range 0.0062 to 131.5 Pa·m³/mol, and K_{OA} values in the range 4.44×10^5 to 2.57×10^9. The K_{OA} values used for the correlation were estimated as the ratio K_{OW}/K_{AW}.

The relationship shows that leaf material behaves like a mixed-phase system consisting of 70% water, 19% air, and 5% octanol. The model overestimates BCF_{LA} values of the training set by as much as a factor of 4. The error is probably due to errors in the training-set data.

Example

Estimate the leaf–air bioconcentration factor of alachlor at 20°C (293 K) using the azalea model, Equation 11.23.

1. Calculate the value of K_{AW}. The Henry's law constant of alachlor is 6.2×10^{-3} Pa-m³/mol (Suntio et al., 1988). As shown in Chapter 8,

$$K_{AW} = H/RT$$

$$= \left(6.2 \times 10^{-3}\ Pa - m^3/mol\right) / \left(8.3145\ Pa - m^3/mol\right) (293\ K)$$

$$= 2.5 \times 10^{-6}$$

2. Calculate the value of K_{OA}. For alachlor, $K_{OW} = 6.3 \times 10^2$ (Suntio et al., 1988).

$$K_{OA} = K_{OW}/K_{AW}$$

$$= \left(6.3 \times 10^2\right) / \left(2.5 \times 10^{-6}\right)$$

$$= 2.5 \times 10^8$$

3. Estimate BCF_{LA} using Equation 11.23.

$$BCF_{LA} = 0.19 + 0.7\ K_{AW} + 0.05\ K_{OA}$$

$$= 0.19 + (0.7)\left(2.5 \times 10^{-6}\right) + 0.05\left(2.5 \times 10^8\right)$$

$$= 1.3 \times 10^7$$

The reported value is 2.82×10^5 (Paterson et al., 1991). Azalea-leaf bioconcentration factors of alachlor, sulfotep, and γ-HCH are poorly estimated with this model.

There are several important limitations to the applicability of the bioconcentration models described in this section for chemicals in plants and vegetables. Uptake may be an active process that is specific for certain chemicals or a passive process that is not chemical specific (Ryan et al., 1988). Active processes and passive processes are not easy to distinguish. Furthermore, the uptake of a specific chemical may be an active process in one plant and a passive process in another (O'Conner et al., 1990). The relationships described in this section apply only to the passive uptake of chemicals.

Equation 11.17 shows that chemical partitioning between plant root and soil solution is directly proportional to the K_{OW} value in the range $-2 < \log K_{OW} < 5$. Highly hydrophobic chemicals presumably accumulate in the lipid found in root membranes and cell walls (Paterson et al., 1990). In a laboratory study of chemical uptake by the aquatic macrophyte *Myriophylum spicatum*, the lipid-normalized plant–water bioconcentration factor was found to be approximately the same as K_{OW} for nine aromatic hydrocarbons, suggesting that chemical uptake is due to partitioning between water and plant lipid (Gobas et al.,1991). Translocation to other parts of the plant appears to reach a maximum when $\log K_{OW}$ is 1.8 (Briggs et al., 1982; Topp et al., 1986). As a result, chemicals with intermediate values of K_{OW} are found throughout the plant, while highly hydrophobic chemicals tend to accumulate in plant roots. Polychlorinated biphenyls are found strongly adsorbed to the outer cell walls in carrots and do not penetrate into the root (Strek and Weber, 1982). Similar considerations apply to the uptake of chemicals by foliage. Water-soluble chemicals may be distributed throughout the plant, while highly hydrophobic chemicals tend to remain sorbed to the waxy cuticle of the leaves (Paterson et al., 1990).

Steady-state concentrations of chemicals may not be reached during the growing season (Topp et al., 1986). In addition, xenobiotics may be metabolized by plants. This is believed to be the case for pentachlorophenol in soybeans and spinach (Casterline et al., 1985). Photodegradation of chemicals sorbed on foliage may also occur. For instance, dioxin sorbed to grass foliage undergoes photodegradation (McCrady and Maggard, 1993). Bioconcentration models can be expected to overestimate the levels of xenobiotics in plants if chemical conversion is significant.

REFERENCES

AQUIRE, 1996. Aquatic Information Retrieval Database. U.S. Environmental Protection Agency, Environmental Research Laboratory — Duluth, 6201 Congdon Blvd., Duluth, MN 55804.

Bacci, E., Calamari, D., Gaggi, C., and Vighi, M., 1990. Bioconcentration of organic chemical vapors in plant leaves: experimental measurements and correlation. *Environ. Sci. Technol.*, **24**, 885–889; Bacci, E., Cerejeira, M.J., Gaggi, C., Chemello, G., Calamari, D., and Vighi, M., 1990. Bioconcentration of organic chemical vapors in plant leaves; the azalea model. *Chemosphere*, **21**, 525–535.

Baes, C.F., 1982. Prediction of radionuclide K_d values from soil-plant concentration ratios. *Trans. Amer. Nuclear Society*, **41**, 53–54.

Banerjee, S. and Baughman, G.L., 1991. Bioconcentration factors and lipid solubility.*Environ. Sci. Technol.*, **25**, 536–539.

Barron, M.G., 1990. Bioconcentration. *Environ. Sci. Technol.*, **24**, 1612–1618.

Baughman, G.L. and Paris, D.F., 1981. Microbial bioconcentration of organic pollutants from aquatic systems — a critical review. In *CRC Critical Reviews in Microbiology*. CRC Press, New York, pp. 205–228.

Bintein, S., Devillers, J., and Karcher, W., 1993. Nonlinear dependence of fish bioconcentration on n-octanol/water partition coefficient. *SAR QSAR Environ. Res.*, **1**, 29–39.

Briggs, G.G., 1981. Theoretical and experimental relationships between soil adsorption, octanol–water partition coefficients, water solubilities, bioconcentration factors, and the parachor. *J. Agric. Food Chem.*, **29**, 1050–1059.

Briggs, G.G., Bromilow, R.H., and Evans, A.A., 1982. Relationships between lipophilicity and root uptake and translocation of nonionized chemicals by barley. *Pesticide Sci.*, **13**, 495–504.

Briggs, G.G., Bromilow, R.H., Evans, A.A., and Williams, M., 1983. Relationships between lipophilicity and the distribution of nonionized chemicals by barley shoots following uptake by the roots. *Pesticide Sci.*, **14**, 492–500.

Bysshe, S.E., 1990. Bioconcentration factor in aquatic organisms. Chapter 5. In *Handbook of Chemical Property Estimation Methods. Environmental Behavior of Organic Compounds*. Lyman, W.J., Reehl, W.F., and Rosenblatt, D.H., Eds. American Chemical Society, Washington, D.C.

Calamari, D., Bacci, E., Focardi, S., Gaggi, C., Morosini, M., and Vighi, M., 1991. Role of plant biomass in the global environmental partitioning of chlorinated hydrocarbons. *Environ. Sci. Technol.*, **25**, 1489–1495.

Casterline, J.L., Barnett, N.M., and Ku, Y., 1985. Uptake, translocation, and transformation of pentachlorophenol in soybean and spinach plants. *Environ. Res.*, **37**, 101–118.

CHEMFATE, 1986. Chemfate database. Syracuse Research Corp., Merrill Lane, Syracuse, NY 13210–4080.

Connell, D.W. and Hawker, D.W., 1988. Use of polynomial expressions to describe the bioconcentration of hydrophobic chemicals by fish. *Ecotoxicol. Environ. Saf.*, **16**, 242–257.

Davies, R.P. and Dobbs, A.J., 1984. The prediction of bioconcentration in fish. *Water Res.*, **18**, 1253–1262.

Dean, J.A., 1992. *Lange's Handbook of Chemistry*, 14th Edition. McGraw-Hill, New York.

Devillers, J., Bintein, S., and Domine, D., 1996. Comparison of BCF models based on log P. *Chemosphere*, **33**, 1047–1065.

ENVIROFATE, 1985. Envirofate database, U.S. Environmental Protection Agency, Environmental Research Laboratory–Duluth, 6201 Congdon Blvd., Duluth, MN 55804.

Esser, H.O. and Moser, P., 1982. An appraisal of problems related to the measurement and evaluation of bioaccumulation. *Ecotoxicol. Environ. Saf.*, **6**, 131–148.

Garten, C.T. and Trabalka, J.R., 1983. Evaluation of models for predicting terrestrial food chain behavior of xenobiotics. *Environ. Sci. Technol.*, **17**, 590–595.

Geyer, H.J., Scheunert, I., Bruggemann, R., Steinberg, C., Korte, F., and Kettrup, A., 1991. QSAR for organic chemical bioconcentration in *Daphnia*, algae, and mussels. In *QSAR in Environmental Toxicology — IV*. Proceedings of the Fourth International Workshop, Veldhoven, The Netherlands, 16–20 September 1990. Hermens, J.L.M. and Opperhuizen, A., Eds. Elsevier, Amsterdam, pp. 387–394.

Gobas, F.A.P.C., McNeil, E.J., Lovett-Doust, L., and Haffner, G.D., 1991. Bioconcentration of chlorinated aromatic hydrocarbons in aquatic macrophytes. *Environ. Sci. Technol.*, **25**, 924–929.

Govers, H., Ruepert, C., and Aiking, H., 1984. Quantitative structure-activity relationships for polycyclic aromatic hydrocarbons: correlations between molecular connectivity, physico-chemical properties, bioconcentration and toxicity in *Daphnia pulex*. *Chemosphere*, **13**, 227-236.

Hamelink, J.L., Waybrant, R.C., and Ball, R.C., 1971. A proposal: exchange equilibria control the degree chlorinated hydrocarbons are biologically magnified in lentic environments. *Trans. Amer. Fish Soc.*, **100**, 207–211.

Harner, T. and Mackay, D., 1995. Measurement of octanol–air partition coefficients for chlorobenzenes, PCBs, and DDT. *Environ. Sci. Technol.*, **29**, 1599–1606.

HSDB, 1996. Hazardous Substance Data Base, Toxicology Data Network (TOXNET), National Library of Medicine, Washington, D.C.

Hawker, D.W. and Connell, D.W., 1986. Bioconcentration of lipophilic compounds by some aquatic organisms. *Ecotoxicol. Environ. Saf.*, **11**, 184–197.

Hope, B.K., 1995. A review of models for estimating terrestrial ecological receptor exposure to chemical contaminants. *Chemosphere,* **30**, 2267–2287.

Isnard, P. and Lambert, S., 1988. Estimating bioconcentration factors from octanol–water partition coefficient and aqueous solubility. *Chemosphere*, **17**, 21–34.

Kenaga, E.E., 1980. Correlation of bioconcentration factors of chemicals in aquatic and terrestrial organisms with their physical and chemical properties. *Environ. Sci. Technol.*, **14**, 553–556.

Kenaga, E.E. and Goring, C.A.I., 1980. Relationship between water solubility, soil sorption, octanol–water partitioning, and concentration of chemicals in biota. In *Aquatic Toxicology.* Proceedings of the Third Annual Symposium on Aquatic Toxicology, ASTM Special Technical Publication 707. Eaton, J.G., Parrish, P.R., and Hendricks, A.C., Eds., pp. 79–129.

Knezovich, J.P., 1994. Chemical and biological factors affecting bioavailability of contaminants in seawater. In *Bioavailability. Physical, Chemical, and Biological Interactions.* Hamelink, J.L., Landrum, P.F., Bergman, H.L., and Benson, W.H., Eds., Lewis Publishers, Boca Raton, FL, pp. 23–30.

Koch, R., 1983. Molecular connectivity index for assessing ecotoxicological behavior of organic compounds. *Toxicol. Environ. Chem.*, **6**, 87–96.

Konemann, H. and van Leeuwen, K., 1980. Toxicokinetics in fish: accumulation and elimination of six chlorobenzenes by guppies. *Chemosphere*, **9**, 3–19.

Lide, D.R., 1994. *CRC Handbook of Chemistry and Physics.* 74th Edition. CRC Press, Boca Raton, FL.

Mackay, D., 1982. Correlation of bioconcentration factors. *Environ. Sci. Technol.,* **16**, 274–278.

Mackay, D., Shiu, W.Y., and Ma, K.C., 1992. *Illustrated Handbook of Physical-Chemical Properties and Environmental Fate for Organic Chemicals.* Volume 1. Monoaromatic Hydrocarbons, Chlorobenzenes, and PCBs. Lewis Publishers, Boca Raton, FL.

McCrady, J.K. and Maggard, S.P., 1993. Uptake and photodegradation of 2,3,7,8-tetrachlorodibenzo-p-dioxin sorbed to grass foliage. *Environ. Sci. Technol.*, **27**, 343–350.

McKone, T.E. and Ryan, P.B., 1989. Human exposure to chemicals through food chains. An uncertainty analysis. *Environ. Sci. Technol.*, **23**, 1154–1163.

McKone, T.E., 1993. The precision of QSAR methods for estimating intermedia transfer factors in exposure assessment. *SAR QSAR Environ. Res.*, **1**, 41–51.

McLachlan, M.S., 1993. Mass balance of polychlorinated biphenyls and other organochlorine compounds in a lactating cow. *J. Agric. Food Chem.*, **41**, 474–480.

Mogghissi, A.A., Marland, R.R., Congel, F.J., and Eckerman, K.F., 1980. Methodology for environmental human exposure and health risk assessment. Chapter 31. In *Dynamics, Exposure, and Hazard Assessment of Toxic Chemicals.* Haque, R., Ed. Ann Arbor Science Publishers, Ann Arbor.

Neely, W.B., Branson, D.R., and Blau, G.E., 1974. Partition coefficient to measure bioconcentration potential of organic chemicals in fish. *Environ. Sci. Technol.*, **8**, 1113–1115.

Nendza, M., 1991. QSARs of bioconcentration: validity assessment of log Pow/log BCF correlations. In *Bioaccumulation in Aquatic Systems.* Nagel, R. and Loskill, R., Eds. VCH, Weinheim, pp. 43–66.

O'Conner, G.A., Lujan, J.R., and Jin, Y., 1990. Adsorption, degradation, and plant availability of 2,4-dinitrophenol in sludge-amended calcerous soils. *J. Environ. Qual.*, **19**, 587–593.

Oliver, B.G., 1984. The relationship between bioconcentration factor in rainbow trout and physical-chemical properties for some halogenated compounds. In *QSAR in Environmental Toxicology*. Kaiser, K.L.E., Ed. D. Reidel Publishing, Dordrecht, Holland, pp. 300–317.

Oliver, B.G. and Niimi, A.J., 1988. Trophodynamic analysis of polychlorinated biphenyl congeners and other chlorinated hydrocarbons in the Lake Ontario ecosystem. *Environ. Sci. Technol.*, **22**, 388–397.

Opperhuizen, A., Serne, P., and van der Steen, J.M.D., 1988. Thermodynamics in fish/water and octan-1-ol/water partitioning of some chlorinated benzenes. *Environ. Sci. Technol.*, **22**, 286–292.

Paterson, S., Mackay, D., Tam, D., and Shiu, W.Y., 1990. Uptake of organic chemicals by plants: a review of process, correlations and models. *Chemosphere*, **21**, 297–331.

Paterson, S., Mackay, D., Bacci, E., and Calamari, D., 1991. Correlation of the equilibrium and kinetics of leaf–air exchange of hydrophobic organic chemicals. *Environ. Sci. Technol.*, **25**, 866–871.

Ryan, J.A., Bell, R.M., Davidson, J.M., and O'Connor, G.A., 1988. Plant uptake of non-ionic organic chemicals from soils. *Chemosphere,* **17**, 2299–2323.

Sabljic, A., Gusten, H., Schonherr, J., and Riederer, M., 1990. Modeling plant uptake of airborne organic chemicals. Plant cuticle/water partitioning and molecular connectivity. *Environ. Sci. Technol.*, **24**, 1321–1326.

Sabljic, A., 1987. The prediction of fish bioconcentration factors of organic pollutants from the molecular connectivity model. *Z. Gesamt. Hyg.*, **33**, 493–496.

Sabljic, A., 1991. Chemical topology and ecotoxicology. *Sci. Total Environ.,* **109/110**, 197–220.

Sabljic, A., Lara, R., and Ernst, W., 1989. Modeling association of highly chlorinated PCBs with marine humic substance. *Chemosphere*, **19**, 1503–1511.

Schonherr, J. and Riederer, M., 1989. Foliar penetration and accumulation of organic chemicals in plant cuticles. *Rev. Environ. Toxicol.*, **108**, 1–70.

Schwarzenbach, R.P., Gschwend, P.M., and Imboden, D.M., 1993. *Environmental Organic Chemistry*. John Wiley & Sons, New York.

Simonich, S.L. and Hites, R.A., 1994. Vegetation-atmosphere partitioning of polycyclic aromatic hydrocarbons. *Environ. Sci. Technol.*, **28**, 939–943.

Strek, H.J. and Weber, J.B., 1982. Behavior of polychlorinated biphenyls (PCBs) in soils and plants. *Environ. Pollut.,* **28**, 291–312.

Suntio, L.R., Shiu, W.Y., and Mackay, D., 1988. Critical review of Henry's law constants for pesticides. *Rev. Environ. Contam. Toxicol.*, **103**, 1–59.

Thomann, R.V., 1989. Bioaccumulation model of organic chemical distribution in aquatic food chains. *Environ. Sci. Technol.*, **23**, 699–707.

Topp, E., Scheunert, I., Attar, A., and Korte, F., 1986. Factors affecting the uptake of ^{14}C-labeled organic chemicals by plants from soil. *Ecotoxicol. Environ. Saf.*, **11**, 219–228.

Topp, E., Scheunert, I., and Korte, F., 1989. Kinetics of uptake of ^{14}C-labeled chlorinated benzenes by plants. *Ecotoxicol. Environ. Saf.,* **17**, 157–166.

Travis, C.C. and Arms, A.D., 1988. Bioconcentration of organics in beef, milk, and vegetation. *Environ. Sci. Technol.*, **22**, 271–274.

Travis, C.C. and Hattemer-Frey, H.A., 1988. Uptake of organics by aerial plant parts: a call for research. *Chemosphere*, **17**, 277–283.

Tulp, M.T.M. and Hutzinger, O., 1978. Some thoughts on the aqueous solubilities and partition coefficients of PCB, and the mathematical correlation between bioaccumulation and physico-chemical properties. *Chemosphere*, **7**, 760–849.

Valsaraj, K.T., 1995. *Elements of Environmental Engineering*. Lewis Publishers, Boca Raton, FL.

Veith, G.D., DeFoe, D.L., and Bergstedt, B.V., 1979. Measuring and estimating the bioconcentration factor of chemicals in fish. *J. Fish Res. Board Can.*, **36**, 1040–1048.

Veith, G.D., Macek, K.J., Petrocelli, S.R., and Carroll, J., 1980. An evaluation of using partition coefficients and water solubility to estimate bioconcentration factors for organic chemicals in fish. In *Aquatic Toxicology*. Proceedings of the Third Annual Symposium on Aquatic Toxicology, ASTM Special Technical Publication 707. Eaton, J.G., Parrish, P.R., and Hendricks, A.C., Eds. American Society for Testing and Materials, Philadelphia, PA, pp. 116–129.

Zaroogian, G.E., Heltshe, J.F., and Johnson, M., 1985. Estimation of bioconcentration in marine species using structure–activity models. *Environ. Toxicol. Chem.*, **4**, 3–12.

CHAPTER **12**

Diffusivity

12.1 INTRODUCTION

Molecular diffusion is the migration of matter along a concentration gradient from a region of high concentration to a region of low concentration. It is the result of random molecular motion, and it produces spontaneous mixing in inhomogeneous systems. Molecular diffusion controls, to a degree, the transport rate of chemicals at phase boundaries and in the slow-moving fluids found in porous media. To a lesser degree, it controls the dispersion of chemicals in turbulent fluids.

Diffusivity values span a narrow range within any particular medium, but they differ greatly from medium to medium. **Diffusivity in air, D_A**, ranges from values as high as 3×10^{-5} m²/s for ammonia to values as small as 3×10^{-6} m²/s for benzo(a)pyrene. Diffusivity values in liquids are 10^3 to 10^4 times smaller. For instance, **diffusivity in water, D_W**, varies from 3×10^{-9} m²/s for ammonia to 3×10^{-10} m²/s for p-aminophenol. By comparison, molecular diffusivities in solids lie below 10^{-14} m²/s. Turbulent diffusivities in air and in water can be greater than 10 m²/s. As a result, chemical transport models are usually not sensitive to error in estimated diffusivity values.

This chapter explores methods of estimating molecular diffusivity in air, water, and soil. Turbulent diffusion is also discussed. This and the following chapters deal with the transport of chemicals in the environment. Calculus is indispensable in dealing with chemical transport. A review of differential and integral calculus is presented in Appendix 7.

12.1.1   Chemical Transport and the Governing Equations

Experiment shows that the net flow rate of chemical mass through a plane lying perpendicular to the direction of transport is proportional both to the area of the plane and to the chemical's concentration gradient measured at the plane. This is stated mathematically by **Fick's first law of diffusion**. For transport in one direction (z), Fick's first law is

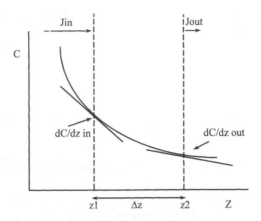

Figure 12.1 As chemical diffuses along the z direction, the mass flux, J, at each point is proportional to –dC/dz. The rate at which chemical concentration changes in a volume element of width Δz is $\delta C/\delta t = (J_{in} - J_{out})/\delta z$.

$$J = \frac{1}{A}\frac{dm}{dt} = -D\frac{dC}{dz} \qquad (12.1)$$

where J (g/m^2·s) is the mass flux, the rate of mass transport per unit area through a surface lying perpendicular to the direction of motion, A (m^2) is the cross-sectional area of the surface, dm/dt (g/s) is the flow rate of chemical mass, C is the concentration of chemical (g/m^3), and dC/dz (g/m^3/m) is the concentration gradient in the direction of transport. The relationship between mass flux and concentration gradient is graphically illustrated in Figure 12.1.

The proportionality factor, D (m^2/s), is called the **diffusivity**. Since J is positive when dC/dz is negative (see Figure 12.1), Equation 12.1 shows that diffusivity must be a positive quantity. Its value depends on the molecular size of the chemical and on the temperature, pressure, viscosity, and local composition of the fluid in which the chemical moves.

Equation 12.1 is useful if the concentration gradient does not change with time, a rare occurrence. In the likely event that the concentration gradient is not fixed, we can still evaluate the rate at which the chemical concentration in soil changes during diffusion. The principle of mass conservation requires that the change of chemical concentration with time in a small volume of space, $\delta C/\delta t$ (g/m^3·s), is equal to the difference between the rate of mass transport into the volume element and the rate of mass transport out of the volume element. The mathematical relationship that expresses this principle is called the **equation of continuity**: $\delta C/\delta t = (J_{in} - J_{out})/\delta z = -\delta J/\delta z$.

Fick's second law of diffusion, also called **the diffusion equation**, relates the change of chemical concentration to the net mass flux in a volume element of space. It is derived by combining Fick's first law and the equation of continuity: $\delta C/\delta t = -\delta J/\delta z = \delta(D\delta C/\delta z)\delta z = (\delta D/\delta z)(\delta C/\delta z) + D\delta^2 C/\delta z^2$. Diffusivity varies with concentration, so if a concentration gradient exists along z, then D must vary with z.

However, D is nearly constant in dilute solutions, $(\delta D/\delta z) \approx 0$, and the diffusion equation is, to a good approximation,

$$\frac{\delta C}{\delta t} = D \frac{\delta^2 C}{\delta z^2} \tag{12.2}$$

where $\delta^2 C/\delta z^2$ (g/m^3/m^2) is the change in concentration gradient with distance along the direction of transport.

12.1.2   Analytical Solutions to the Diffusion Equation

Solutions to the diffusion equation describe the distribution of chemicals in space as a function of time. The odds are that you will use numerical solutions obtained with the aid of a digital computer. Still, it is instructive to examine some analytical solutions. In order to solve Equation 12.2, we must specify the shape of the region from which the chemical diffuses and any restrictions placed on its concentration and concentration gradient. For example, lets say that a fixed amount of chemical found in a thin layer of soil with thickness = 1 (m) at depth $z = 0$ in the column diffuses into the soil both above and below it. The function $C(z,t)$ which describes the concentration of chemical at penetration depth z and time t must satisfy the boundary conditions $C(z,0) = C_0$ if $z = 0$ and $C(z,0) = 0$ if $z \neq 0$. The solution is (Crank, 1956)

$$C(z,t) = \frac{C_0 l}{2\sqrt{\pi Dt}} e^{-z^2/4Dt} \tag{12.3}$$

where C_0 is the initial concentration at $z = 0$ and $t = 0$.

Example

Using Equation 12.3, calculate the ratio C/C_0 vs. z for a chemical diffusing in water with $D = 1 \times 10^{-9}$ m^2/s and $\Delta z = 0.1$ m. The results are given in Figure 12.2 for several values of Dt.

If the contaminated layer is at the top of the column and chemical diffuses downward into the soil only, then the value of $C(z,t)$ is

$$C(z,t) = \frac{C_0 l}{\sqrt{\pi Dt}} e^{-z^2/4Dt} \tag{12.4}$$

Finally, suppose that chemical diffuses from a thick slab of soil with uniform concentration C_0 into a clean slab of soil where the initial concentration of chemical is zero. The boundary conditions are $C(z,0) = C_0$ for $z < 0$ at $t = 0$ and $C(z,0) = 0$ for $z > 0$ at $t = 0$. If the chemical is continuously applied to or slowly depleted from the contaminated layer so that its concentration there is roughly constant, then

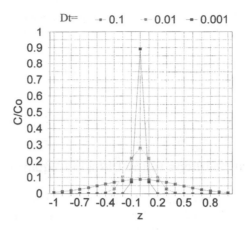

Figure 12.2 Concentration vs. position for a chemical diffusing from a thin layer of water located at z = 0 m. Δz = 0.1 m.

$$C(z,t) = \frac{C_0}{2}\left[1 - erf\left(z/2\sqrt{Dt}\right) = \frac{C_0}{2} erfc\left(z/2\sqrt{Dt}\right)\right]$$ (12.5)

where erf(z) is the error function and erfc(z) is the complementary error function of z. The complementary error function is discussed in Appendix 6. More solutions to the diffusion equation are given in Chapters 13 and 14.

Example

Using Equation 12.5, calculate C/C₀ vs. z for a chemical diffusing in water with D = 10⁻⁹ m²/s. The results are given in Figure 12.3 for several values of the product Dt.

12.1.3 Turbulent Diffusion

Fick's laws of diffusion, Equations 12.1 and 12.2, are good models of transport in gases and liquids that are not turbulent, but they are not useful as models of transport in turbulent fluids. Turbulence, random currents (eddies) induced by the shear stress that accompanies the relative motion of fluids, is almost always the dominant control of chemical transport in open air and water. Since eddy diffusion is due to the random motion of packets of fluid, we model it in much the same way that we model diffusion resulting from random molecular motion. In one dimension, the mean turbulent flux, J_{turb}, is related to the concentration gradient of a chemical:

$$J_{turb} = -D_T \frac{dC}{dz}$$ (12.6)

where the proportionality factor D_T is called the turbulent or eddy diffusivity.

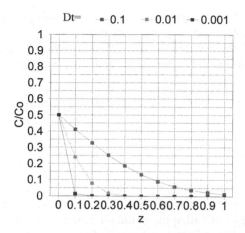

Figure 12.3   Concentration vs. position for a chemical diffusing from a thick layer of water. The boundary between the source and the receptor is located at z = 0 m.

Unlike D, D_T is never constant. Its value is scale dependent and, moreover, D_T is anisotropic as a result. Since vertical transport distances are frequently limited by thermal stratification and are usually small compared to horizontal distances anyway, the vertical eddy diffusivity is often smaller than the horizontal eddy diffusivity. In open ocean water, vertical eddy diffusivity values range from 10^{-5} to 1 m²/s, while horizontal eddy diffusivity values span the range from 10^{-2} to 10^4 m²/s.* In air, vertical eddy diffusivity values range from 10 to 100 m²/s. Horizontal eddy diffusivity in air is no doubt larger, but the question is academic since advection almost always controls horizontal transport in air. Since molecular and turbulent diffusion have the same general form, molecular and turbulent diffusivities are usually lumped together into a general mass-transfer coefficient. More information on turbulent diffusion is given in Chapter 14.

12.1.4   Advection and Dispersion in Porous Media

Chemicals are transported in soil by advection and dispersion in addition to molecular diffusion. **Advection** is *solute transport due to the bulk flow of gas or liquid*. The flow may be caused by gravity, by a pressure or temperature gradient, by a change in the height of the water table in soil, or by the friction of wind moving over open water. The mass flux due to advection, J_{AD} (g/m²·s), is the product of the solute concentration and average linear velocity of the fluid.

The relationship between the average linear velocity of soil gas, V_G, and the volume discharge of gas, Q_G (m³/s), per unit area, A (m²), is given by **Darcy's law**:

$$V_G = \frac{Q_G}{aA} = -\frac{1}{aA}\frac{k_A}{\mu_G}\left(\frac{dP_G}{dz}\right) \tag{12.7}$$

* Okubo (1971) showed that D_t is proportional to $F^{4/3}$ where F is the fetch of open ocean water for 10 m < F < 1000 km. Fetch is the surface distance across the body of water in the direction of wind flow.

where k_A (m^2) is soil–gas permeability, μ_G (g/m·s) is soil–gas viscosity, and dP_G/dz (g/m^2·s^2) is the vertical pressure gradient of gas.*

Darcy's law relates the average linear velocity of soil liquid, V_L (g/s), to the volume discharge of liquid, Q_L (m^3/s), per unit area, A (m^2):

$$V_L = \frac{Q_L}{wA} = -\frac{1}{wA}\frac{k_W}{\mu_L}\left(\frac{dP_L}{dz} + \rho_L g\frac{dh}{dz}\right) \tag{12.8}$$

where w (m$^3_{wat}$/m$^3_{soil}$) is the volumetric liquid content of soil, k_W (m^2) is soil–water permeability, μ_L (g/m·s) is soil–liquid viscosity, dP_L/dz (g/m^2·s^2) is the vertical pressure gradient of liquid, ρ_L (g/m^3) is liquid density, g (m/s^2) = 9.81 m/s^2 is the acceleration of gravity, and dh/dz (m/m) is the vertical head gradient of liquid.

The term $k_W\rho_L g/\mu_L$ is called the hydraulic conductivity, K (m/s). The value of K ranges from 1×10^{-10} m/s for sandstone to over 1×10^{-1} m/s for gravel (Freeze and Cherry, 1979). Above pore-water velocities of D_{SW}/d, where D_{SW} (m^2/s) is diffusivity in soil water and d (m) is grain diameter, or above about 2×10^{-5} m/s, advection is the dominant mode of transport in the saturated zone of soil, and diffusion may be ignored (Tucker and Nelken, 1990).

Hydrodynamic dispersion, *spreading due to local variability of the pore velocity in porous media*, is accounted for by adding a dispersive flux term, $J_{HD} = -D_{HD}\delta C/\delta z$, to Equation 12.1. The hydrodynamic dispersion coefficient is $D_{HD} = \lambda V_E + D_{SW}$, where λ (m) is the longitudinal dispersivity (mixing length), V_E (m/s) is the effective (average) advection velocity in soil, and D_{SW} (m^2/s) is the diffusivity in soil water. The value of λ is scale dependent. Molecular diffusion and hydrodynamic dispersion have the same general mathematical form, and the proportionality constants are usually lumped together.

12.2 ESTIMATION METHODS

12.2.1 Diffusivity in Air, D_A

Kinetic theory shows that a chemical's diffusivity in air is related to ambient temperature, T (K), pressure, P (atm), the chemical's molar mass, M (g/mol), and the chemical's molar volume, V (m^3/mol). Specifically, D_A is shown to be proportional to $T^{3/2} \cdot P^{-1} \cdot M^{-1/2} \cdot V^{-2/3}$. While theoretical predictions of the magnitude of diffusivity in air are not sufficiently accurate for our purposes, the theoretical relationships do serve as the basis of several useful semi-empirical models. Predictions of the semi-empirical models agree with experimental values to within 5 or 10% (Reid et al., 1987; Tucker and Nelken, 1990), and the relationships may be extrapolated beyond the range of the training data.

---

* To account for gravity flow of a gas that is heavier than air, Equation 12.7 is modified by adding $-k_W\rho_G g/a\mu_L)$ (dh/dz) to the right-hand side, where ρ_G (g/m^3) is gas density, g (m/s^2) = 9.807 m/s^2 is the acceleration of gravity, and dh/dz (m/m) is the vertical head gradient of soil vapor.

Table 12.1 Atomic Diffusion Volumes

Atom/Feature	V_i cm³/mol[a]
C	15.9
H	2.31
O	6.11
N	4.54
S	22.9
F	14.7
Cl	21.0
Br	21.9
I	29.8
Ring (aromatic/heterocyclic)	−18.30

[a] The diffusion volumes of C, H, O, and rings were determined with a larger training set than the diffusion volumes of the other features.

From Fuller, E.N., Ensley, K., and Giddings, J.C., 1969. *J. Phys. Chem.*, **73**, 3679–3685.

Fuller et al. (1966, 1969) describe the following semi-empirical relationship:

$$D_A = \frac{1.00 \times 10^{-3} T^{1.75}}{P} \frac{\left[(1/M_A) + (1/M_C)\right]^{1/2}}{\left[\left(\sum V_A\right)^{1/3} + \left(\sum V_C\right)^{1/3}\right]^2}$$

(12.9)

where D_A = chemical diffusivity in air, cm²/s
T = temperature, K
P = atmospheric pressure, bar
M_A = average molar mass of air, 28.97 g/mol
M_C = molar mass of chemical, g/mol
$\sum V_A$ = estimated diffusion volume of air, 19.7 cm³/mol
$\sum V_C$ = estimated diffusion volume of chemical, cm³/mol

The molar diffusion volumes used in Equation 12.9 are estimated from fragment contributions (Fuller et al., 1969).

$$\sum V_C = \sum_i n_i \cdot v(i)$$

(12.10)

where $\sum V_C$ = estimated diffusion volume of chemical, cm³/mol
n_i = number of atoms/features of type i
$v(i)$ = atomic diffusion volume of feature i, cm³/mol

The atomic diffusion volumes, v_i, are given in Table 12.1. They were evaluated with a regression analysis of 512 measured values of the diffusivity of diverse chemicals in air and other gases. The values for C, H, O, and rings are based on a much larger training set and are more reliable than the values given for the other features. The training set is listed in Table 12.2.

Fuller et al. report an average absolute error of 4.2% for the method. Reid et al. (1987) report similar results with a small test set of diverse chemicals. Shen (1981) used the method to estimate the diffusivities in air of 41 chemicals of various types over a range of temperatures. His estimates are consistent with measured values.

Example

Estimate the diffusivity in air of carbon tetrachloride, CCl_4, at 25°C and 1 atm pressure. The chemical's molar mass is 153.8 g/mol.

1. Estimate the molar diffusion volume using Equation 12.10. The atomic diffusion volumes are given in Table 12.1.

$$V_C = V(C) + 4V(Cl)$$

$$= 15.9 + 4(21.0)$$

$$= 99.9 \ cm^3/mol$$

2. Calculate the diffusivity in air using Equation 12.9.

$$D_A = (0.00101) (298)^{1.75} [(1/28.97) + (1/153.8)]^{1/2} / (1) (19.7^{1/3} + 99.9^{1/3})^2$$

$$= 0.81 \ cm^2/s$$

The measured diffusivity of carbon tetrachloride in air is 0.078 cm²/s at 25°C and 1 atm (USEPA, 1989). The estimate is in error by 3.9%.

More complex methods of estimating diffusivity have been proposed. While some may be more widely applicable, none appear to be as accurate as Fuller's relationship (Reid et al., 1987; Tucker and Nelken, 1990).

The diffusivities of closely related molecules should be related by:

$$D_A(1) = D_A(2) \left(\frac{M_2}{M_1} \right)^{1/2} \tag{12.11}$$

where $D_A(i)$ = diffusivity of chemical i in air, cm²/s
  M_i = molar mass of chemical i, g/mol

Shen (1981) describes a method in which the diffusivity of a chemical is estimated using Equation 12.11 and the experimental value of a reference compound having similar diffusion volume and molar mass.

Table 12.2 Training Set for the Atomic Diffusion Volumes, Table 12.1

Chemical	Note	Chemical	Note
		Hydrocarbons	
Methane	a, c, f	n-Octane	a, b, c, d, e
Ethane	a, d	2,2,4-Trimethylpentane	a, b, c, d, e
Ethylene	d	n-Decane	a, d
n-Butane	a, d	2,3,3-Trimethylheptane	a, d
Isobutane	d	n-Dodecane	a, d
n-Hexane	a, c, d, e	Benzene	a, b, c, d, f
2,3-Dimethylbutane	a, b, d, e	Toluene	f
Cyclohexane	a, b, d, e	Biphenyl	f
Methylcyclopropane	a, b, d, e	Pyridine	a, b, d
n-Heptane	a, b, d, e	Piperidine	a, b, d
2,4-Dimethylpentane	a, b, c, d, e	Aniline	f
Cyclohexane	a	Thiophene	a, b, d
Methylcyclopentane	a	Tetrahydrothiophene	a, b, d
		Alcohols	
Methanol	c	2-Butanol	f
Ethanol	c, f	1-Pentanol	c
1-Propanol	c	2-Pentanol	f
2-Propanol	c, f	Hexanol	c
1-Butanol	c, f		
		Halogenated Hydrocarbons	
Difluoromethane	c	1-Bromopropane	c
1,1-Difluoroethane	c	2-Bromopropane	c
1-Fluorohexane	c	1-Bromobutane	c
Fluorobenzene	c	2-Bromobutane	c
Hexafluorobenzene	c	1-Bromohexane	c
4-Fluorotoluene	c	2-Bromohexane	c
Dichloromethane	c	3-Bromohexane	c
Trichloromethane	c	Bromobenzene	c
1,2-Dichloroethane	c	2-Bromo-1-chloropropane	c
1-Chloropropane	c	Iodomethane	c
1-Chlorobutane	c	Iodoethane	c
2-Chlorobutane	c	1-Iodopropane	c
1-Chloropentane	c	2-Iodopropane	c
Chlorobenzene	c	1-Iodobutane	c
Dibromomethane	c	2-Iodobutane	c
Bromoethane	c		
		Other	
NH_3	a, b, c, d, e, f	Phosgene	f
SF_6	a, b	Chloropicrin	f
HCN	f	Ethyl acetate	f
Cyanogen chloride	f	Nitrobenzene	c, f

[a] Diffusivity measured in H_2.	[d] Diffusivity measured in N_2.
[b] Diffusivity measured in O_2.	[e] Diffusivity measured in Ar.
[c] Diffusivity measured in He.	[f] Diffusivity measured in air.

From Fuller, E.N., Schettler, P.D., and Geddings, J.C., 1966. *Ind. Eng. Chem.*, **58**, No. 5, 18–27; Fuller, E.N., Ensley, K., and Giddings, J.C., 1969. *J. Phys. Chem.*, **73**, 3679–3685.

Example

Estimate the diffusivity in air of trichlorofluoromethane (Freon-11) at 25°C and 1 bar from the diffusivity of carbon tetrachloride. The molar mass of CCl_3F is 137.37 g/mol.

1. The diffusion volume of CCl_3F is estimated using Equation 12.10.

$$V_C = V(C) + 3V(Cl) + V(F)$$

$$= 15.9 + 3(21.0) + 14.7 \ \ cm^3/mol$$

$$= 93.6 \ \ cm^3/mol$$

The diffusion volume of CCL_4 is 99.9 cm³/mol. Its molar mass is 153.8 g/mol. As expected, the diffusion volume and molar mass of CCl_3F are close to the values of the reference chemical, CCl_4.
2. The diffusivity in air is estimated using Equation 12.11 and the experimental value of the diffusivity of CCl_4, 0.078 cm²/s.

$$D_A(CCl_3F) = D_A(CCl_4)(M_{CCl_4}/M_{CCl_3F})^{1/2}$$

$$= (0.078)(153.82/137.37)^{1/2} \ \ cm^2/s$$

$$= 0.083 \ \ cm^2/s$$

The measured value is 0.087 cm²/s (USEPA, 1989). The estimate is in error by 4.6%.

A chemical's diffusivity in air depends upon the ambient temperature and pressure as shown in Equation 12.9. The diffusivities of a chemical at different temperatures and pressures are related by (Hamaker, 1972):

$$\frac{D_A(T_2)}{D_A(T_1)} = \frac{P(T_1)}{P(T_2)}\left(\frac{T_2}{T_1}\right)^n \qquad (12.12)$$

where $D_A(T_i)$ = chemical diffusivity at temperature T_i, cm²/s
  $P(T_i)$ = ambient pressure at temperature T_i, bar
  T_i = ambient temperature, K
  n = empirical coefficient, dimensionless

Fuller et al. (1966, 1969) report the best value of n to be 1.75. This is close to the theoretical value of 1.5.

12.2.2   Diffusivity in Water, D_W

The **Stokes–Einstein equation** describes the theoretical relationship between a chemical's diffusivity in water and its molecular size. The relationship is

$$D_W = \frac{RT}{6\pi\eta r} \tag{12.13}$$

where D_W = chemical diffusivity in water, cm^2/s
      k   = Boltzmann constant, erg/K
      T   = temperature, K
      η = solution viscosity, poise
      r   = solute hydrodynamic radius, cm

As with gases, theoretical predictions of diffusivity in water are not accurate, but the Stokes–Einstein relationship serves as the basis of useful semi-empirical estimation methods. It shows that D_W is inversely proportional to solvent viscosity and molecular hydrodynamic radius. The hydrodynamic radius is not readily measured, but it should be proportional to $V^{2/3}$, where V (cm^3/mol) is the chemical's molecular volume.

Hayduk and Laudie (1974) report the following semi-empirical relationship:

$$D_W = \frac{13.26 \times 10^{-5}}{\eta^{1.4}V^{0.589}} \tag{12.14}$$

where D_W = chemical diffusivity in water, cm^2/s
      η = solution viscosity, centipoise (10^{-2} g/cm·s)
      V   = chemical molar volume, cm^3/mol

The molar volume is equal to the chemical's molar mass divided by its liquid density:

$$V = \frac{M}{\rho_l} \tag{12.15}$$

where   V   = chemical molar volume, cm^3/mol
      M = chemical molar mass, g/mol
      ρ_l = liquid density, g/cm^3

In the event that liquid density is not determined, molar volume can be estimated using Le Bas (1915) additivity constants

$$V = \sum_i n_{iB}(i) \tag{12.16}$$

Table 12.3 Le Bas Atomic and Group Molar Volume

Atom/Group	$V_B(i)$ (cm³/mol)
C	14.8
H	3.7
O (except as noted below)	7.4
In methyl esters and ethers	9.1
In ethyl esters and ethers	9.9
In higher esters and ethers	11.0
In acids	12.0
Joined to S, P, or N	8.3
N, double bonded	15.6
In primary amines	10.5
In secondary amines	12.0
Br	27
Cl	24.6
F	8.7
I	37
S	25.6
Ring, three-membered	−6.0
Four-membered	−8.5
Five-membered	−11.5
Six-membered	−15.0
Naphthalene	−30.0
Anthracene	−47.5

From Reid, R.C., Prausnitz, J.M., and Poling, B.E., 1987. *The Properties of Gases and Liquids.* Fourth Edition. McGraw-Hill, New York.

where V   = chemical molar volume, cm³/mol
      n_i = number of atoms/features of type i
     $v_B(i)$ = Le Bas volume of feature i, cm³/mol

The Le Bas constants are given in Table 12.3.

The relationship is based on the measured diffusivities of 87 diverse chemicals including alkanes, alcohols, aldehydes, ketones, amides, carboxylic acids, amino acids, halogenated alkanes, and aromatic hydrocarbons. The chemicals are listed in Table 12.4. Hayduk and Laudie's model produces an average absolute predictive error of 5.8% when applied to the training set. Le Bas molar volumes were used in the study.

Other methods of estimating diffusivity in water are reviewed by Reid et al. (1987) and by Tucker and Nelken (1990). The methods are not as general as Hayduk and Laudie's, do not produce more accurate estimates, and are more complicated in some cases.

Le Bas volumes are good estimates of molar volumes. Reid et al. (1987) compared the molar volumes of 32 diverse chemicals with their Le Bas volumes. They report an average absolute error of 4.0%. Hayduk and Laudie (1974) report a 5.8% error with the training set of 87 compounds listed in Table 12.4.

Table 12.4   Molar Volumes, V (cm³/mol), and Diffusivities in Water, $D_W \times 10^5$ (cm²/s), at 25°C[a]

Chemical	V LeBas	V Actual	D_W
CH_4	29.6	37.7	1.67
CH_3C1	47.5	50.6	1.49
C_2H_3C1	62.3	64.5	1.34
C_2H_6	51.8	53.5	1.38
C_2H_4	44.4	49.4	1.55
C_3H_8	74.0	74.5	1.16
C_3H_6	66.6	69.0	1.44
C_4H_{10}	96.2	96.6	0.97
Pentane	118	118	0.97
Cyclopentane	99.5	97.9	1.04
Cyclohexane	118	117	0.90
Methylcyclopentane	112	120	0.93
Benzene	96.0	96.5	1.09
Toluene	118	118	0.95
Ethylbenzene	140	140	6.90
Methanol	37.0	42.5	1.66
Ethanol	59.2	62.6	1.24
Propanol	81.4	81.8	1.12
Isopropanol	81.4	81.8	1.08
Butanol	104	101	0.98
Isobutanol	104	101	0.93
n-Butanol	104	102	0.97
Benzyl alcohol	126	12l	0.93
Ethylene glycol	66.6	63.5	1.16
Triethylene glycol	170	163	0.76
Propylene glycol	88.8	84.5	1.00
Glycerol	96.2	85.1	0.93
Ethane-hex-diol	200	197	0.64
Acetone	74.0	77.5	1.28
Ethyl acetate	111	106	1.12
Furfural	95.7	97.5.	1.12
Urea	58.0	ND	1.38
Formamide	43.8	46.6	1.67
Acetamide	66.0	67.1	1.32
Propionamide	88.2	87.4	1.20
Butyramide	110	109	1.07
Isobutyramide	110	103	1.02
Diethylarnine	112	109	1.11
Aniline	110	106	1.05
Formic acid	41.6	41.1	1.52
Acetic acid	63.8	64.1	1.19
Propionic acid	86.0	85.3	1.01
Butyric acid	108	108	0.92
Isobutyric acid	108	109	0.95
Valeric acid	130	129	0.82
Trimethylacetic acid	130	127	0.82
Chloracetic acid	81.7	79.6	1.04
Glycolic acid	71.2	ND	0.98
Caproic acid	153	ND	0.78

Table 12.4 (continued)   Molar Volumes, V (cm³/mol), and Diffusivities in Water, $D_W \times 10^5$ (cm²/s), at 25°C[a]

Chemical	V LeBas	V Actual	D_W
Succinic acid	120	ND	0.86
Glutaric acid	142	ND	0.79
Adipic acid	165	ND	0.74
Pimelic acid	187	ND	0.71
Glycine	78.0	ND	1.06
Alanine	100	ND	0.91
Serine	108	ND	0.88
Aminobutyric acid	122	ND	0.83
Valine	145	ND	0.77
Leucine	167	ND	0.73
Proline	127	ND	0.88
Hydroxyproline	135	ND	0.83
Hisitidine	168	ND	0.73
Phenylalanine	189	ND	0.705
Tryptophan	223	ND	0.660
Diglycine	138	ND	0.791
Triglycine	198	ND	0.665
Glycyl-leucine	227	ND	0.623
Leucyl-glycine	227	ND	0.613
Leucyl-glycyl-glycine	287	ND	0.551
Amino benzoic acid	144	ND	0.840
Dextrose	166	ND	0.675
Sucrose	325	ND	0.524
Raffinose	480	ND	0.434

[a] ND: not determined. See Hayduk and Laudie (1974) for data sources.

From Hayduk, W. and Laudie, H., 1974. *AIChE J.*, **20**, 611–615.

Example

Estimate the diffusivity of vinyl chloride, C_2H_3Cl, in water at 25°C.

1. Estimate the molar volume using Equation 12.16 and the Le Bas molar volumes given in Table 12.3.

$$V = 2V_B(C) + 3V_B(H) + V_B(Cl)$$

$$= 2(14.8) + 3(3.7) + (24.6) \ cm^3/mol$$

$$= 65.3 \ cm^3/mol$$

2. The viscosity of water at 25°C is 0.8904 cp (Lide, 1994). The diffusivity in water is estimated using Equation 12.14.

$$D_W = 13.26 \times 10^{-5} / (0.8904)^{1.4} (65.3)^{0.589}$$

$$= 1.33 \times 10^{-5} \ cm^2/s$$

Reported values of the diffusivity of vinyl chloride in water at 25°C range from 1.30 to 1.39×10^{-5} cm²/s (Hayduk and Laudie, 1974). The estimated diffusivity differs from the mean of measured values, 3.4×10^{-5} cm²/s, by 1%.

The diffusivity in water of closely related chemicals should be related by:

$$D_W(1) = D_W(2) \left[\frac{V(2)}{V(1)} \right]^{0.6} \tag{12.17}$$

where $D_W(i)$ = diffusivity of chemical i in water, cm²/s
$\quad\quad V(i)$ = molar volume of chemical i, cm³/mol

The diffusivity of a chemical in water may be estimated using Equation 12.17 and the experimental value for a reference compound having similar molar volume and molar mass.

Example

Estimate the diffusivity of trichloroethylene, TCA(C_2HCl_3), in water at 25°C from that of trichloroethane.

1. Estimate the molar volumes of the two compounds using Le Bas molar volumes.

$$V(TCA) = 2V_B(C) + 3V_B(H) + 3V_B(Cl)$$

$$= 2(14.8) + 3(3.7) + 3(24.6) \ cm^3/mol$$

$$= 84.9 \ cm^3/mol$$

$$V(TCE) = 2V_B(C) + 1V_B(H) + 3V_B(Cl)$$

$$= 2(14.8) + (3.7) + 3(24.6) \ cm^3/mol$$

$$= 77.5 \ cm^3/mol$$

The molar mass of C_2HCl_3 is 131.4 g/mol. That of $C_2H_3Cl_3$ is 133.4 g/mol. The molar volumes and masses of the two compounds are similar.

2. Estimate the diffusivity in water using Equation 12.17. The diffusivity of 1,1,2-trichloroethane, $C_2H_3Cl_3$, is 8.8×10^{-6} cm²/s (USEPA, 1989).

$$D_W(TCE) = D_W(TCA)\left[V_B(TCA)/V_B(TCE)\right]^{0.6}$$

$$= 8.8 \times 10^{-6} \; (84.9/77.5)^{0.6} \; cm^2/s$$

$$= 9.6 \times 10^{-6} \; cm^2/s$$

The measured diffusivity of trichloroethylene in water is 9.1×10^{-6} cm²/s at 25°C (USEPA, 1989). The error of the estimate is 5.2%.

As the Stokes–Einstein equation shows, chemical diffusivity in water varies with temperature. Average groundwater temperatures in the U.S. range from 4 to 25°C. Surface water temperatures span a wider range. Values of D_W estimated with Hayduk and Laudie's relationship, Equation 12.11, are corrected for temperature because the value of the solvent viscosity, η, is temperature dependent. The viscosity of water varies from 1.567 cp at 4°C to 0.8904 cp at 25°C (Lide, 1994), resulting in a factor of 1.76 increase in estimated diffusivity over the temperature range from 4 to 25°C using Equation 12.11. However, diffusivity in water actually increases by a factor of 5.5 between 4 and 25°C, and the Hayduk–Laudie relationship appears to account poorly for the temperature dependence of D_W.

When comparing diffusivities in water over a large range of temperature, the following correlation can be used (Reid et al., 1987):

$$\frac{D_W(T_2)}{D_W(T_1)} = \left(\frac{374.1 - T_1}{374.1 - T_2}\right)^6 \tag{12.18}$$

where $D_W(T_i)$ = diffusivity in water at temperature T_i, cm²/s
         T_i = temperature i, K

Equation 12.18 produces an average absolute error of about 9%.

Diffusivity in water is sensitive to the presence of other chemicals, speciation, and pressure. These factors are discussed by Reid et al. (1987).

12.2.3   Diffusivity in Soil, D_E

Chemicals diffuse through the air and water contained in the pore structure of soils. Since soil particles block the linear motion of diffusing molecules, the flow area is reduced, the average path length of diffusing chemicals is increased, and diffusivities in soil air and soil water are smaller than diffusivities in open air and open water. To account for this, correction factors are used in models of diffusion flux in soils as shown in Equation 12.19.

$$D_{SA} = at_A D_A \tag{12.19}$$

where D_{SA} = apparent diffusivity in soil air, cm²/s
         a   = volumetric air content of soil, cm^3_{air}/cm^3_{soil}

t_A = tortuosity (dimensionless)
D_A = diffusivity in air, cm²/s

The volumetric air content, the ratio of air volume to total volume of soil, corrects for the reduced flow area in the soil column. Tortuosity, t_A, corrects for the average increase in path length as molecules in soil air diffuse around soil particles. For diffusion in air, tortuosity has been estimated in several ways (Weeks, 1982). However, the Millington and Quirk (1961) model is most commonly used:

$$t_A = \frac{a^{7/3}}{P_T^2}$$ (12.20)

where t_A = soil air tortuosity (dimensionless)
  a = volumetric air content of soil, cm^3_{air}/cm^3_{soil}
  P_T = total soil porosity (dimensionless)

Combining Equations 12.19 and 12.20, the apparent diffusivity in soil air is $D_{SA} = (a^{10/3}/P_T^2)D_A$. In extremely dry soils, the volumetric air content, a, is approximately equal to total porosity, P_T, and $D_{SA} \approx P_T^{4/3}D_A$.
The total soil porosity, P_T, may be estimated with the following relationship:

$$P_T = 1 - \frac{d_b}{d_m}$$ (12.21)

where P_T = total soil porosity, dimensionless
  d_b = soil bulk density, g/cm³
  d_m = mineral density, g/cm³

Soil mineral density, d_m, varies from 2.65 g/cm³ for clays to 2.75 g/cm³ for sand. In the absence of measured values, a value of 2.65 g/cm³ is usually assumed (Shen, 1981). Soil bulk density, d_b, lies between 1.0 and 2.0 g/cm³. For dry, uncompacted soils, a value of 1.2 g/cm³ is assumed for d_b, resulting in a value of 0.55 for P. P may decrease to as little as 0.35 for compacted soils.
The Millington and Quirk model has been tested and compared with other soil air tortuosity models. It has been found to produce good estimates of pesticide diffusivities over a wide range of water content up to saturation (Jury, 1986) and to be superior to other models in estimating the diffusion of gasoline hydrocarbons in sand columns with moisture content near the field capacity (Baehr, 1987).
Diffusivity in soil water is estimated using

$$D_{SW} = wt_w D_W$$ (12.22)

where D_{SW} = apparent diffusivity in soil water, cm²/s
  w = volumetric water content of soil, cm^3_{water}/cm^3_{soil}
  t_W = soil water tortuosity (dimensionless)

The volumetric water content, the ratio of pore water volume to total volume of soil, accounts for the reduced flow volume in the soil column due to the presence of solids. A Millington–Quirk model has been used for t_W in unsaturated soils, i.e., $t_W = w^{7/3}/P_T^2$ and $D_{SW} = (w^{10/3}P_T^2)D_W$ (Jury et al., 1983). The model's estimates agree with measurements made of apparent diffusivity of pesticides in soils over a large range of water content. Jury et al. (1984) summarize the evidence. It isn't clear that the model applies to diffusivity in saturated soils, however. Measured values of the tortuosity factor, wt_W, vary from 0.01 to 0.5 in saturated porous rock and unconsolidated soils (Freeze and Cherry, 1979).

The average effective diffusivity of a chemical in unsaturated soil depends upon the distribution of the chemical between soil air and soil water and on the soil water content. Jury et al. (1983) suggests the following averaging method:

$$D_E = \frac{D_{SA}K_{AW} + D_{SW}}{d_b K_D + w + a K_{AW}}$$ (12.23)

where D_E = effective diffusivity in soil, cm²/s
  D_{SA} = apparent diffusivity in soil air, cm²/s
  K_{AW} = air–water partition coefficient (dimensionless)
  D_{SW} = apparent diffusivity in soil water, cm²/s
  d_b = soil bulk density, g/cm³
  a   = volumetric air content of soil, cm^3_{AIR}/cm^3_{SOIL}
  W   = volumetric water content of soil, cm^3_{WATER}/cm^3_{SOIL}
  K_D = solid-liquid distribution coefficient, cm³/g
       = $f_{OC}K_{OC}$
  f_{OC} = organic carbon fraction of soil (dimensionless)
  K_{OC} = organic carbon distribution coefficient, cm³/g

The term $1/(d_b K_D + w + a K_{AW})$ is the fraction of solute in the liquid phase, and $K_{AW}/(d_b K_D + w + a K_{AW})$ is the fraction of solute in the vapor phase.

Example

Estimate the average effective diffusivity of trichloroethylene in soil at 20°C. For trichloroethylene, D_A is 0.079 cm²/s, D_W is 9.1×10^{-6} cm²/s, K_{AW} is 0.38, and K_{OC} is 130 cm³/g (USEPA, 1989). The soil bulk density is assumed to be 1.2 g/cm³, and the mineral density is 2.65 g/cm³, $P_T = 0.55$. The volumetric water content is 0.20.

1. Since $P_T = a + w$, the volumetric air content is

$$a = P_T - w$$

$$= 0.55 - 0.20$$

$$= 0.35$$

2. The apparent diffusivity in soil air is estimated using Equations 12.19 and 12.20.

$$D_{SA} = at_A D_A \text{ and } t_A = a^{7/3}/P_T^2$$

$$D_{SA} = \left(a^{10/3}/P_T^2\right) D_A$$

$$= \left(0.35^{10/3}/0.55^2\right) 0.079 \text{ cm}^2/s$$

$$= 0.0079 \text{ cm}^2/s$$

3. The apparent diffusivity in soil water is estimated for transport in the vadose zone using Equation 12.23 and a Millington-Quirk tortuosity model.

$$D_{SW} = wt_W D_W \text{ and } t_W = w^{7/3}/D_T^2$$

$$D_{SW} = \left(w^{10/3}/P^2\right) D_W$$

$$= \left(0.20^{10/3}/0.55^2\right) 9.1 \times 10^{-6} \text{ cm}^2/s$$

$$= 1.4 \times 10^{-7} \text{ cm}^2/s$$

4. The average effective diffusivity lies between 7.9×10^{-3} and 1.4×10^{-7} cm^2/s. For trichloroethylene, $K_{AW} = 0.38$ and $K_{OC} = 130$ cm^3/g. If f_{OC} is 0.025, then $K_D = 3.25$ cm^3/g. We will assume $w = 0.20$ and $d_b = 1.2$ g/cm^3. Using Equation 12.23,

$$D_E = \left(D_{SA} K_{AW} + D_{SW}\right)/\left(d_b K_D + w + a K_{AW}\right)$$

$$D_E = \left[(0.0079)(0.0091) + 1.4 \times 10^{-7}\right)\right]$$

$$\Big/\left[\left(1.2 \text{ g/cm}^3\right)\left(3.25 \text{ cm}^3/g\right) + 0.20 + (0.35)(0.38)\right]$$

$$= 1.7 \times 10^{-5} \text{ cm}^2/s$$

Polar compounds and ionic species are retarded in the soil column by electric charges on the surface of soil particles. $D_{SW} = rwt_W D_W$, where r is the retardation factor, which is typically 0.2 for clay, 0.5 for silts, and 1.0 for sand.

REFERENCES

Abramowitz, M. and Stegun, I.A., Eds., 1964. *Handbook of Mathematical Functions with Formulas, Graphs, and Mathematical Tables.* U.S. Government Printing Office, Washington, D.C.

Baehr, A.L., 1987. Selective transport of hydrocarbons in the unsaturated zone due to aqueous and vapor phase partitioning. *Water Resources Res.,* **23**, 1926–1938.

Beyer, W.H., Ed., 1991. *CRC Standard Mathematical Tables and Formulae.* CRC Press, Boca Raton, FL.

Crank, J., 1956. *The Mathematics of Diffusion.* Oxford University Press, New York.

Freeze, R.A. and Cherry, J.A., 1979. *Groundwater.* Prentice-Hall, Englewood Cliffs, NJ.

Fuller, E.N., Schettler, P.D., and Giddings, J.C., 1966. A new method for prediction of binary gas-phase diffusion coefficients. *Ind. Eng. Chem.*, **58**, No. 5, 18–27.

Fuller, E.N., Ensley, K., and Giddings, J.C., 1969. Diffusion of halogenated hydrocarbons in helium. The effect of structure on collision cross sections. *J. Phys. Chem.*, **73**, 3679–3685.

Hamaker, J.W., 1972. Diffusion and Volatilization. Chapter 5. In *Organic Chemicals in the Soil Environment.* Volume 3. Goring, C.A.I. and Hamaker, J.W., Eds., Marcel Dekker, New York.

Hayduk, W. and Laudie, H., 1974. Prediction of diffusion coefficients for non-electrolytes in dilute aqueous solutions. *AIChE J.*, **20**, 611–615.

Jury, W.A., Spencer, W.F., and Farmer, W.J., 1983. Behavior assessment model for trace organics in soil. I. Model description. *J. Environ. Qual.*, **12**, 558–564.

Jury, W.A., Spencer, W.F., and Farmer, W.J., 1984. Behavior assessment model for trace organics in soil. IV. Review of experimental evidence. *J. Environ. Qual.*, **13**, 580–586.

Jury, W.A., 1986. Volatilization from soil. In *Vadose Zone Modeling of Organic Pollutants.* Hern, S.C. and Melancon, S.M., Eds. Lewis Publishers, Chelsea, MI, pp. 159–190.

Le Bas, G., 1915. *The Molecular Volumes of Liquid Chemical Compounds.* Longmans, Green, New York.

Lide, D.R., 1994. *CRC Handbook of Chemistry and Physics.* 74th Edition. CRC Press, Boca Raton, FL.

Luckner, L. and Schestakow, W.M., 1991. *Migration Processes in the Soil and Groundwater Zone.* Lewis Publishers, Boca Raton, FL.

Lyman, W.J., Reidy, P.J., and Levy, B., 1992. *Mobility and Degradation of Organic Contaminants in Subsurface Environments.* C.K. Smoley, Chelsea, MI.

Mayer, R., Letey, J., and Farmer, W.J., 1974. Models for predicting volatilization of soil-incorporated pesticides. *Soil. Sci. Soc. Amer. Proc.*, **38**, 563–568.

Millington, R.J. and Quirk, P.J., 1961. Permeability of porous solids. *Trans. Faraday Soc.*, **57**, 1200–1207.

Okubo, A., 1971. Oceanic diffusion diagrams. *Deep Sea Res.*, **18**, 789–802.

Reid, R.C., Prausnitz, J.M., and Poling, B.E., 1987. *The Properties of Gases and Liquids.* Fourth Edition. McGraw-Hill, New York.

Shen, T.T., 1981. Estimating hazardous air emissions from disposal sites. *Pollut. Eng.*, **13**, 31–34.

Thomas, R.G., 1990. Volatilization from Soil. Chapter 16. In *Handbook of Chemical Property Estimation Methods. Environmental Behavior of Organic Compounds.* Lyman, W.J., Reehl, W.F., and Rosenblatt, D.H., Eds. American Chemical Society, Washington, D.C.

Tucker, W.A. and Nelken, L.H., 1990. Diffusion coefficients in air and water. Chapter 17. In *Handbook of Chemical Property Estimation Methods.* Lyman, W.J., Reehl, W.F., and Rosenblatt, D.H., Eds. American Chemical Society, Washington, D.C.

U.S. EPA, 1989. *Hazardous Waste Treatment, Storage and Disposal Facilities (TSDF) — Air Emission Models.* EPA-450/3-87-026. U.S. Environmental Protection Agency, Office of Air and Radiation, Office of Air Quality Planning and Standards, Research Triangle Park, NC 27711.

Weeks, E.P., 1982. Use of atmospheric fluorocarbons F-11 and F-12 to determine the diffusion parameters of the unsaturated zone in the southern high plains of Texas. *Water Resources Res.*, **18**, 1365–1378.

CHAPTER 13

Volatilization from Soils

13.1 INTRODUCTION

This chapter deals with the spontaneous transport of organic chemicals from soil to air. It describes methods of estimating emission rates, residence times, and residual concentrations of a chemical from the time it is applied to soil until such time as equilibrium concentration levels are reached in soil and in air. In the following, we will assume that we are dealing with inert chemicals, but keep in mind that residence times in soil are long enough that chemical conversion processes such as biodegradation probably compete with volatilization.

Some comprehensive sources of information on volatilization are available. A general model of volatilization from soils was proposed by Jury et al. (1983). The model also describes advection and chemical degradation in the soil column. The subject was reviewed by Jury (1986). Some analytical models found useful for special purposes are reviewed by Thomas (1990). These special-purpose analytical models are often encountered as components of hybrid soil-transport models.

Calculus is indispensable in describing chemical transport in soils. A review of differential and integral calculus is presented in Appendix 7.

13.2 CHEMICAL TRANSPORT BETWEEN SOIL AND AIR

13.2.1 A Model Soil Column

Imagine the soil column divided into several horizontal layers as shown in Figure 13.1, one that is uniformly contaminated with chemical and one or more layers that initially are clean. The contaminated layer may be located anywhere in the column. In the contaminated layer, chemicals are distributed in soil air, soil water, and adsorbed on soil particles. The amount and type of mobile phase present are determined by soil structure, soil porosity, and soil–water content.

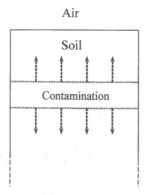

Figure 13.1   The model soil column is composed of several horizontal layers that are homo-geneous and isotropic. One layer is uniformly contaminated with chemical.

The total concentration of chemical in soil, C_T (g/m^3), is the weighted sum of the chemical concentrations found in each phase:

$$C_T = aC_G + wC_L + d_bC_S \qquad (13.1)$$

where a (m^3 air/m^3 soil) is the volumetric air content of soil, C_G (g/m^3 soil air) is the concentration of chemical in soil air, w (m^3 water/m^3 soil) is the volumetric water content of soil, C_L (g/m^3 soil solution) is the concentration of chemical in soil water, d_b (g/m^3 soil) is the bulk density of soil, and C_S (g solute/g soil) is the concentration of chemical adsorbed on soil solids. The sum of the volumetric air content and the volumetric water content of soil is equal to the specific soil pore volume or total soil porosity, P_T (m^3 pore space/m^3 soil); P_T = a + w.

Chemicals move slowly in soil, and it seems reasonable to assume that the concentrations of chemical in soil air, in soil water, and adsorbed on soil particles are equilibrium values. The ratio of the vapor concentration to solution concentration of chemical is given by the air–water partition coefficient, $K_{AW} = C_G/C_L$. (See Chapter 8.) The ratio of the sorbate to solution concentration is given by the soil–water distribution ratio, K_d (m^3/g) = C_S/C_L. (See Chapter 10.) We assume that the sorption isotherm is linear, and the sorption coefficient, K_p, equals K_d.* Since organic solutes are sorbed preferentially on organic matter in soil, $K_d = f_{OC}K_{OC}$, where f_{OC} (g organic carbon/g soil) is the mass fraction of organic carbon in soil and K_{OC} (m^3/g) is the organic carbon–water partition coefficient. Polar and ionic solutes are adsorbed primarily on mineral matter. The presence of clay in the soil column promotes exchange and adsorption of polar chemicals, retarding their solution levels.

We can derive some useful relationships relating the total concentration of a chemical in soil to the chemical's concentration in soil–air, soil–water, and on soil surfaces. Substituting $K_{AW}C_L$ for C_G and K_dC_L for C_S in Equation 13.1 produces

---

* Sorption isotherms are approximately linear at low solute concentrations.

$$C_T = \left(d_b K_D + w + a K_H\right) C_L \tag{13.2}$$

Since $C_L = C_G/K_{AW}$, Equation 13.2 can be written

$$C_T = \left(\frac{d_b K_D}{K_H} + \frac{w}{K_H} + a\right) C_G \tag{13.3}$$

Moreover, $C_L = C_S/K_d$, and Equation 13.2 can also be written

$$C_T = \left(d_b + \frac{w}{K_D} + \frac{a K_H}{K_D}\right) C_S \tag{13.4}$$

13.2.2 Transport in the Soil Column

Movement of the pesticides lindane and dieldrin in the soil column is almost entirely due to diffusion in soil air, diffusion in soil water, and advection* with soil water (Spencer and Cliath, 1973). There is reason to believe that all volatile chemicals behave similarly. Vapor diffusion in soil–air is faster than liquid diffusion in soil–water by a factor of about 10^4. Both are much faster than diffusion on soil surfaces by a factor of at least 10^9.

Furthermore, advection with soil–water is the dominant mode of mass transport in soil so long as the fluid is in motion. It will be important under almost all soil conditions if the concentration of chemical in soil water is high. Advection is moderated by atmospheric humidity. When there is little or no evaporation of water, there is little transport of water to the surface, and only molecular diffusion is important.

We are most interested in vertical transport of chemicals in the soil column, and we will assume for the sake of simplicity that horizontal transport is insignificant and can be ignored. **Fick's first law of diffusion** for transport in the vertical direction, z, is

$$J_S = \frac{1}{A}\frac{dm}{dt} = -D_S\frac{dC}{dz} \tag{13.5}$$

where J_S (g/m^2·s) is the vertical flux of chemical in soil (the rate of mass transport per unit area through a horizontal layer of the soil column), A (m^2) is the cross-sectional area of the horizontal layer, dm/dt (g/s) is the flow of mass per unit time, D_S is the apparent diffusivity of chemical in soil (m^2/s), C is the mass concentration of chemical (g/m^3), and dC/dz (g/m^3/m) is the concentration gradient of chemical in the direction of transport. (See Section 12.1.1.)

---

* **Advection** is solute transport due to the bulk flow of gas or liquid. It may be caused by gravity, capillary action, a pressure gradient, or a change in the height of the water table.

The diffusion equation, derived from the continuity equation and Fick's second law of diffusion, relates the change of chemical concentration in the soil column to the mass flux. *At low concentrations*, D_S is nearly constant, $(dD_S/dz) \approx 0$, and the diffusion equation is

$$\frac{\delta C}{\delta t} = D_S \frac{\delta^2 C}{\delta z^2} \qquad (13.6)$$

where $\delta C/\delta t$ (g/m³·s) is the time rate of change of chemical concentration in soil, and $\delta^2 C/\delta z^2$ (g/m³/m²) is the change in concentration gradient with distance along the direction of transport. (Some analytical solutions to the diffusion equation are given in Chapter 12.)

Crank (1956), Mayer et al. (1974), and Thibodeaux (1979) examined the problem of a chemical diffusing directly into air from a slab of soil with thickness l (m) lying at the surface of the soil column. Solutions to the diffusion equation were obtained assuming that volatilization is limited by diffusion in soil air. We will assume, also, that the chemical is uniformly applied to the contaminated slab with an initial concentration of C_0 throughout and that the concentration of chemical at the surface of the slab is always zero. In mathematical notation with z = 0 at the bottom of the contaminated layer, the initial and boundary conditions are $C(z,0) = C_0$ for $0 \le z \le 1$, $C(1,t) = 0$, and $\delta C/\delta z = 0$ for z = 0. One solution to the diffusion equation is (Thibodeaux, 1979)

$$C(z,t) = \frac{4 C_0}{\pi} \sum_{n=0}^{\infty} \frac{(-1)^n}{(2n+1)} e^{-D_{SA}(2n+1)^2 \pi^2 t/4l^2} \cdot \cos \frac{(2n+1)\pi z}{2l} \qquad (13.7)$$

Here, n is an index integer. Its value extends from 0 to ∞.*

In addition, the flux at the surface of the slab, J_0 (g/m²s), was shown to be

$$J_0 = \frac{2 D_{SA} C_0}{l} \sum_{n=0}^{\infty} e^{\frac{-D_{SA}(2n+1)^2 \pi^2 t}{4l^2}} \qquad (13.8)$$

Depending on the boundary conditions, other solutions may be found that lead to simpler computational methods. For example, if the contaminated soil layer is thick, the residence time short, or the diffusivity low, Mayer et al. (1974) find

$$J_0 = \left(\frac{D_S}{\pi t}\right)^{\frac{1}{2}} C_0 \qquad (13.9)$$

---

* In this and in succeeding equations, the summation may converge on a value when n is a small number. (See Appendix A6.) The number of terms needed in the summation can be calculated by setting the exponential term to 10^{-99} and solving for n.

Equations 13.7, 13.8, and 13.9 are valid so long as the concentration of chemical at the lower boundary remains fixed. Solutions to the diffusion equation grow more complex if chemical moves across both upper and lower boundaries of the contaminated soil layer.

Chemicals are transported to the surface of the soil column by advection in soil air and soil water. For example, low humidity and high wind speed cause drying of surface soils. Chemical may then be transported to the surface in soil solution as a result of capillary forces. This is called the **wick effect**. Another example is the transport of a chemical by the convective sweep of gases generated in municipal or sanitary landfills. If chemical is transported to the surface by advection in liquid soil water as well as by diffusion in soil air and soil water, the flux of chemical is given by an expanded version of Equation 13.5:

$$J_S = D_{SA} \frac{\delta C_G}{\delta Z} - D_{SW} \frac{\delta C_L}{\delta Z} + V_L C_L \qquad (13.10)$$

where D_{SA} (m^2/s) is the chemical's apparent diffusivity in soil air, D_{SW} (m^2/s) is the chemical's apparent diffusivity in soil water, and V_L (m^3/m^2/s) is the volumetric flux of water or the average linear velocity of liquid water in soil. The evaluation of D_{SA} and D_{SW} is described in Section 12.2.3.

The relative importance of partitioning as opposed to chemical transport in the soil column is made clear if, using Equations 13.2 and 13.3, we write Equation 13.10 as

$$J_S = -D_E \frac{\delta C_T}{\delta Z} + V_E C_T \qquad (13.11)$$

where D_E (m^2/s) = $(D_{SA}K_{AW} + D_{SW})/(d_b K_d + w + a K_{AW})$ is the total effective diffusivity of chemical in soil and V_E (m/s) = $V_L/(d_b K_d + w + a K_{AW})$ is the effective advection velocity of chemical in soil solution. Above pore–water velocities of 2×10^{-5} m/s, advection is the dominant mode of transport in the saturated zone of soil, and diffusion can be ignored (Tucker and Nelken, 1990).

The **generalized diffusion equation** is derived by adding a term describing advective transport to the diffusion equation. It is

$$\frac{\delta C_T}{\delta t} = D_E \frac{\delta^2 C_T}{\delta z^2} - V_E \frac{\delta C_T}{\delta z} \qquad (13.12)$$

Analytical solutions to the generalized diffusion equation are obtained if we assume that, in the soil column, properties are homogeneous and isotropic, chemicals are distributed uniformly, and the vertical flow of water is constant and uniform.

If conversion processes also remove chemical from the soil column, the reaction rate law must be added to the diffusion equation producing **the material-balance**

equation; $\delta C_T/\delta t = D_E \delta^2 C_T/\delta z^2 - V_E \delta C/\delta z - kC_T$, where k (s^{-1}) is a first-order kinetic rate constant for the chemical reaction.

At high water flux or in structured soils in which substantial variation in water velocity occurs, **hydrodynamic dispersion** of solute, *spreading due to the local variability of pore velocity in porous media*, is an important mode of chemical transport. It is accounted for by adding a dispersive flux term, $J_{HD} = -D_{HD} \delta C/\delta z$, to Fick's law. D_{HD} (m^2/s) $= \lambda V_E + D_{SW}$, where λ (m) is the longitudinal dispersivity (mixing length). Since molecular diffusion and hydrodynamic dispersion have the same general form, the proportionality constants are usually lumped together. The equation obtained by including hydrodynamic dispersion in the material balance model is called the **advection-dispersion equation** for porous media. Solutions to the advection-dispersion equation are used to describe contaminant transport in groundwater. The subject is discussed in detail by Luckner and Schestakow (1991).

As we have seen, analytical solutions to the equations that govern chemical transport are obtained if we assume that soil properties are homogeneous, isotropic, and constant. But soil properties are usually inhomogeneous, anisotropic, and transient, and analytical solutions to the governing equations have limited applicability. They are used to model long-term, regional-scale transport of chemicals in soil. Analytical models are also used to compare the relative susceptibility of chemicals to volatilization, leaching, and biodegradation. They are not used to model short-term or field-scale behavior because these depend on the detailed distribution of solute in the soil column and on the timing with which chemical and water are introduced.

In order to model contaminant transport under inhomogeneous, anisotropic, and transient conditions, numerical solutions to the governing equations are used. Numerical models employ sets of linear algebraic equations that approximate the differential form of the governing equation for small volume elements of the soil column over discrete time steps. The amount of input data needed is an obstacle to the application of numerical models to study large-scale transport over long time periods, and hybrid models, collections of submodels that simulate important hydrological and chemical processes, are typically used for the purpose.

Numerical models are classified as being either lumped or distributed. In a typical lumped model, the soil column is treated as a stack of horizontal layers, each layer having homogeneous, isotropic properties and uniform distribution of chemical. In a distributed model (finite difference, finite element, method of characteristics), transport and conversion are described at selected points in the column.

13.2.3 Transport through the Soil–Air Boundary Layer

A stagnant boundary layer of air lies in contact with the surface of the soil column. Frictional drag opposes air flow near the surface. Wind speed drops and turbulent diffusion is damped. While advection and, to a lesser degree, turbulent eddy diffusion control the transport of water vapor and chemicals in the bulk of the air column, molecular diffusion is an important control of transport in the stagnant boundary layer.

The rate of volatilization may be limited by diffusion in the stagnant boundary layer (Jury et al., 1984; Spencer et al., 1988). The flux of chemical through the boundary layer is given by Fick's first law:

$$J_S = -\frac{D_A}{h}\left[C_G(h) - C_G(0)\right] \tag{13.13}$$

where D_A (m²/s) is the chemical's diffusivity in air, h (m) is the boundary-layer thickness, $C_G(h)$ (g/m³) is the chemical vapor concentration at the upper limit of the boundary layer, and $C_G(0)$ (g/m³) is the chemical vapor concentration in soil air which we assume to be the same throughout the soil column. Equation 13.13 is the discrete version of Equation 13.5 with dC/dz replaced by its average value in the boundary layer, $[C_G(h) - C_G(0)]/h$.

In order to use Equation 13.13, we need to know the depth of the stagnant air–soil boundary layer. Unfortunately, its geometry is a complex function of wind speed, fetch, and surface roughness, and we know little about it.* One way around this is to combine D_A and h to form a first-order exchange constant, the boundary-layer mass-transfer coefficient, k_g (m/s) = D_A/h, which is evaluated empirically.

The speed of transport through the soil column to the surface relative to transport through the stagnant boundary layer at the surface is determined mainly by the magnitude of the chemical's air–water partition coefficient, K_{AW} (Spencer et al., 1988). Jury et al. (1984) suggest that if K_{AW} is 2.5×10^{-3} or greater, the chemical is transported primarily in soil air, and the vertical flux is limited by the rate of movement through the soil column to the surface as described by Equation 13.5.

If K_{AW} is 2.5×10^{-7} or less, the chemical is transported primarily in soil water, and the vertical flux is limited by diffusion through the boundary layer as described by Equation 13.13. Since advection to the surface is faster than diffusion through the boundary layer, the flux must increase with time as the chemical accumulates near the surface. Therefore, moderately volatile, water-soluble compounds should exhibit enhanced volatilization rates over time. This pattern of behavior was demonstrated for the pesticide prometon, $K_{AW} = 1.0 \times 10^{-7}$ (Spencer et al., 1988).

Glotfelty et al. (1984) report that volatilization rates are very sensitive to solar heating and to the degree of incorporation of chemical into the soil. Solar heating of surface soils increases the vapor pressure and diffusivity of surface contaminants, increasing their volatility. The evaporation of soil water is also increased, increasing the rate of advection. Large diurnal variation in the vertical flux of chemicals in surface soils coupled with the diurnal variation of solar radiation can be expected. Diurnal variation in the volatilization of pesticides applied to surface soils was demonstrated for trifluralin, heptachlor, chlordane, lindane, and dacthal.

---

* The effective viscosity of fluids increases as eddy size decreases, causing a marked decline in the frequency of occurrence of eddys with diameter below about 1 mm in air. So, the upper limit of the thickness of the air–soil boundary layer over bare soil should be roughly 1 mm. Spencer et al. (1988) report values ranging from 0.22 to 1.2 mm, based on laboratory studies of the volatilization of the pesticide prometon.

Even shallow incorporation of chemical into soil greatly reduces the volatilization rate. This situation is too complex to describe with a simple analytical model. Empirical models are often used to describe the volatilization of chemicals from surface soils.

For intermediate values of K_{AW}, $2.5 \times 10^{-7} < K_{AW} < 2.5 \times 10^{-3}$, the chemical may be transported by vapor diffusion in dry soils and by advection in soil water in wet soils. The volatilization rate is limited by the sum of resistances to transport in the soil column and in the air–soil boundary layer. The ratio of the volatilization rate to the concentration of chemical in soil is given by the overall mass transfer coefficient, a proportionality constant which can be derived as discussed in Section 14.2. This approach is taken in modeling volatilization from water, but it has not been widely used in modeling volatilization from soil.

13.3 METHODS OF ESTIMATING THE RATE OF VOLATILIZATION FROM SOIL

The results presented so far form the basis of estimation methods that are widely used alone and as components of hybrid computer models. Almost all apply only when the chemical's vertical flux is limited by its rate of transport through the soil column. The work of Spencer et al. (1988) shows that this should be true if K_{AW} is much greater than 2.65×10^{-5}, e.g., if $K_{AW} > 2.65 \times 10^{-3}$. The methods and some useful applications are as follows:

13.3.1 Volatilization of a chemical through a soil cover in the absence of advection.
    (1) **Farmer's Equation**. Assumes an infinite source, steady-state, vapor-phase diffusion through the soil cover limits volatilization. Used for chemicals found in soil liquid beneath a clean soil cover, e.g., in a buried sludge cell or contaminated shallow aquifer.
    (2) **Transient Model**. Assumes an infinite source, diffusion through the soil column controls volatilization. Used for chemicals adsorbed on unsaturated soil beneath a clean soil cover, e.g., in a covered landfill or old chemical spill.

13.3.2 Volatilization of chemical from contaminated soil with no soil cover in the absence of advection.
    (1) **Landfarming Equation**. Assumes diffusion through the air–liquid boundary layer limits volatilization. Used for chemical in soil water or oil at surface of soil column, e.g., in a surface spill or in saturated soil.
    (2) **Time-Average Model**. Assumes an infinite source diffusion in soil air and soil–air partitioning limits volatilization. Used for chemical on unsaturated soil at surface, e.g., old spill site or uncovered landfill.

13.3.3 Volatilization limited by advection.
    (1) **Constant Advection Model**. Assumes an infinite source, steady-state advection limits volatilization. Used for chemical in wet soil.

13.3.4 Volatilization of chemical found at the surface of a soil column.
    (1) **Dow Kinetic Rate Model**. Relates first-order volatilization rate constant to chemical's vapor pressure, solubility, and soil adsorption coefficient. Used for chemicals applied to or accumulated at the soil surface.

13.3.1 Volatilization through a Soil Cover

(1) Farmer's Equation

Farmer et al. (1980) used Fick's first law, Equation 13.5, to model the volatilization of hexachlorobenzene wastes from a uniformly contaminated soil column overlaid by a clean, dry soil cover. It is assumed that the mass of chemical is not depleted during volatilization. Soil humidity is assumed to be low enough that advection is negligible and volatilization is controlled only by the diffusive transport of chemical through soil air.

Farmer's equation is derived from Equation 13.5 by replacing the term dC/dz with its average value across the soil cover, $[C_G(atm) - C_G]/l_C$. Here $C_G(atm)$ (g/m^3) is the concentration of chemical in the atmosphere at the surface of the soil column, C_G (g/m^3) is the average concentration of chemical in soil air below the soil cover, and l_C (m) is the depth of soil cover. The discrete version of Fick's law is $J_S = -(D_{SA}/l_C)[C_G(atm) - C_G]$. The emission rate of a volatile chemical from a soil column of surface area A is $E = J_S \cdot A$. If the concentration of chemical in the atmosphere over the soil column is negligible, $C_G(atm) \approx 0$ and

$$E = \frac{D_{SA}}{1} A C_G \qquad (13.14)$$

where E   = emission rate of volatile chemical, g/s
 D_{SA} = apparent diffusivity in soil air, m^2/s
 l_C = depth of dry soil cover, m
 A   = surface area of soil column, m^2
 C_G = average chemical concentration in soil air, g/m^3

The calculation of apparent diffusivity in soil air, D_{SA}, is described in Section 12.2.3.

Example

Estimate the emission rate of chloroform from a closed landfill using Farmer's model. Compare the estimate with the emission rate due to barometric pumping. Assume that the chloroform is dissolved in liquid organic waste at a mole fraction, X_{CCl_4}, of 0.5. The molar mass of chloroform, MW_{CCl_4}, is 119.4 g/mol. Its diffusivity in air at 15°C is about 1.0×10^{-5} m^2/s, and its saturation vapor pressure, P_{sat}, is 0.162 atm at 15°C (USEPA, 1989). The properties of the landfill are given in Table 13.1. They are median values taken from a survey of closed landfills and are characteristic of a landfill with a clay cap (USEPA, 1989).

1. Calculate the apparent diffusivity of chloroform in soil air, D_{SA}. We use the Millington and Quirk (1961) model for tortuosity given by Equations 12.18 and 12.19, Chapter 12:

$$D_{SA} = at_A D_A$$

Table 13.1 Closed Landfill Characteristics

Parameter	Symbol	Value
Waste bed thickness	l	5.0 m
Surface area	A	1.5×10^4 m^2
Cap thickness	l_c	1.0 m
Cap total porosity	P_T	0.40 m^3/m^3
Waste air porosity	a	0.25 m^3/m^3
Waste temperature	T	15°C
Average atmospheric pressure	P	1013 mbar
Daily pressure drop	ΔP	4.0 mbar

From U.S. EPA, 1989.

where D_{SA} = effective diffusivity in soil air, m^2/s

$\quad$ a $\quad$ = volumetric air content of soil, m^3_{air}/m^3_{soil}

$\quad t_A \quad$ = tortuosity (dimensionless)

$\qquad$ = $a^{7/3}/P_T^2$

$\quad D_A \quad$ = diffusivity in air, m^2/s

$\quad P_T \quad$ = total soil porosity, m^3/m^3

Combining the terms above, the apparent diffusivity of chloroform in soil air is $D_{SA} = (a^{10/3}/P_T^2) \cdot D_A$, but since the cap is dry, $a \approx P_T$ and

$$D_{SA} \approx P_T^{4/3} \cdot D_A$$

$$= 0.40^{4/3} \left(1.0 \times 10^{-5} \, m^2/s \right)$$

$$= 2.9 \times 10^{-6} \ m^2/s$$

2. Estimate the concentration of chloroform in soil gas C_G(soil). Since the chloroform is dissolved in organic liquid, we estimate the solute partial pressure over the mixture using Raoult's law*

$$P_{CCl_4} = X_{CCl_4} P_{sat}$$

$$= 0.50 (0.162 \ atm)$$

$$= 0.081 \ atm$$

The concentration of chloroform in soil gas is estimated using **the ideal gas equation**:

$$P_{CCl_4} V = n_{CCl_4} RT$$

where P_{CCl_4} $\quad$ = partial pressure of CCl$_4$, atm

$\quad$ V $\qquad$ = gas volume, m^3

---

* If the chemical is incorporated uniformly into soil, the equilibrium concentration in soil air is given by Equation 13.3: C_G(soil) = $K_{AW}C_0/(p_b K_d + w + aK_{AW})$.

n_{CCl_4} = quantity of CCl_4, mol
= m/MW_{CCl_4}
m = mass of chemical, g
MW_{CCl_4} = molar mass of CCl_4, g/mol
R = 0.08206 l·atm/mol·K
T = temperature, K

Since $C_G = m/V$, the gas equation can be rearranged:

$$C_G = P_{CCl_4} \cdot MW_{CCl_4}/RT$$

$$= (0.081 \text{ atm})(119.4 \text{ g/mol})/(0.08206 \text{ l·atm/mol·K})(288 \text{ K})$$

$$= 0.41 \text{ g/m}^3$$

3. Calculate the average emission rate using Equation 13.14.

$$E = D_{SA} \cdot A \cdot C_G/l_C$$

$$= (2.9 \times 10^{-6} \text{ m}^2/\text{s})(1.5 \times 10^4 \text{ m}^2)(0.41 \text{ g/m}^3)/(1 \text{ m})$$

$$= 0.018 \text{ g/s}$$

4. A chemical waste landfill would probably be vented. It is instructive to compare the emission rate given by Farmer's model with the average daily emission rate due to barometric pumping. The total volume of gas in the landfill is

$$V_T = 1 \cdot A \cdot a$$

$$= (5.0 \text{ m})(1.5 \times 10^4 \text{ m}^2)(0.25 \text{ m}^3/\text{m}^3)$$

$$= 1.9 \times 10^4 \text{ m}^3$$

5. To calculate the volume of gas exiting the vent due to barometric pumping, we use the combined gas equation, $P_i V_i/T_i = P_f V_f/T_f$, where P_i, V_i, and T_i are the initial values of soil–gas pressure, volume, and temperature, respectively, and P_f, V_f, and T_f are the final values. Rearranging the combined gas equation, $V_f = V_i(P_i/P_f)(T_f/T_i)$. The volume of gas exiting the vent is $V_B = V_f - V_i = V_i(P_i/P_f)(T_f/T_i) - V_i$. If we let $V_i = V_T$, $P_i = P$, $P_f = P + \Delta P$, and $T_i = T_f$,

$$V_B = V_T \cdot [P/(P - \Delta P)] - V_T$$

$$= (1.9 \times 10^4 \text{ m}^3)(1013 \text{ mbar})/(1013 - 4 \text{ mbar}) - 1.9 \times 10^4 \text{ m}^3$$

$$= 74 \text{ m}^3$$

The mass of CCl_4 exiting is

$$m_B = C_G(\text{soil}) \cdot V_B$$

$$= \left(0.41 \ \text{g/m}^3\right)\left(74 \ \text{m}^3\right)$$

$$= 30 \ \text{g}$$

The average emission rate due to barometric pumping over a 24-h $(8.64 \times 10^4 \ \text{s})$ period is

$$E_B = m_B/t$$

$$= 30 \ \text{g}/8.64 \times 10^4 \ \text{s}$$

$$= 3.5 \times 10^{-4} \ \text{g/s}$$

Farmer's equation provides an estimate of the maximum diffusive flux through a layer of dry soil. If the soil has a significant moisture level, the emission rate is reduced. For example, Table 13.1 shows that the average value of air porosity in a clay landfill cap is 0.08 m^3/m^3 (USEPA, 1989). Using this value of a in the tortuosity relationship above yields an apparent diffusivity that is 17 times smaller than the dry-soil value. Furthermore, advection may be important in moist soil, and a simple diffusion model such as Farmer's may not be applicable.

(2) Transient Model

The transient model is used to estimate the volatilization rate of a chemical present in unsaturated soil beneath a dry cover. Assume that the soil cover extends to a depth l_C and is initially uncontaminated. The contaminated soil extends to a depth 1, and initially $C = C_0$ for $l_C < z < 1$. The concentration of chemical in air at the surface of the soil cover is zero, $C = 0$ at $z = 0$. Advection is negligible, and volatilization is limited by diffusion of chemical vapor in the contaminated soil.

The problem is similar to that examined by Thibodeaux (1979), and the flux at the surface of the slab is given by a relationship similar to Equation 13.8 (USEPA, 1992). The emission rate, $E = J_0 \cdot A$, is

$$E = \frac{2D_{SA}}{l} A C_0 \sum_{n=0}^{\infty} e^{-D_{SA}(2n+1)^2 \pi^2 t/4l^2} \cos\left[\frac{(2n+1)\pi l_C}{2l}\right] \qquad (13.15)$$

where E   = emission rate of chemical at time t, $\text{g/m}^2 \cdot \text{s}$
       D_{SA} = apparent diffusivity of chemical in soil air, m^2/s
       1   = depth to bottom of contaminated layer, m
       A   = surface area, m^2
       C_0 = initial concentration of chemical in soil, g/m^3

Table 13.2 Properties of the Soil Column

Parameter	Symbol	Value
Porosity	P_T	0.55 m³/m³
Bulk density	p_b	1200 kg/m³
Organic carbon content	f_{OC}	0.025
Water content	w	0.20
Soil cover depth	l_C	1 m
Contamination depth	l	5 m
Area of soil column	A	1.5×10^4 m²

From U.S. EPA, 1989.

t   = time, s
l_C = depth of soil cover, m
n   = index integer, $1 < n < \infty$

Example

Estimate the instantaneous volatilization rate of trichloroethylene from unsaturated soil found beneath a dry soil cover using the transient model. Let $C_0 = 5 \times 10^4$ g/m³. For trichloroethylene, D_A is 7.9×10^{-6} m²/s, K_{AW} is 0.38, and K_{OC} is 0.130 m³/kg (USEPA, 1989). The soil properties are given in Table 13.2. The time elapsed since application is assumed to be 3.2×10^7 s (1 year).

1. Estimate the apparent diffusivity of trichloroethylene in soil air, D_{SA}. As was shown in the previous example, the apparent diffusivity in dry soil is

$$D_{SA} \approx P_T^{4/3} \cdot D_A$$

$$= (0.55)^{4/3} \left(7.9 \times 10^{-6} \text{ m}^2/\text{s}\right)$$

$$= 3.6 \times 10^{-6} \text{ m}^2/\text{s}$$

2. Determine the number of terms needed in the summation of Equation 13.15 by setting the exponential term equal to 10^{-99} and solving for n.

$$-D_E(2n+1)^2 \cdot \pi^2 \cdot t/4l^2 = \ln 10^{-99}$$

$$(2n+1)^2 = -4 \cdot l^2 \cdot \ln 10^{-99}/D_E \cdot \pi^2 \cdot t$$

$$= 4(5 \text{ m})^2 (228)/\left(3.6 \times 10^{-6} \text{ m}^2/\text{s}\right)\pi^2 \left(3.0 \times 10^7 \text{ s}\right)$$

$$= 21$$

$$n = 1.8 \approx 2$$

3. Estimate the emission rate using Equation 13.15. Since the summation converges at the second term:

$$E \approx \frac{2D_{SA}}{1} AC_0 \left[e^{-D_{SA}\pi^2 t/4l^2} \cos\left(\frac{\pi l_C}{21}\right) \right]$$

$$\approx 2\left[\left(3.6 \times 10^{-6} \text{ m}^2/\text{s}\right)\left(1.5 \times 10^4 \text{ m}^2\right)\left(5 \times 10^4 \text{ g/m}^3\right)/5 \text{ m} \right]$$

$$X\, e - \left[\left(3.56 \times 10^{-6} \text{ m}^2/\text{s}\right) (1)^2 (3.14)^2 \left(3.0 \times 10^7 \text{ s}\right)/(4)(5 \text{ m})^2 \right]$$

$$X \cos(\pi \cdot 1 \text{ m}/2 \cdot 5 \text{ m})$$

$$= \left(1.08 \times 10^3 \text{ g/s}\right)\left(2.64 \times 10^{-5}\right)(0.951)$$

$$= 2.7 \times 10^{-2} \text{ g/s}$$

For most problems of interest, transient model estimates are most readily obtained using a digital computer. The average flux is estimated by numerically integrating Equation 13.15 over the time interval of interest.

13.3.2 Volatilization in the Absence of a Soil Cover

(1) Landfarming Equation

The landfarming equation describes the steady-state volatilization of a chemical that is spilled or otherwise incorporated into surface soil. The chemical initially coats the soil particles in the contaminated layer of the column. We assume that the soil pore spaces are connected with the surface, temperature is constant, chemical equilibrium is established across all phases, and capillary rise does not affect the volatilization rate. As the surface layers of the liquid are lost, a dry surface layer of soil forms, its depth increasing with time. The depth of the contaminated layer decreases with time. The concentration of chemical in soil liquid and soil air remains constant, however. If diffusion in soil air of the dry surface layer controls the volatilization rate, Thibodeaux and Huang (1982) show:

$$E = \frac{2D_{SA}C_G A}{l_C + \sqrt{l_C^2 + \left(2D_{SA}C_G t/C_T\right)}}$$ (13.16)

where E   = average volatilization rate over time t, g/s
  D_{SA} = apparent diffusivity in soil air, m²/s
  C_G = chemical concentration in soil air, g/m³
  C_T = total concentration of chemical in soil, g/m³
  A   = surface area of soil column, m²

Table 13.3 Properties of Contaminated Soil

Parameter	Symbol	Value
Porosity	P_T	0.50 m³/m³
Bulk density	d_b	1350 kg/m³
Organic carbon fraction	f_{OC}	0.0125
Water content	w	0.30 m³/m³
Water evaporation rate	V_L	6.0×10^{-8} m/s
Surface cover depth	l_C	0.020 m
Area of soil column	A	2.5×10^4 m²

From U.S. EPA, 1989.

l_C = depth of dry surface layer at sampling time, m
t = elapsed time since sampling, s

Methods for estimating D_{SA} are discussed in Chapter 12. When chemical is applied to or spilled on the soil, $l_C = 0$ and the relationship takes a form similar to that of Equation 13.9. For more details, see Thibodeaux and Huang (1982) and Farino et al. (1983).

Example

Estimate the volatilization rate of benzene from a partially dry surface spill. We will estimate the volatilization rate for an elapsed time of 1 year (3.15×10^7 s) after sampling the soil column. At the time of sampling, the bulk concentration of benzene, C_T, in soil beneath the dry surface layer was found to be 3.00×10^3 g/m³. The average temperature was 20°C. For benzene, $D_A = 8.8 \times 10^{-5}$ m²/s, $K_{AW} = 0.23$ at 20°C, $K_{OC} = 83 \times 10^{-3}$ m³/kg (USEPA, 1989). The properties of the contaminated soil column recorded at the sampling time are given in Table 13.3.

1. Estimate the apparent diffusivity of benzene in soil air at the surface. As was shown in a previous example, the effective diffusivity in dry soil is

$$D_{SA} = P_T^{4/3} D_A$$

$$= (0.50)^{4/3} \left(8.8 \times 10^{-5} \text{ m}^2/\text{s}\right)$$

$$= 3.5 \times 10^{-8} \text{ m}^2/\text{s}$$

2. Estimate K_d.

$$K_d = f_{OC} \cdot K_{OC}$$

$$= 0.0125 \left(83 \times 10^{-3} \text{ m}^3/\text{kg}\right)$$

$$= 1.04 \times 10^{-3} \text{ m}^3/\text{kg}$$

3. Estimate the concentration of benzene in soil air using Equation 13.3.

$$C_G = H \cdot C_T / (d_b \cdot K_d + w + a \cdot H)$$

$$= (0.23)(3.0 \times 10^3 \ g/m^3)$$

$$/ [(1350 \ kg/m^3)(1.04 \times 10^{-3} \ m^3/kg) + 0.30 + 0.20(0.23)]$$

$$= 4.0 \times 10^2 \ g/m^3$$

4. The emission rate is estimated using Equation 13.16.

$$E = 2 \cdot D_{SA} \cdot C_G \cdot A / \left\{1 + \left[1^2 + (2 \cdot D_{SA} \cdot C_G \cdot t/C_T)\right]^{1/2}\right\}$$

$$= 2(3.5 \times 10^{-8} \ m^2/s)(4.0 \times 10^2 \ g/m^3)(25,000 \ m^2)/(0.020 \ m) + \left[(0.020 \ m)^2\right.$$

$$+ \ 2(3.5 \times 10^{-8} \ m^2/s)(4.0 \times 10^2 \ g/m^3)(3.15 \times 10^7 \ s)/(3.0 \times 10^3 \ g/m^3)\right]^{1/2}\Big\}$$

$$= (7.0 \times 10^{-1} \ g \cdot m/s)/(0.54 \ m)$$

$$= 1.2 \ g/s$$

The landfarming equation is useful to estimate short-term emission rates. Eventually, the chemical will become depleted, and soil pore spaces will no longer be coated with contaminants. A long-term model such as the time-average model is a good choice for estimating emission rates in such circumstances.

(2) Time-Average Model

Mayer et al. (1974) examined several special situations in which the volatilization rate is controlled only by vapor-phase diffusion in the soil column. We assume that the soil column is uniformly contaminated with $C(0,t) = 0$ and no chemical is transported across the lower boundary at the base of contamination. If the contaminated layer of soil is thick, the residence time short, or the diffusivity low so that $1^2/D_{SA}t < 14.4$, the chemical flux at the surface of the slab is given by Equation 13.9. The volatilization rate, $E = J_0 \cdot A$, is

$$E = \left(\frac{D_{SA}}{\pi t}\right)^{\frac{1}{2}} AC_0 \tag{13.17}$$

where E   = Emission rate of chemical at time t, g/s
        D_{SA} = apparent diffusivity of chemical in soil air, m²/s
        A   = surface area of the soil column, m²

C_0 = initial concentration of chemical in soil, g/m^3
t   = elapsed time, s

The results are similar to those reported By Jury et al. (1980), Hamaker (1972), and others (USEPA, 1986).

Example

Estimate the average volatilization rate of trichloroethylene diffusing from unsaturated soil with no soil cover using the time-average model. $D_A = 7.9 \times 10^{-6}$ m^2/s (USEPA, 1989). Except that there is no soil cover, the soil properties are given in Table 13.2. Let $C_0 = 5 \times 10^4$ g/m^3 and $t = 3.2 \times 10^7$ s (1 year).

1. The apparent diffusivity of trichloroethylene in soil air was estimated in a previous example. (See the transient model example in Section 13.3.1 [2].) It is 3.6×10^{-6} m^2/s.
2. The average volatilization rate is estimated using Equation 13.17.

$$E = \left(D_{SA}/\pi \cdot t\right)^{1/2} \cdot A \cdot C_0$$

$$= \left[\left(3.6 \times 10^{-6}\ m^2/s\right)/(3.14)\left(3.2 \times 10^7\ s\right)\right]^{1/2}\left(1.5 \times 10^4\ m^2\right)\left(5 \times 10^4\ g/m^3\right)$$

$$= 1.4 \times 10^2\ g/s$$

13.3.3   Volatilization Limited by Advection

(1) Constant Advection Model

For chemicals with $K_{AW} > 2.5 \times 10^{-5}$, volatilization rates are sensitive to the movement of water in the soil column. Above pore water velocities of 2×10^{-5} m/s, advection dominates transport, and Equation 13.10 becomes $J_S \approx V_L \cdot C_L$. We assume that the contaminated layer is infinite, advection is constant, and evapotranspiration of water controls advection. Then V_L is the volume-average evaporation rate. Since $E = \Lambda \cdot J_S$,

$$E = AV_LC_L \tag{13.18}$$

where E = emission rate, g/s
       A = surface area of soil column, m^3
       V_L = volume-average evaporation rate, m/s
       C_L = concentration in soil water, g/m^3

A similar equation describes the emission rate of a chemical transported by the convective sweep of gases generated in municipal or sanitary landfills (Thibodeaux, 1981).

$$E = AV_G C_G \qquad (13.19)$$

where E = emission rate, g/s
      A = area, m²
      V_G = velocity of soil gas, m/s
      C_G = average concentration in soil air, g/m³

Thibodeaux (1981) suggested that V_G has an average value of 1.63×10^{-5} m/s in sanitary landfills.

Example

Estimate the volatilization rate of benzene due to advection from a partially dry surface spill. At the time of sampling, the bulk concentration of benzene, C_T, in soil beneath a dry surface layer was found to be 3.00×10^3 g/m³. The average temperature was 20°C. For benzene, $K_{AW} = 0.23$ at 20°C, $K_{OC} = 83 \times 10^{-3}$ m³/kg (USEPA, 1989). The measured properties of the soil column at the sampling time are given in Table 13.3.

1. Estimate the concentration of benzene in soil water with Equation 13.2.

$$C_L = C_T / (d_b \cdot K_d + w + a \cdot H)$$

$$= (3.0 \times 10^3 \ g/m^3) / [(1350 \ kg/m^3)(1.04 \times 10^{-3} \ m^3/kg) + 0.30 + 0.20(0.23)]$$

$$= 1.7 \times 10^3 \ g/m^3$$

2. The emission rate is estimated with Equation 13.18.

$$E = A \cdot V_L \cdot C_L$$

$$= (25,000 \ m^2)(6.0 \times 10^{-8} \ m/s)(1.7 \times 10^3 \ g/m^3)$$

$$= 2.6 \ g/s$$

Evaporation rates are very sensitive to wind speed, humidity, and solar heating. Large diurnal and day-to-day variations can be expected. For most purposes, advection estimates would be made with hybrid computer models.

13.3.4   Volatilization of a Chemical Accumulated at the Soil Surface

(1) Dow Kinetic Rate Model

Workers at the Dow Chemical Company developed an empirical volatilization model for chemicals applied to the surface soil. They assumed volatilization to be

Table 13.4 Chemical Properties and Volatilization Rate Constants for Pesticides Applied to Surface Soil

Chemical	P^s, mmHg	K_{OC}	S, mg/l	k_v, day^{-1}
Atrazine	8.5×10^{-7}	150	24	1.0×10^{-2}
Carbofuran	2.0×10^{-6}	10	415	2.1×10^{-2}
Chlorpyrifos	1.9×10^{-5}	13,000	2	3×10^{-2}
DDT	2.7×10^{-7}	240,000	1.2×10^{-3}	4.1×10^{-2}
Dinoseb	5×10^{-5}	700	50	6.3×10^{-2}
Diuron	1.9×10^{-6}	150	42	1.3×10^{-3}
Lindane	8×10^{-5}	2,500	7	2×10^{-1}
Nitrapyrin	3×10^{-3}	600	4	6×10^{1}
Trifluralin	1.3×10^{-4}	6,000	6×10^{-1}	2

From Thomas, R.G., 1990. Volatilization from soil. Chapter 16. In *Handbook of Chemical Property Estimation Methods. Environmental Behavior of Organic Compounds.* Lyman, W.J., Reehl, W.F., and Rosenblatt, D.H., Eds. American Chemical Society, Washington, D.C.

a first-order kinetic rate process and reported a correlation between the rate constant and some properties that control partitioning of chemical between soil air, soil water, and soil surfaces. The Dow model is (Thomas, 1990)

$$C = C_0 e^{-k_v \cdot t} \tag{13.20}$$

where C(t) = concentration of chemical at time t, g/m^3
  C_0 = initial concentration of chemical, g/m^3
  k_v = first-order kinetic rate constant
    = 4.4×10^7 ($P^s/K_{OC} \cdot S$) day^{-1}
  P^s = saturation vapor pressure of chemical, mmHg
  K_{OC} = organic carbon partition function, dimensionless
  S   = aqueous solubility of chemical, mg/L
  t   = elapsed time, day

The expression is based on measurements of the volatilization of the nine pesticides listed in Table 13.4. The estimated rate constants were compared with laboratory measurements for all nine chemicals and with field measurements for chlordane, heptachlor, lindane, and trifluralin. The estimated and measured rate constants agreed within a factor of four (Thomas, 1990). However, volatilization from surface soil varies greatly with soil properties, solar insolation, and meteorological conditions. Estimates made with the Dow model should be considered semiquantitative at best.

Example

Estimate k_v and the volatilization half-life of chlordane applied to the surface of a soil column. For chlordane, $P^s = 4.6 \times 10^{-4}$ mmHg, $K_{OC} = 24,600$, and S = 0.1 mg/l (Howard, 1991). The value of K_{OC} is estimated from the chemical's octanol–water partition function. Since the values for chlordane fall within the range of values used to derive the Dow relationship, we are justified in proceeding.

1. Estimate the value of k_V for chlordane. The value of k_V used in Equation 13.20 is estimated from the vapor pressure, organic-carbon partition coefficient, and solubility of the chemical. It is

$$k_V = 4.4 \times 10^7 \left(P^S / K_{OC} \cdot S \right) \ day^{-1}$$

$$= 4.4 \times 10^7 \left(4.6 \times 10^{-4} \right) / (24,600) \ (0.1) \ day^{-1}$$

$$= 8.2 \ day^{-1}$$

2. The half-life, $t_{1/2}$, is the time required for the concentration of chemical to drop to half its initial value. Using Equation 13.20,

$$c = c_0 \ e^{-kvt} \ \text{or} \ \ln(c/c_0) = -k_v t$$

Rearranging,

$$t = -\ln(c/c_0)/k_v$$

$$t_{1/2} = -\ln(0.5)/8.23 \ day = 0.693/8.23 \ day$$

$$= 0.084 \ day = 2.0 \ h$$

Field experiments reported by Glotfelty et al. (1984) show that, under some ambient conditions, the half-life of chlordane is as long as several days.

The first-order decay of chemical concentration assumed in deriving Equation 13.20 is not observed in the field. Glotfelty et al. (1984) report the following regression formula for the volatilization of chemicals applied to the surface of bare, moist soil:

$$\log m = \log m_0 - kt^{1/2} \tag{13.21}$$

where m = mass of chemical in soil at time t, g/m^2
       m_0 = initial mass of chemical in soil, g/m^2
       k   = regression parameter, $h^{-1/2}$
       t   = elapsed period of daylight, h

Since volatilization of surface-applied chemicals is reported to be significant during daylight hours only, the time period used is hours of daylight elapsed since application.

In one experiment, the regression parameter k was found to range from $0.03 \pm 0.02 \ h^{-1}$ for dacthal to $0.18 \pm 0.11 \ h^{-1}$ for trifluralin. The results were obtained for pesticides applied to moist soil. Volatilization rates from dry soils are expected to be lower due to enhanced adsorption of chemicals on soil surfaces.

13.4 MODEL SENSITIVITY AND METHOD ERROR

The analytical models discussed in this chapter produce order-of-magnitude estimates of volatilization rates. Volatilization competes with leaching, washoff, and chemical conversion in soil. It is often not possible to estimate the rates of the competing processes accurately. In such cases, volatilization models need not be highly accurate, and the order-of-magnitude estimates are satisfactory.

Our model of the soil column is a useful but limited approximation. At high water flow rates, equilibrium partitioning may not be achieved in soil. At very low humidity levels, small changes in soil moisture may greatly alter chemical partitioning. Also, capillary tension may affect partitioning. Hysteresis and nonlinear adsorption are not included in the model but are frequently observed. Most volatilization models do not take account of these factors.

The volatilization rate will depend upon such factors as rainfall, soil management practices, and degree of incorporation of chemical into soil. Furthermore, it is rarely the case that soil column properties are homogeneous and isotropic and the chemical distribution uniform. For these reasons, analytical models are not used to estimate short-term, field-scale behavior.

Volatilization estimates are not very sensitive to the choice of diffusivity value. Diffusivities in air and water span a narrow range of values near 1×10^{-5} and 1×10^{-9} m^2/s, respectively. Also, the temperature dependence of volatilization is accounted for indirectly in the temperature dependence of input parameters such as diffusivity.

REFERENCES

Crank, J., 1956. *The Mathematics of Diffusion*. Clarendon Press, Oxford.

Farino, W., Spawn, P., Jasinski, M., and Murphy, B., 1983. Evaluation and selection of models for estimating air emissions from hazardous waste treatment, storage, and disposal facilities. Revised Draft Final Report. GCA Corporation, GCA Technology Division, Bedford, MA, prepared for U.S. Environmental Protection Agency, Office of Solid Waste, Land Disposal Branch. Contract No. 68-02-3168.

Farmer, W. J., Yang, M. S., Letey, J., and Spencer, W. F., 1980. Hexachlorobenzene: its vapor pressure and vapor phase diffusion in soil. *Soil Sci. Soc. Amer. J.,* **44**, 676–680.

Glotfelty, D.E., Taylor, A.W., Turner, B.C., and Zoller, W.H., 1984. Volatilization of surface-applied pesticides from fallow soil. *J. Agric. Food Chem.*, **32**, 638–643.

Hamaker, J.W., 1972. Diffusion and volatilization. Chapter 5. In *Organic Chemicals in the Soil Environment. Volume 3*. Goring, C.A.I. and Hamaker, J.W., Eds. Marcel Dekker, New York.

Howard, P.H., 1991. *Handbook of Fate and Exposure Data. Volume III. Pesticides*. Lewis Publishers, Chelsea, MI.

Jury, W.A., 1986. Volatilization from soil. In *Vadose Zone Modeling of Organic Pollutants*. Hern, S.C. and Melancon, S.M., Eds. Lewis Publishers, Chelsea, MI, pp. 159–190.

Jury, W. A., Grover, R., Spencer, W.F., and Farmer, W.J., 1980. Modeling vapor losses of soil-incorporated triallate. *Soil Sci. Soc. Amer. J.*, **44**, 676–680.

Jury, W.A., Spencer, W.F., and Farmer, W.J., 1983. Behavior and assessment model for trace organics in soil. I. Model description. *J. Environ. Qual.*, **12**, 558–564.

Jury, W.A., Farmer, W.J., and Spencer, W.F., 1984. Behavior and assessment model for trace organics in soil. II. Chemical classification and parameter sensitivity. *J. Environ. Qual.*, **13**, 567–572.

Jury, W.A., Spencer, W.F., and Farmer, W.J., 1984. Behavior and assessment model for trace organics in soil. III. Application of screening model. *J. Environ. Qual.*, **13**, 573–579.

Jury, W.A., Spencer, W.F., and Farmer, W.J., 1984. Behavior and assessment model for trace organics in soil. IV. Review of experimental evidence. *J. Environ. Qual.*, **13**, 580–586.

Luckner, L. and Schestakow, W.M., 1991. *Migration Processes in the Soil and Groundwater Zone.* Lewis Publishers, Boca Raton, FL.

Mayer, R., Letey, J., and Farmer, W.J., 1974. Models for predicting volatilization of soil-incorporated pesticides. *Soil Sci. Soc. Amer. Proc.*, **38**, 563–568.

Millington, R.J. and Quirk, P.J., 1961. Permeability of porous solids. *Trans. Faraday Soc.*, **57**, 1200–1207.

Spencer, W.F., Cliath, M.M., Jury, W.A., and Zhang, L.-Z., 1988. Volatilization of organic chemicals from soil as related to their Henry's law constants. *J. Environ. Qual.*, **17**, 504–509.

Spencer, W. F. and Cliath, M.M., 1973. Pesticide volatilization as related to water loss from soil. *J. Environ. Qual.*, **2**, 284–289.

Thibodeaux, L.J., 1981. Estimating the air emissions of chemicals from hazardous waste landfills. *J. Hazardous Mater.*, **4**, 235-244.

Thibodeaux, L.J., 1979. *Chemodynamics.* John Wiley & Sons, New York.

Thibodeaux, L.J. and Huang, S.T., 1982. Landfarming of petroleum wastes: modeling the air emission problem. *Environ. Progress*, **1**, 42–46.

Thomas, R.G., 1990. Volatilization from soil. Chapter 16. In *Handbook of Chemical Property Estimation Methods. Environmental Behavior of Organic Compounds.* Lyman, W.J., Reehl, W.F., and Rosenblatt, D.H., Eds. American Chemical Society, Washington, D.C.

Tucker, W.A. and Nelken, L.H., 1990. Diffusion coefficients in air and water. Chapter 17. In *Handbook of Chemical Property Estimation Methods.* Lyman, W.J., Reehl, W.F., and Rosenblatt, D.H., Eds., American Chemical Society, Washington, D.C.

U.S. EPA, 1986. Development of Advisory Levels for Polychlorinated Biphenyls (PCBs) Cleanup. EPA/600/6-86/002, NTIS PB86-232774. Office of Health and Environmental Assessment, Exposure Assessment Group, U.S. Environmental Protection Agency, Washington, D.C., 20460.

U.S. EPA, 1989. GCA physical/chemical database. In Hazardous Waste Treatment, Storage and Disposal Facilities (TSDF) — Air Emission Models. EPA-450/3-87-026. U.S. Environmental Protection Agency, Office of Air and Radiation, Office of Air Quality Planning and Standards, Research Triangle Park, NC 27711.

U.S. EPA, 1992. MMSOILS. Multimedia Contaminant Fate, Transport, and Exposure Model. Documentation and User's Manual. Office of Health and Environmental Assessment, Exposure Assessment Group, U.S. Environmental Protection Agency, Washington, D.C. 20460.

CHAPTER **14**

Volatilization from Water

14.1 INTRODUCTION

This chapter deals with nonequilibrium transport of organic chemicals into air from open bodies of water: open tanks, impoundments, lakes, streams, rivers, estuaries, bays, harbors, and oceans. It describes methods of estimating emission rates, residence times, and residual concentrations of solutes that have not achieved equilibrium partitioning between air and water. Both well-mixed and plug-flow models are described. Equilibrium partitioning of a chemical between air and water is discussed in Chapter 8.

Several important reviews of this subject have been published. Schwarzenbach et al. (1993) give a useful overview of volatilization from open water. The literature prior to 1981 was reviewed by Mackay (1981) and by Thomas (1990).

Volatilization is one of several competitive processes that must be considered when estimating a solute's residence time and concentration in water. Wet and dry deposition of atmospheric aerosols and gases adds chemicals to the water column. Adsorption by solids and seepage into the soil column remove chemicals. These processes are known to affect the concentration of solutes such as polycyclic aromatic hydrocarbons and SO_2 in large bodies of water (Baker and Eisenreich, 1990). Sorption is discussed in Chapter 10.

14.2 A MODEL OF INTERFACIAL CHEMICAL TRANSPORT

The **two-film model** of interfacial chemical transport was first proposed by Whitman (1923). It has been applied to atmosphere–ocean exchange of gases by Liss and Slater (1974) and to air–water exchange of organic chemicals by Mackay and Leinonen (1975), Smith et al. (1980, 1981), and Rathbun and Tai (1982), among others.

Imagine the water column divided into two parts as shown in Figure 14.1, a surface film or boundary layer and the bulk of the column. Turbulent diffusion and advection* are the major controls of mass transport in the bulk of the water column. Both are damped at the phase boundary, however, as eddy currents no longer move toward the air–water interface but spread along the water's surface instead (Hunt, 1984). As a result, molecular diffusion is an important control of mass transport in the surface layer. (A similar phenomenon is observed at the air–soil boundary layer. See Chapter 13.)

Fick's first law, discussed in Section 12.1.1, describes mass transport by molecular diffusion in the surface layer. The flux of chemical, J (g/m²·s),** in the layer toward the air–water interface is, according to Fick's law,

$$J = -\frac{D_W}{d}\left[C_l(0) - C_l(d)\right] \tag{14.1}$$

where D_W (m²/s) is the solute diffusivity in water, d (m) is the surface layer thickness, $C_l(0)$ (g/m³) is the solute concentration in the liquid at the air–water interface, and $C_l(d)$ (g/m³) is the solute's concentration at the interface between the surface film and the bulk of the water column. The flux is positive when the concentration gradient, $[C_l(0) - C_l(d)]/d$, is negative.

As with the stagnant air–soil boundary layer, the geometry of the air–water boundary layer and the processes controlling transport within it are not known precisely,*** and we won't attempt to evaluate d and D_W separately. Instead, we combine them, creating the liquid-film mass-transfer coefficient, $k_l = D_W/d$ (m/s), which we evaluate empirically. Using this approach, our model accommodates all modes of mass transport in the surface layer. Note that k_l has units of velocity and is often called the liquid-film mass-transfer velocity.

Mass transport in the bulk of the water column is controlled by eddy diffusion induced by wind, current, and the motion of living creatures that dwell there (bioturbation). In this layer, $J = -[(D_W + D_e)/d][C_l(d) - C_l(Z)]$, where D_e is the eddy diffusivity (usually much greater the D_W) and Z (m) is the average depth of the water column. D_e is not identical for all directions of transport. In rivers, D_e for horizontal motion commonly exceeds that for vertical motion by an order of magnitude. Furthermore, D_e for horizontal motion is scale dependent and can have different values for cross-stream and downstream transport.

In deep lakes, a thermocline may develop during summer months, further stratifying the water column into a warm upper layer of 10 to 20 m depth (the epilimnion) and a cool lower layer (the hypolimnion). Eddy diffusion is damped in the thermocline. (In salt water, a stable salinity gradient may develop, damping eddy diffusion,

---

* **Advection** is the process in which solute is transported due to the bulk flow of gas or liquid. The flow may be caused by gravity, a pressure gradient, a temperature gradient, or the friction of wind moving over open water.

** The *diffusion flux* is the rate of transport of chemical across a surface of unit area that lies perpendicular to the direction of mass flow. It is a vector quantity pointing in the direction of transport with a magnitude of $J = E/A$, where E (g/s) is the time rate of transport of chemical mass and A (m²) is the cross-sectional area.

*** The effective viscosity of fluids increases as eddy size decreases. This results in a marked decline in the frequency of occurrence of eddys below a diameter of about 0.1 mm in water and 1 mm in air, the upper limits for the depth of the surface films in water and air, respectively.

when the upper layer is less saline than the lower layer.) Transport of chemical through the water–solids boundary layer, the hypolimnion, the thermocline, the epilimnion, and the water surface layer is controlled by the concentration gradient of chemical in each layer. The magnitude of the first-order exchange constant in each layer depends upon the depth and the relative importance of molecular and eddy diffusion in each.

Frictional drag opposes air flow and causes wind speed to decrease near the air–water interface. We imagine the air column to be divided like the water column, into a nonturbulent surface film or boundary layer, where molecular diffusion is an important control of mass transport and the bulk of the air column where eddy diffusion controls transport. This is illustrated in Figure 14.1. On occasion, the air column may be divided by temperature inversions into additional layers. The chemical vapor flux away from the air–water interface through the air boundary layer is

$$ J = -\frac{D_A}{h}\left[C_g(h) - C_g(0)\right] \tag{14.2} $$

where D_A (m²/s) is the chemical's diffusivity in air, h (m) is the boundary layer thickness, $C_g(0)$ (g/m³) is the concentration of chemical vapor at the air–water interface, and $C_g(h)$ (g/m³) is the chemical vapor concentration at the interface between the surface film and bulk of the air column. The air-film mass-transfer coefficient for chemical in the air boundary layer is $k_g = D_A/h$ (m/s). Like k_l, it will be evaluated empirically.

To simplify the task of modeling, we assume that turbulent mixing is efficient and chemicals are homogeneously distributed in the bulk layers of the air and water columns, i.e., $C_l(d)$ is equal to C_l, the uniform mass concentration of chemical in the bulk layer of the water column, and $C_g(h)$ is equal to C_g, the uniform chemical concentration in the bulk layer of the air column. See Figure 14.1. If large-concentration gradients do occur outside of the boundary layer, the model should be amended to include the additional mass-transfer coefficients. This issue is discussed in the next section.

A relationship between C_g, C_l, and the volatilization rate is obtained as follows. Assume that volatilization is controlled solely by mass transport across the air–water boundary layer. Diffusion across other boundary layers found in the water column is assumed to be fast by comparison. Furthermore, assume that a steady state has been reached and that transport of chemical through the water boundary layer equals that through the air boundary layer. The emission rate, E (g/s), of chemical transported across either boundary layer at the interface is then

$$ E = JA = -k_l A\left[C_l(0)\right] = k_g A\left[C_g - C_g(0)\right] \tag{14.3} $$

We can safely assume that equilibrium exists at the air–water interface,* and Henry's law is obeyed. The ratio of the equilibrium concentrations of gaseous and

---

* Effusion rates in air and water are such that molecules collide with the air–water interface at the rate of about 10^{27} collisions/m²·s. This is more than enough to ensure that equilibrium is established rapidly at the interface.

dissolved chemical at the interface, $C_g(0)/C_l(0)$, is the solute's air–water partition coefficient, K_{AW} (see Figure 14.1). This is discussed in Chapter 8. Equation 14.3 can be used to show*

$$E = kA\left(C_l - \frac{C_g}{K_{AW}} \right) \tag{14.4}$$

where k (m/s) is the overall mass-transfer coefficient.

The two-film model neglects situations in which turbulence within the water column grows so great that bulk fluid replaces surface water. This might occur, for instance, in turbulent river water. Eddy-structure models have been proposed to describe the situation. The **surface renewal model** is a representative example (Danckwerts, 1951). It describes a water column in which 'packets' of bulk fluid are transported to the surface, remain there for brief periods of time while volatilization takes place, and are subsequently replaced by other 'packets' of bulk material. The flux of chemical across the air–water boundary initially increases as the surface is renewed and declines as a stagnant boundary layer develops again. The chemical flux becomes periodic as the process is repeated.

Since equilibrium is rapidly established as 'packets' arrive at the air–water interface, chemical transport from water to air is limited by the surface renewal rate, r_w (s^{-1}). Equation 14.3 can be modified to model this situation by setting k_l equal to $(D_w r_w)^{1/2}$.** While eddy-structure models differ in their details, all predict that k_l is proportional to $D_w^{1/2}$. This has been verified for turbulent water columns (Dickey et al., 1984; Ledwell, 1984).

As with the two-film model, we know little about the transport process at an agitated air–water boundary, and so the transfer coefficients must still be determined empirically. However, the foregoing leads us to expect that the relationship between k_l and D_w is not fixed. The important point is that $k_l = D_w^n$, where the exponent, n, decreases from 1 to 0.5 as the turbulence increases in the water column.

If solute is lost from the water column by volatilization alone, the instantaneous time rate of change of residual solute concentration, dc_l/dt (g/m^3·s), is related to the emission rate by $E = -V \cdot dC_l/dt$, where V (m^3) is the volume of the water column. Substituting this term for E in Equation 14.4 and rearranging, we find:

$$\frac{dC_l}{dt} = -k\frac{A}{V}\left(C_l - \frac{C_g}{K_{AW}} \right) \tag{14.5}$$

Note that A/V, the ratio of surface area to volume of the body of water, is equal to 1/Z, where Z (m) is the average depth of the water column. Also, most chemicals

---

* Replacing $C_g(0)$ with $K_{AW}C_l(0)$ in Equation 14.3 and solving for $C_l(0)$, we find that $C_l(0) = (k_lC_l + k_gC_g)/(k_l + k_gK_{AW})$. Equation 14.4 is derived by substituting the expression for $C_l(0)$ into Equation 14.3 and setting $k = k_lk_gK_{AW}/(k_l + k_gK_{AW})$, i.e., $1/k = 1/k_l + 1/k_gK_{AW}$.
** The mass transferred in time t is proportional to the diffusion penetration depth (given by the Einstein–Smoluchowsky equation) multiplied by the concentration gradient. For the liquid film, it is proportional to $(D_w t)^{1/2}[C_l - C_l(0)]$.

are present in air at trace concentration levels, and advection and dispersion in the air column usually are efficient at transporting chemical vapor away from the air–water interface. Therefore, C_g is usually small enough to be neglected. Substituting $1/Z$ for A/V and setting C_g equal to 0 in Equation 14.5 yield the first-order differential equation:

$$\frac{dC_l}{dt} = -\frac{k}{Z}C_l \qquad (14.6)$$

with solution:

$$C_l(t) = C_l(0)e^{-\frac{kt}{Z}} \qquad (14.7)$$

where $C_l(t)$ (g/m³) is the residual solute concentration at time t and $C_l(0)$ (g/m³) is the initial concentration of chemical in the bulk water column.

The concentration half-life of solute, $t_{1/2}$ (s), is found by setting $C_l(t)$ equal to $0.5C_t(0)$ in Equation 14.7. We find that $t_{1/2} = 0.693\ Z/k = 0.693\ Z(1/k_l + 1/k_gK_{AW})$; the half-life is proportional to Z and inversely proportional to k_l, k_g, and K_{AW}. As the value of K_{AW} approaches 0.1, $t_{1/2}$ approaches an asymptotic value of $0.693\ Z/k_l$.

14.3 METHODS OF ESTIMATING THE RATE OF VOLATILIZATION FROM WATER

The choice of procedure for estimating volatilization rates and residual solute concentrations depends upon whether the water column can be treated as well mixed due to back flow or if a plug-flow model is more appropriate. The well-mixed model is used to describe volatilization from lakes, impoundments, and tanks. Streams, rivers, and impoundments with surface films (which inhibit mixing) are described with a plug-flow model.

14.3.1 Volatilization from a Well-Mixed System

In a well-mixed system, the chemical in the water column is mixed by molecular and turbulent diffusion to produce a solution of uniform concentration. The system may have an effluent outflow (a lake or a storage impoundment) or it may not (a tank or a disposal impoundment). Volatilization is assumed to be the only significant process by which solute is removed from solution.

Estimation methods for well-mixed bodies of water are based on Equation 14.4. As noted above, advection and dispersion in the air column are usually efficient at dispersing chemical, and C_g is often negligible. Setting C_g equal to 0, Equation 14.4 becomes

$$E = kAC_l \qquad (14.8)$$

where E = emission rate from the liquid surface, g/s
      k = overall mass-transfer coefficient, m/s
      A = liquid surface area, m^2
      C_1 = aqueous concentration of chemical, g/m^3

Methods of estimating the model parameters are described below. If more than one volatile chemical is present in the system, little error is incurred (less than 10%) by applying Equation 14.8 separately to each solute (Mackay and Yeun, 1983).

(1) Estimation of k

In deriving Equation 14.4, we defined the overall mass-transfer coefficient with the relationship:*

$$\frac{1}{k} = \frac{1}{k_l} + \frac{1}{k_g \cdot K_{AW}}$$

(14.9)

where k   = overall mass-transfer coefficient, m/s
      k_l = liquid-film mass-transfer coefficient, m/s
      k_g = gas-film mass-transfer coefficient, m/s
      K_{AW} = air–water partition coefficient

In open water, the k_l/k_g ratio is observed to vary from 0.003 to 0.06, depending on wind speed. A value of 0.01 is typical. For volatile, sparingly soluble chemicals, K_{AW} is typically greater than 0.1, and 1/k is roughly equal to $1/k_l$. Volatilization is said to be liquid-film controlled.** For such chemicals, the value of k_l can be estimated by comparing D_l for the compound of interest with that of a reference compound which behaves similarly. Molecular oxygen is often chosen as the reference compound because adsorption of O_2 is liquid-film controlled and a large amount of data is available on oxygen reaeration rates.

Resistance to transport through the stagnant air layer is the same whether the underlying layer consists of soil or water. Spencer et al. (1988) found that the effective air-boundary layer thickness over soils varied from 2.2×10^{-4} to 1.5×10^{-3} m for mixtures of lindane and prometon, and the value of the air-film mass-transfer coefficient, $k_g = D_A/h$, lies in the range from 3×10^{-3} to 3×10^{-2} m/s (Jury et al., 1983). The values are similar to those observed for stagnant air boundary layers found over quiescent water surfaces. However, resistance to transport is higher

---

* In the two-film model, we imagine that the overall resistance to volatilization at the liquid surface, $R_v = V/Ak$ (s), is the sum of two resistances: resistance in the liquid interfacial film, $R_l = V/Ak_l$ (s), and resistance in the gas interfacial film, $R_g = V/Ak_g K_{AW}$ (s). An alternate derivation of Equation 14.9 involves substituting the equivalent terms given above for R_v, R_l and R_g in the resistance equation, $R_v = R_l + R_g$, and multiplying through by A/V.
** The volatilization of solutes with Henry's law constants greater than about 5×10^{-3} atm·m³/mol is liquid-film controlled. This includes most chemicals listed on federal and state volatile priority pollutant lists.

in soils than in water because of the enhanced tortuosity and chemical adsorption in soils.

For nonvolatile, water-soluble contaminants such as herbicides and many pesticides, K_{AW} is typically less than 0.001, and $1/k$ is roughly equal to $1/k_g K_{AW}$. The value of k_g can be estimated by comparing D_A of the compound of interest to that of a reference compound having similar behavior. Water is usually chosen as the reference compound because its vaporization can be limited only by resistance in the gas phase.

It is important to estimate the value of k_l for volatile solutes accurately. Volatile solutes are rapidly removed from the water column by volatilization alone, and errors in the value of k_l can lead to large errors in predicted half-life and emission rates. It is usually not as important to accurately estimate the value of k_g for nonvolatile solutes. Nonvolatile solutes are often removed from the water column by a variety of biotic and abiotic processes in addition to volatilization. Accurate rate constants for these processes are not typically available, however, and good accuracy in the value of k_g will not significantly enhance the accuracy of predicted half-lives and residual solute concentrations.

For compounds of intermediate volatility, both k_l and k_g must be evaluated. There are two alternative approaches. In one, the magnitude of k_l and k_g are estimated by comparing diffusion coefficients of the chemical of interest with those of a compound that behaves similarly such as diethyl ether (Lunney et al., 1985) or benzene (Mackay and Yeun, 1983). This approach is demonstrated below for the volatilization of solutes from lakes and impoundments. The second approach involves comparing diffusion coefficients of the compound of interest with that of oxygen to evaluate k_l and with that of water to evaluate k_g. This method was reviewed by Rathbun (1990) and is demonstrated below for the volatilization of solutes from streams and estuaries.

The various methods of evaluating k_l and k_g are described below. Procedures for estimating the air–water partition coefficient, K_{AW}, are described in Chapter 8.

(2) Estimation of k_l

Lakes, Impoundments, Estuaries, and Oceans — Wind induces turbulent mixing in large, open bodies of water. When the water flow velocity is small or zero, air flow controls the rate of volatilization.

Three wind-speed regimes have been identified within the environmental range of 0 to 20 m/s* in which different physical processes control gas exchange between water and air. Based on published laboratory and field studies using various tracers, Liss and Merlivat (1986) identify a smooth surface regime when the wind speed at an altitude of 10 m, U_{10} (m/s), lies between 0 and 5 m/s, a capillary wave regime when U_{10} lies between 5 to 10 m/s, and a turbulent regime characterized by breaking waves, spray in the air column, and the injection of air bubbles into the water column when U_{10} exceeds 10 m/s. Gas transfer is enhanced in progressing from the smooth surface to the turbulent surface regime. Cohen et al. (1978), Mackay and Yeun

---

* The normal range of wind speed is 0 to 10 m/s. Wind gusts do exceed 15 m/s. Violent weather may produce winds in excess of 20 m/s.

(1983), and Lunney et al. (1985) propose similar schemes based on laboratory studies of the volatilization of organic chemicals.

Turbulence is scale dependent, and we expect volatile components to be purged faster from shallow water columns than from deep water columns. Mackay (1981), Mackay and Yeun (1983), and Wanninkhof et al. (1985) point out that transfer rates measured in tanks are higher than those measured in lake and ocean water. Lunney et al. (1985) conclude that the differences are probably due to failure to compare fetch* to depth ratios in tank and environmental studies. Three fetch/depth regimes were identified for which k_l must be evaluated differently. They are F/Z < 14, 14 ≤ F/Z ≤ 50, and F/Z > 50. Relationships for F/Z < 14 apply to deep bodies of water such as deep lakes and also to most laboratory simulators. Those for F/Z > 14 apply to shallow bodies of water, such as storage and disposal impoundments, and very wide bodies of water.

It was suggested that the difference between laboratory simulators and open water is also due to differences in wave development. Jahne et al. (1984) show that k_l depends greatly on surface turbulence as measured by capillary wave slope. Capillary wave slope depends on wind fetch, surface contamination, water current, and so on. Wanninkhof et al. (1985) point out that laboratory measurements show a stronger dependence of k_l on wind speed than do field measurements. They suggest this is due to the production of short, steep waves in the short fetch of laboratory wind-wave tanks that are unlike the long, shallow waves commonly found in open water.

Formulas to calculate the liquid-film mass-transfer coefficient for slow-moving water were proposed by Liss and Merlivat (1986), Mackay and Yeun (1983), and Lunney et al. (1985). Their recommendations, categorized by type of water column, fetch-to-depth ratio, and wind speed, are given in Table 14.1.** The relationships have been tested in studies of volatilization from open water, but not extensively. Recent field measurements of gas exchange in lake and ocean water have been made with ^{14}C, [222]Rn, [3]He, and O_2 (Emerson, 1975; Torgersen et al., 1982; Broecker and Peng, 1984), with dichlorobenzene and tetrachloroethylene (Schwarzenbach et al., 1979), with SF_6 (Wanninkhof et al., 1985, 1987; Upstill–Goddard et al., 1990), and with a dual tracer technique employing both SF_6 and [3]He (Watson et al., 1991). The results are generally in agreement with predictions of the models listed in Table 14.1. Lunney et al. (1985) assert that, when fetch-to-depth ratio is properly accounted for, their model predictions are in good agreement with measurements of k_l for benzene made in open water.

Note that Equation 14.10 shows k_l to be independent of wind speed when U_{10} lies below 5 m/s. Environmental measurements show that k_l increases slightly with increasing wind speed in this regime, as shown in Equations 14.13 and 14.15.

Rivers and Streams — In fast-moving rivers and streams, water flow is an important control of volatilization. Field measurements of gas exchange in rivers have been done with SF_6, and with organic solutes such as 1,1,2-trichlorotrifluoro-ethane, trichloroethene, trichloromethane, and tribromomethane (Neely et al., 1976;

---

* *Fetch* is the distance across the surface of a body of water in the direction of wind flow.
** Some models proposed by Lunney et al. for volatilization from bodies of water with F/Z < 14 are not included here, since they are not validated with environmental measurements.

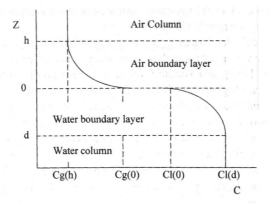

Figure 14.1   The model air–water interface is composed of a water boundary layer of depth
d and an air boundary layer of height h. Molecular diffusion controls chemical
transport within the boundary layers. Beyond the boundary layers, eddy diffusion
controls transport and produces a uniform chemical concentration. At z = 0,
$C_g(0)/C_l(0) = K_{AW}$.

Rathbun, 1990; Cirpka et al., 1993; Wilcock, 1984; Duran and Hemond, 1984). They
show that regardless of mixing conditions, liquid-vapor mass-transfer rates of organic
chemicals are roughly proportional to oxygen reaeration rates (Thomann and Mueller, 1987; Thomas, 1990). Therefore, the magnitude of k_l for an organic solute in a
river or stream can be estimated by comparison with the oxygen mass-transfer
coefficient, K_{O2} (m/s), using

$$k_l = \phi K_{O2} \qquad (14.18)$$

where ϕ is a dimensionless proportionality factor.

Liss and Slater (1974) assert that k_l is proportional to $M^{-1/2}$, where M (g/mol) is
the molar mass of the organic solute, and $\phi = (M_{O2}/M)^{1/2}$, where M_{O2} is the molar
mass of O_2 (32.0 g/mol). Rathbun (1990), evaluating data for 40 volatile organic
solutes, reports that setting $\phi = (M_{O2}/M)^{1/2}$ produced a 14.5% average predictive
error. Mackay and Yeun (1983) suggest that k_l is proportional to $D_W^{1/2}$, where D_W is
the solute diffusivity, and $\phi = (D_W/D_{O2})^{1/2}$, where D_{O2} is the diffusivity of oxygen in
water. Rathbun (1990) found that regression models based indirectly on diffusivity
were more accurate than those based on molar mass.* On the basis of this work,
the following relationship is likely to produce minimum predictive error:

$$k_l = \left(\frac{D_W}{D_{O2}}\right)^{\frac{1}{2}} K_{O2} \qquad (14.19)$$

---

* Regression analysis of data on 50 volatile and semivolatile chemicals shows that DW is proportional
to $M^{-0.433}$. Regression models in which ϕ is proportional to $M^{-0.433/2}$ produce predictive errors of less than
10% (Rathbun, 1990).

Table 14.1   Semi-Empirical Models for Calculating Liquid-Film Mass-Transfer Coefficients

Shallow Impoundments[a]

When $U_{10} < 5$ m/s,

$$K_l = 2.788 \times 10^{-6} \left(\frac{D_w}{D_{eth}}\right)^{\frac{2}{3}}$$

(14.10)

When $14 \leq F/Z \leq 51.2$ and $U_{10} \geq 5$ m/s,

$$k_l = \left[2.605 \times 10^{-9} \left(\frac{F}{Z}\right) + 1.277 \times 10^{-7}\right] U_{10}^2 \left(\frac{D_w}{D_{eth}}\right)^{\frac{2}{3}}$$

(14.11)

When $F/Z > 51.2$ and $U_{10} \geq 5$ m/s,

$$k_l = 2.611 \times 10^{-7} \, U_{10}^2 \left(\frac{D_w}{D_{eth}}\right)^{\frac{2}{3}}$$

(14.12)

Lakes[b]

When $U_{10} < 5$ m/s,

$$k_l = 1.0 \times 10^{-6} + 144 \times 10^{-4} \, U^{*2.2} \, Sc_L^{-0.5}$$

(14.13)

When $U_{10} \geq 5$ m/s,

$$k_l = 1.0 \times 10^{-6} + 34.1 \times 10^{-4} \, U^* \, Sc_L^{-0.5}$$

(14.14)

Ocean, Estuaries[c]

When $U_{10} < 3.6$ m/s,

$$k_l = 4.7 \times 10^{-7} \, U_{10} \left(\frac{Sc_L}{600}\right)^{-\frac{2}{3}}$$

(14.15)

When 3.6 m/s[c] $\leq U_{10} \leq 13$ m/s,

$$k_l 2.78 \times 10^{-6} \left(2.85 U_{10} - 9.65\right) \left(\frac{Sc_L}{600}\right)^{-0.5}$$

(14.16)

When $U_{10} > 13$ m/s,

$$k_l = 2.78 \times 10^{-6} \left(5.9 U_{10} - 49.3\right) \left(\frac{Sc_L}{600}\right)^{-0.5}$$

(14.17)

**Table 14.1 (continued)   Semi-Empirical Models for Calculating
Liquid-Film Mass-Transfer Coefficients**

where k_l = liquid-film mass-transfer coefficient, m/s
  D_w = diffusivity of chemical in water, cm²/s
  D_{eth} = diffusivity of ethyl ether in water, cm²/s
    = 8.5×10^{-6} cm²/s at 25°C
  F/Z = fetch-to-depth ratio of system
  U_{10} = wind speed at 10 m above surface, m/s
  Sc_L = liquid side Schmidt number = $\mu_L/\rho_L D_w$
  μ_L = viscosity of water, g/cm·s
  ρ_L = density of water, g/cm³

[a] Based on laboratory measurements with diethyl ether (Lunney, Springer, and Thibodeaux, 1985).
[b] Based on laboratory measurements with benzene, toluene, 1,2-dichloropropane, 1,2-dibromoethane, chlorobenzene, carbon tetrachloride, 2-pentanone, 2-heptanone, 1-pentanol, 2-methyl-1-propanol, and 1-butanol in a wind-wave tank with F/Z = 10 and correlated with environmental measurements on CO_2, O_2 and ^{222}Rn (Mackay and Yeun, 1983).
[c] Based on laboratory and field measurements with CO_2, ^{222}Rn, SF_6, and ^3He (Liss and Merlivat, 1986). The relationships incorporate the value of Sc_L for the reference compound CO_2 of 600 at 20°C.

where k_l = liquid-film mass-transfer coefficient, m/s

  D_W = diffusivity of chemical in water, m²/s

  D_{O2} = diffusivity of O_2 in water, m²/s

    = 2.090×10^{-9} m²/s at 20°C

  K_{O2} = oxygen mass transfer coefficient, m/s

The oxygen mass transfer coefficient is related to the widely studied oxygen reaeration rate constant, K_A (s⁻¹) = $K_{O2}A/V$, where A/V = 1/Z is the average depth of the water column.* Methods of measuring and estimating the oxygen reaeration rate were reviewed by Thomas (1990) and Thomann and Mueller (1987).

Turbulence and eddy transport to the surface are proportional to stream velocity, U_{curr} (m/s), and inversely proportional to average stream depth, Z. For streams and rivers where surface wind does not contribute significantly to mixing, O'Conner and Dobbins (1958) recommend setting the surface renewal velocity, r_w, equal to U_{curr}/Z, resulting in

$$K_A = \frac{K_{O2}}{Z} = \frac{\left(D_{O2}U_{CURR}\right)^{\frac{1}{2}}}{Z^{\frac{3}{2}}} \tag{14.20}$$

* The rate of change of dissolved-oxygen level due to reaeration is

$$\frac{dC_{O2}}{dt} = K_A \left(C_{O2eq} - C_{O2}\right)$$

where C_{O2eq} (g/m³) is the equilibrium solute concentration and K_A (s⁻¹) is the reaeration rate constant.

where K_A = oxygen reaeration rate constant, s^{-1}
       K_{O2} = oxygen mass-transfer coefficient, m/s
       D_{O2} = diffusivity of oxygen in water
                = 2.090×10^{-9} m^2/h at 20°C
       U_{CURR} = average velocity of stream flow, m/s
       Z = average stream depth, m

Equation 14.20 satisfactorily predicts reaeration rates in large streams, tidal rivers, and estuaries, but it underpredicts oxygen reaeration rates in small streams. When this relationship is applied to reaeration in tidal rivers and estuaries, U_{CURR} is taken as the mean tidal velocity over one complete tide cycle, and Z is the average depth during the cycle.

Substituting K_{O2} in Equation 14.19 with the relationship given in Equation 14.20 gives*

$$ k_l = \left(\frac{D_W U_{CURR}}{Z} \right)^{\frac{1}{2}} $$

(14.21)

where k_l = liquid-film mass-transfer coefficient, m/s
       D_W = diffusivity of chemical in water, m^2/s
       U_{CURR} = average velocity of stream flow, m/s
       Z = average stream depth, m

As with Equation 14.20, we expect Equation 14.21 to accurately predict the value of the liquid-film mass-transfer coefficient with the exception of small streams. Schwarzenbach et al. (1993) find this to be the case. They show that model estimates agree with experiment results within a factor of three. Several empirical models have been proposed for estimating oxygen reaeration rates in small rivers and streams. A few of those cited by Thomas (1990), Thomann and Mueller (1987), and St. John et al. (1993) are listed in Table 14.2.

As with all regression models, the relations given in Table 14.2 are valid only for the range of variables used in deriving them. Since predictions of the various models may differ by a factor of 10 or more, it is important to use a model that is appropriate for the range of conditions observed in the water column being studied.

Most of the models listed in Table 14.2 predict that K_A approaches zero as the depth of the water column increases. K_A cannot decrease without limit, however. Measured values range from 1×10^{-4} s^{-1} for rivers to about 1×10^{-6} s^{-1} for ponds and lakes (Smith et al., 1980). Mackay and Yeun (1983) estimate the minimum still-air value of k_l to be 1×10^{-6} m/s. This is consistent with a minimum value of K_A of roughly 1×10^{-6} s^{-1}.

The magnitude of k_l for lakes and estuaries is observed to lie in the range of 6.5×10^{-6} to 3.5×10^{-5} m/s. The diffusivity of most volatile chemicals in water is

---

* It was noted above that the surface renewal model requires $k_l = (D_W r_W)^{1/2}$. Equation 14.21 is obtained by setting r_W equal to U_{CURR}/Z.

Table 14.2 Empirical Models for Estimating Oxygen Reaeration Rates in Small Rivers and Streams

Flow Velocity, Depth Models

$K_A = 4.56 \times 10^{-5} \, U_{CURR}^{0.50} \, Z^{-1.50}$	O'Conner-Dobbins[a]
$K_A = 6.17 \times 10^{-5} \, U_{CURR}^{0.67} \, Z^{-1.85}$	Owens et al.[b]
$K_A = 6.69 \times 10^{-5} \, U_{CURR} \, Z^{-1.33}$	Langbein-Durum
$K_A = 6.17 \times 10^{-5} \, U_{CURR} \, Z^{-1.67}$	Churchill et al.[c]
$K_A = 6.47 \times 10^{-5} \, U_{CURR}^{0.61} \, Z^{-1.69}$	Bennett-Rathbun
$K_A = 6.47 \times 10^{-5} \, U_{CURR} \, Z^{-1.5}$	Isaacs-Grundy

Flow Velocity, Slope Models

$K_A = 3.61 \times 10^{-1} \, U_{CURR} \, S$	Tsivoglou-Wallace[d]
$K_A = 1.76 \times 10^{-5} \, U_{CURR} \, S$	Tsivoglou-Wallace[e]
$K_A = 1.96 \times 10^{-1} \, U_{CURR} \, S$	Shinadala-Truax
$K_A = 3.75 \times 10^{-4} \, U_{CURR}0.413 \, S^{0.273} \, Z^{-1.408}$	Bennett-Rathbun
$K_A = 2.83 \times 10^{-7} \, U_{CURR}^{2.695} \, S^{-0.823} \, Z^{-3.085}$	Churchill et al.[c]

[a] $0.15 < U_{CURR} < 0.49$ m/s, $0.31 < Z < 9.1$ m, $0.015 \leq K_a \leq 3.7$ hr^{-1}.
[b] $0.030 < U_{CURR} < 1.5$ m/s, $0.12 < Z < 3.4$ m.
[c] $0.55 < U_{CURR} < 1.5$ m/s, $0.61 < Z < 3.4$ m.
[d] $0.03 < Q < 0.28$ m^3/s, $Z_{avg} = 0.75$ m, $U_{avg} = 0.12$ m/s, T = 20°C.
[e] $0.28 < Q < 8.5$ m^3/s, $Z_{avg} = 0.75$ m, $U_{avg} = 0.12$ m/s, T = 20°C.

where K_A = oxygen reaeration rate coefficient, s^{-1}
U_{CURR} = stream flow velocity, m/s
Z = average stream depth, m
S = stream-bed slope, m drop/m run

roughly 1×10^{-9} m^2/s. Substituting this value of D_W in Equation 14.21 and setting k_1 equal to 6.5×10^{-6} m/s yields the minimum value of U_{curr}/Z of 0.04 s^{-1} below which gas exchange is not affected significantly. Similar reasoning led Schwarzenbach et al. (1993) to conclude that current is important in gas exchange provided $U_{curr}/Z > 0.03 \text{s}^{-1}$.

The diffusivity of oxygen in water is temperature dependent. The relationship is $D_{O2} = 2.090 \times 10^{-9}(1.037)^{T-20}$ (m^2/s) (Dobbins, 1964). The temperature-dependence of the reaeration rate constant is similar. The relationship is $K_{A(T)} = K_{A(20)}\theta^{T-20}$, where $K_{A(T)}$ is the reaeration rate constant at ambient temperature T (°C). The adjustable parameter θ ranges from 1.005 to 1.030, depending on mixing conditions (Thomann and Mueller, 1987). A value of 1.024 is commonly used (Elmore and West, 1961).

When estimating K_{O2} for oxygen and k_1 for volatile solutes in rivers, Thomas suggests supplementing Equations 14.20 and 14.21 with a wind speed term of $e^{0.526(U10 - 1.9)}$, when U_{10} lies between 1.9 and 5 m/s. However, the wind speed correction for volatilization rates from rivers is not reliable (Frexes et al., 1984) and is rarely used.

(3) Estimation of k_g

Mackay and Yeun (1983) studied the evaporation of 11 compounds of varying Henry's law constant from a wind-wave tank with F/Z = 10. The compounds were benzene, toluene, 1,2-dichloropropane, 1,2-dibromoethane, chlorobenzene, carbon tetrachloride, 2-pentanone, 2-heptanone, 1-pentanol, 2-methyl-1-propanol, and 1-butanol. The data were correlated with rates of CO_2, O_2, and ^{222}Rn transfer to air from large open bodies of water. They recommended using the following relationship to predict environmental volatilization rates:

$$k_g = 1.0 \times 10^{-3} + 46.2 \times 10^{-5} (6.1 + 0.63 U_{10})^{0.5} U_{10} Sc_G^{-0.67} \qquad (14.22)$$

where k_g = gas-film mass-transfer coefficient, m/s
         U_{10} = wind speed, m/s
         Sc_G = Schmidt number on gas side, $Sc_G = \mu_G/\rho_G \cdot D_A$
         μ_G = viscosity of air, g/cm·s
         ρ_G = density of air, g/cm³
         D_A = diffusivity of chemical in air, cm²/s

The value of k_g for rivers and streams is often estimated by comparison with the water transfer coefficient, K_{H_2O} (m/s):

$$k_g = \psi K_{H_2O} \qquad (14.23)$$

where k_g = gas-film mass-transfer coefficient, m/s
         K_{H_2O} = water transfer velocity, m/s
         ψ = proportionality factor

It is assumed that the relationship holds regardless of mixing conditions in the air column.

As noted earlier, k_g is proportional to D_A^n, where n lies between 0.5 to 1. Smith et al. (1980) estimate n as 0.61. Mackay and Yeun (1983) suggest that k_g is proportional to $D_A^{0.67}$, and $\psi = (D_A/D_{H_2O})^{0.67}$. Liss and Slater (1974) suggest and Rathbun (1990) confirms that k_g is proportional to $M^{-0.5}$.

The value of K_{H_2O} is determined by measuring the evaporation rate of water. Several studies of K_{H_2O} vs. wind speed have been done in the laboratory as well as in the field. Observed values range from 0.003 to 0.3 m/s. These are reviewed by Schwarzenbach et al. (1993) who note the similarity of the published empirical correlations. A few of those cited by Schwarzenbach et al. are listed in Table 14.3.

As always, empirical relationships are valid only for the range of variables used to derive them. However, the gas-film transfer coefficients need not be estimated as carefully as the liquid-film coefficients. Schwarzenbach et al. suggested the following approximate relationship for general use:

$$K_{H_2O} = 0.003 + 0.002 U_{10} \qquad (14.24)$$

where K_{H_2O} = water transfer velocity, m/s
         U_{10} = wind speed at 10 m altitude, m/s

Table 14.3 Empirical Models for Estimating Evaporation Rates in Open Water

$K_{H_2O} = 1.5 \times 10^{-3}U_{10}$	Sverdrup et al.
$K_{H_2O} = 5.9 \times 10^{-3} + 2.3 \times 10^{-3}U_{10}$	Penman
$K_{H_2O} = 3.5 \times 10^{-3} + 1.2 \times 10^{-3}U_{10}$	Rohwer
$K_{H_2O} = 5 \times 10^{-5} + 2.1 \times 10^{-3}U_{10}$	Liss
$K_{H_2O} = 5 \times 10^{-3} + 1.85 \times 10^{-3}U_{10}$	Munnich et al.
$K_{H_2O} = 6.5 \times 10^{-4}(6.1 + 0.63U_{10})^{0.5}U_{10}$	Mackay and Yeun

where K_{H_2O} = water transfer velocity, m/s
U_{10} = wind speed at 10 m altitude, m/s

Note that the transfer velocity is estimated to be 0.003 m/s at $U_{10} = 0$. Saturated air is lighter than dry air, and convective motion produces transport away from the water surface. The value of 0.003 m/s is the minimum value of K_{H_2O}, and it corresponds to a 1-mm-thick layer of stagnant air.

A general model for predicting k_g is obtained by combining Equations 14.23 and 14.24 and setting $\psi = (D_A/D_{H_2O})^{0.67}$:

$$k_g = \left(0.003 + 0.002U_{10}\right)\left(\frac{D_{H_2O}}{D_A}\right)^{0.67} \qquad (14.25)$$

where k_g = gas-film mass-transfer coefficient, m/s
D_{H_2O} = diffusivity of water in air, m²/s
= 2.6×10^{-5} m²/s at 20°C
D_A = diffusivity of chemical in air, m²/s
U_{10} = wind speed at 10 m altitude, m/s

If average stream velocity is comparable to wind speed, Thomas (1990) recommends adding the two. He suggests the following relationship for k_g in which $\psi = (M_{H_2O}/M)^{0.5}$.

$$k_g = 3.16 \times 10^{-3}\left(U_{10} + U_{curr}\right)\left(M_{H_2O}/M\right)^{0.5} \qquad (14.26)$$

where k_g = gas-film mass-transfer coefficient, m/s
U_{10} = wind speed at 10 m altitude, m/s
U_{curr} = average stream velocity, m/s
M_{H_2O} = molar mass of water, 18.0 g/mol
M = molar mass of chemical, g/mol

Since the value of D_A is sensitive to temperature, K_{H_2O} is also temperature dependent. The relationship is $K_{H_2O(T)} = [T/298]^{1.005}K_{H_2O(298)}$, where $K_{H_2O(T)}$ is the rate constant for water volatilization at T (K) (Hwang, 1982). The range of ambient temperature encountered in the field is not usually wide, and so the temperature corrections to K_{H_2O} are small (on the order of 2%) and are rarely used.

(4) Estimation of C_l

At **steady state** in a **well-mixed system**, such as a lake or disposal impoundment in which volatilization is the only significant removal process, material balance requires that $QC_0 = kAC_1 + QC_1$, where Q (m³/s) is the volumetric flow rate, C_0 (g/m³) is the concentration of organic chemical in influent, and A (m²) is the surface area of the water column. Rearranging this expression, we obtain:

$$C_l = \frac{QC_0}{kA + Q} \qquad (14.27)$$

where C_1 = steady-state chemical concentration, g/m³
       Q   = volumetric flow rate, m³/s
       C_0 = concentration of solute in influent, g/m³
       k   = overall mass-transfer coefficient, m/s
       A   = liquid surface area, m²

Example

Calculate the volatilization rate of benzene from a well-mixed storage impoundment. The concentration of benzene in the influent is assumed to be 10 g/m³. The storage impoundment characteristics used are average values taken from a national survey of storage impoundments (Westat Corp., 1984). The data are given in Table 14.4.

Table 14.4 Average Characteristics of Storage Impoundments

Area, A	1500 m²
Depth, Z	1.8 m
Volume, V	2700 m³
Retention time, t	20 days
Flow, Q	0.00156 m³/s
Temperature, T	25°C
U_{10}	5 m/s
Initial benzene concentration, c_0	10 g/m³
Henry's law constant, H	5.5×10^{-3} m³·atm/g·mol
Diffusivity of benzene (in air), D_a	8.8×10^{-2} cm²/s
Diffusivity of benzene (in water), D_w	9.8×10^{-6} cm²/s
Diffusivity of ether (in water), D_{eth}	8.5×10^{-6} cm²/s
Viscosity of air, μ_G	1.81×10^{-4} g/cm·s
Density of air, ρ_G	1.2×10^{-3} g/cm³

1. Calculate the fetch to depth ratio, F/Z. The fetch is the linear distance across the impoundment. We will use the impoundment's effective diameter, d_{eff}.

$$d_{eff} = 2(A/\pi)^{0.5}$$

$$= 2(1500 \text{ m}^2/\pi)^{0.5} = 44 \text{ m}$$

$$F/Z = d_{eff}/\text{depth}$$

$$= 44 \text{ m}/1.8 \text{ m} = 24$$

2. Calculate the liquid-film mass-transfer coefficient, k_l. Since F/Z is 24 and U_{10} is 5.0 m/s, k_l is calculated using Equation 14.11 given in Table 14.1.

$$k_l = \left[2.605 \times 10^{-9}(F/Z) + 1.277 \times 10^{-7}\right]U_{10}^2 \left(D_W/D_{eth}\right)^{0.67} \text{ m/s}$$

$$= \left[2.605 \times 10^{-9}(24.3) + 1.277 \times 10^{-7}\right](5.0 \text{ m/s})^2 \cdot \left(9.8 \times 10^{-6}/8.5 \times 10^{-6}\right)^{0.67} \text{ m/s}$$

$$= 5.3 \times 10^{-6} \text{ m/s}$$

3. Calculate the gas-film mass-transfer coefficient using Equation 14.22. The Schmidt number (gas side) for benzene in air is calculated first.

$$Sc_G = \mu_G/\rho_G \cdot D_A$$

$$= (1.81 \times 10^{-4} \text{ g/cm} \cdot \text{s})/(1.2 \times 10^{-3} \text{ g/cm}^3) \cdot (8.8 \times 10^{-2} \text{ cm}^2/\text{s})$$

$$= 1.7$$

$$k_g = 1.0 \times 10^{-3} + 46.2 \times 10^{-5}(6.1 + 0.63U_{10})^{0.5} U_{10} Sc_G^{-0.67}$$

$$= 1.0 \times 10^{-3} + 46.2 \times 10^{-5}[6.1 + 0.63(5)]^{0.5} (5) (1.71)^{-0.67}$$

$$= 5.9 \times 10^{-3} \text{ m/s}$$

4. Calculate the air–water partition coefficient, K_{AW}, for benzene at 25°C. The Henry's law constant is given in Table 14.4. The two are related by:

$$K_{AW} = H/R(273+T)$$

$$= (5.5 \times 10^{-3} \text{ m}^3 \cdot \text{atm/g} \cdot \text{mol})/(8.2 \times 10^{-5} \text{ m}^3 \cdot \text{atm/mol} \cdot \text{K}) (273+25 \text{ K})$$

$$= 0.225$$

5. Calculate the overall mass-transfer coefficient, k, using Equation 14.9.

$$1/k = (1/k_1) + (1/K_H \cdot k_g)$$

$$= 1/(5.3 \times 10^{-6}\,\text{m/s}) + 1/[(0.225)(5.9 \times 10^{-3}\,\text{m/s})]$$

$$= 1.9 \times 10^5\,\text{s/m}$$

$$k = 5.3 \times 10^{-6}\,\text{m/s}$$

6. Calculate the steady-state concentration of benzene in the impoundment using the material balance for a well-mixed system with Equation 14.27. The volumetric flow rate, Q, is calculated first.

$$Q = (V/t)$$

$$= (2700\,\text{m}^3)/(20\,\text{days}) \cdot (24\,\text{hr/day}) \cdot (3600\ \text{s/hr})$$

$$= 1.6 \times 10^{-3}\ \text{m}^3/\text{s}$$

$$C_1 = QC_0/(kA + Q)$$

$$= (1.6 \times 10^{-3}\ \text{m}^3/\text{s}) \cdot (10\ \text{g/m}^3)/[(5.3 \times 10^{-6}\ \text{m/s}) \cdot 1500\ \text{m}^2 + 1.6 \times 10^{-3}\ \text{m}^3/\text{s}]$$

$$= 1.7\ \text{g/m}^3$$

7. Calculate the emission rate with Equation 14.8.

$$E = kAC_1$$

$$= (5.3 \times 10^{-6}\ \text{m/s}) \cdot (1500\ \text{m}^2) \cdot (1.7\ \text{g/m}^3)$$

$$= 1.4 \times 10^{-2}\ \text{g/s}$$

14.3.2 Volatilization from a Plug-Flow System

Plug flow occurs in tanks and impoundments that are designed to prevent back-mixing, as well as in streams and rivers. The concentration of solute in water decreases with time and distance downstream as the chemical volatilizes. The ratio $C_1(t)/C_1(0)$, the fraction of chemical remaining in the plug at time t, was given in Equation 14.7 as $e^{-kt/z}$. The fraction of chemical volatilized at time t is $1 - C_1(t)/C_1(0) = 1 - e^{-kt/z}$, and the average emission rate is

$$E = \left(1 - e^{-kt/Z}\right) QC_l(0) \tag{14.28}$$

where E = emission rate from liquid surface, g/s
   k = overall mass-transfer coefficient, m/s
   t = residence time, s
   Z = average liquid depth, m
   Q = volumetric flow rate, m³/s
   $C_l(0)$ = initial solute concentration, g/m³

In applying Equation 14.28, k is evaluated in the same manner as for well-mixed systems.

The residence time of volatile solute in the stream equals X/U_{curr}, where U_{curr} is the average stream velocity and X is the downstream transport distance. The aqueous concentration of chemical at any point downstream of the source, C_X, is

$$C_X = C_0 e^{-\frac{kX}{U_{curr}Z}} \tag{14.29}$$

where C_X = concentration at downstream distance x, m
   C_0 = influent chemical concentration, g/m³
   X = downstream distance from the source, m
   k = overall mass-transfer coefficient, m/s
   U_{curr} = average stream velocity, m/s
   Z = average stream depth, m

Example

Estimate the volatilization rate and residual concentration of methylene chloride in a contaminated stream using the parameters given in Table 14.5.

The average discharge of methylene chloride from contaminated groundwater into a stream is 4.4 g/day. The flow rate of the stream is (5.0 m)(0.40 m)(0.30 m/s) (86,400 s/day) = 5.2×10^4 m³/day. After complete mixing, the initial methylene chloride concentration, C_0, is (4.4 g/day)/(5.2×10^4 m³/day) = 8.5×10^{-5} g/m³. We will calculate the rate at which methylene chloride volatilizes from the first 1600-m stretch of stream below the mixing zone assuming that there is no dilution from downstream tributaries.

1. Calculate the liquid-film mass-transfer coefficient using Equation 14.21.

$$k_1 = \left(D_w U_{curr}/Z\right)^{1/2} \text{ m/s}$$

$$= \left[\left(1.17 \times 10^{-9} \text{ m}^2/s\right)\left(0.030 \text{ m/s}\right)/\left(0.40 \text{ m}\right)\right]^{1/2} \text{ m/s}$$

$$= 9.4 \times 10^{-6} \text{ m/s}$$

Table 14.5   Characteristics of a Contaminated Stream

CH_2Cl_2 discharge rate = 4.4 g/day
$D_w = 1.17 \times 10^{-9}$ m²/s
$D_A = 1.01 \times 10^{-5}$ m²/s
H = 3.0×10^{-3} m³·atm/mol
Average stream width, W = 5.0 m
Average stream depth, Z = 0.40 m
$U_{curr} = 0.30$ m/s
$U_{10} = 5$ m/s
Ambient temperature, T = 20°C (293 K)

2. Calculate the gas-film mass-transfer coefficient using Equation 14.25.

$$k_g = \left(0.003 + 0.002 U_{10}\right)\left(D_{H_2O}/D_A\right)^{0.67} \text{ m/s}$$

$$= \left(0.003 + 0.002 \times 5 \text{ m/s}\right)\left(2.6 \times 10^{-5}/1.01 \times 10^{-5}\right)^{0.67} \text{ m/s}$$

$$= 2.5 \times 10^{-2} \text{ m/s}$$

3. Calculate the air–water partition coefficient of methylene chloride, K_{AW}, at 20°C. The Henry's law constant, H, is 3.0×10^{-3} m³·atm/mol.

$$K_{AW} = H/R\left(273 + T\right)$$

$$= \left(3.0 \times 10^{-3} \text{ m}^3 \cdot \text{atm/mol}\right)\Big/\left[\left(8.2 \times 10^{-5} \text{ m}^3 \cdot \text{atm/g} \cdot \text{mol} \cdot \text{K}\right)\left(273 + 20 \text{ K}\right)\right]$$

$$= 0.12$$

4. Calculate the overall mass-transfer coefficient using Equation 14.9.

$$1/k = \left(1/k_1\right) + \left(1/K_{AW} \cdot k_g\right)$$

$$= 1/\left(9.4 \times 10^{-6} \text{ m/s}\right) + 1/\left(0.12\right)\left(2.5 \times 10^{-2} \text{ m/s}\right)$$

$$= 1.1 \times 10^5 \text{ s/m}$$

$$= 9.4 \times 10^{-6} \text{ m/s}$$

5. Estimate the emission rate along the 1600-m length of stream. The volume, V, is (1600)(5.0)(0.40) = 3200 m³. First, calculate the volumetric flow rate, Q, and then the residence time, t.

$$Q = U_{curr} \cdot W \cdot Z \ \ m^3/s$$

$$= (0.030 \ m/s)(5.0 \ m)(0.40 \ m) = 0.60 \ m^3/s$$

$$t = V/Q \ s$$

$$= (3200 \ m^3)/(0.60 \ m^3/s) = 5.3 \times 10^3 \ s$$

$$kt/Z = (9.4 \times 10^{-6} \ m/s)(5.3 \times 10^3 \ s)/(0.40 \ m) = 0.13$$

$$E = (1 - e^{-kt/Z})QC_1(0)$$

$$= (1 - e^{-0.13})(0.60 \ m^3/s)(8.5 \times 10^{-5} \ g/m^3) \ g/s$$

$$= 6.0 \times 10^{-6} \ g/s$$

6. The aqueous methylene chloride concentration at a downstream distance of 1600 m from the source is estimated with Equation 14.29.

$$kX/U_{curr}Z = (9.410^{-6} \ m/s)(1600 \ m)/(0.30 \ m/s)(0.40 \ m)$$

$$= 0.13$$

$$C_X = C_0 \ e^{(-Kx/U_{curr}Z)} \ g/m^3$$

$$= (8.5 \times 10^{-5} \ g/m^3) e^{-0.13} = 7.5 \times 10^{-5} \ g/m^3$$

14.3.3 Volatilization from Mechanically Aerated Aquatic Systems

Numerical calculations suggest that hydraulic controls (dams, weirs, rapids, and waterfalls) can be significant point sources of volatile chemicals. In fact, volatilization of chemicals from hydraulic controls may be the dominant pathway into the air column. McLachlan et al. (1990) presented a two-film model which produces order-of-magnitude estimates of the volatilization of organic chemicals from waterfalls. The model predicts that the extent of volatilization is most sensitive to and directly proportional to the air–water partition coefficient and to the flow rate of oxygen into the plunge pool. Compounds such as chloroform, chlorinated benzenes, polychlorinated biphenyls, and mirex volatilized efficiently at hydraulic controls, but not compounds like 2,3,7,8-TCDD, γ-HCH, and large polycyclic aromatic hydrocarbons. The model was not validated.

Cirpka et al. (1993) studied the volatilization of SF$_6$, 1,1,2-trichlorotrifluoro-ethane, trichloroethylene, trichloromethane, and tribromomethane (compounds that exhibit a wide range of air–water partition coefficients) at river cascades. They showed that where entrained air bubbles are produced in large quantities, the two-film model is not sufficient to describe gas exchange for compounds with large air–water partition coefficients ($K_{AW} > 0.3$). A model was presented that combined the two-film model for gas exchange at the free surface of the water column with gas exchange in entrained air bubbles.

Methods of estimating k_l for mechanically aerated ponds, based on studies of volatilization in pulp and paper mill aeration basins, were described by Hwang (1982). Reported methods for estimating k_g for mechanically aerated ponds are based on studies of the volatilization of ammonia from aqueous sulfuric acid solutions (GCA Corp., 1985).

14.4 MODEL SENSITIVITY AND METHOD ERROR

The transfer of chemicals between water and air has been extensively studied in the laboratory and in the field. Several models for the process have been proposed. All produce comparable results when predicting the behavior of highly volatile chemicals. The choice of model does make a difference when estimating the properties of semivolatile chemicals, however.

The sensitivity of the volatilization model depends on the magnitude of the air–water partition coefficient. If the chemical is hydrophilic and k_l is much larger than $k_g \cdot K_{AW}$, the system is "gas-phase limited," and model estimates will be sensitive to error in the values of k_g and K_{AW}, primarily. However, biotic and abiotic processes other than volatilization contribute to removal of semivolatile, hydrophilic solutes from the water column. Chemicals may also leave the system with effluent or seepage. Accurate mass transfer coefficients may not be available for these processes.

Temperature corrections to k_l and k_g are described above. While they do not greatly change emission estimates for volatile solutes, they can significantly alter emission estimates for semivolatile solutes. Since the temperature dependence of K_{AW} is relatively large, it is important to use temperature-corrected values.

Diffusivities in air and water span a narrow range of values near 1×10^{-5} and 1×10^{-9} m²/s, respectively. Regardless of the nature of the solute, estimates of k_l, k_g, and E are not very sensitive to the choice of diffusivity.

Some variation is found in the literature value of any chemical property reported by more than one laboratory, and significant error can be entailed in estimating the Henry's law constant from published data on saturation vapor pressure and solubility. For instance, Suntio et al. (1987) report that published values of the saturation vapor pressure of lindane vary by three orders of magnitude and reported solubility values vary by one order of magnitude. Depending on which data are used, estimates of Henry's law constants for lindane could vary by four orders of magnitude. Henry's law constants for polychlorinated biphenyls and polycyclic aromatic hydrocarbons calculated this way are reported to be no better than order-of-magnitude estimates (Baker and Eisenreich, 1990).

Our models do not account for the spatial and temporal variability of natural aquatic systems. In deriving them, we assume that aqueous solutions of chemical are homogeneous except in the interfacial zones. Obviously, model estimates will be in error if the chemical is not uniformly distributed in the water column. The problem is particularly acute when applying plug-flow models to volatilization from streams and rivers. Influent can travel large distances before complete mixing takes place, and mixing zones can extend for hundreds of meters. In this case, estimates of short-term emission rates may be greatly in error. In view of all this, volatilization estimates made with the methods described here may agree with observation by no more than a factor of two (Thomas, 1990).

The two-film volatilization model may be modified to include additional controls of contaminant transport in the system. For example, volatilization is inhibited by the presence of a surfactant or oil film at the surface.* Most toxic chemicals on state and federal priority lists are nonionic and so miscible with oils. A rigorous model of such a system would include resistance to mass transfer at the oil film, requiring estimates of mass transfer coefficients and equilibrium partitioning at the water–oil and oil–air interface. Mass-transfer and partition coefficients for the oil phase are difficult to estimate, however. In their absence, it is convenient to assume that volatilization is controlled by the gas-phase resistance at the oil layer, and $k = k_g \cdot K_{EQ}$, where K_{EQ} is the gas–oil partition coefficient.

Henry's law is not applicable to gas–liquid partitioning of organic chemicals in oil films. Raoult's law is used to evaluate K_{EQ}. We saw in Chapter 6 that, for a miscible component of an oil film,

$$K_{EQ} = \frac{P_j^S \rho_a M_l}{P_0 \rho_l M_a}$$
(14.30)

where K_{EQ} = gas–liquid partition coefficient,
  P_j^S = saturation vapor pressure of component j, Pa
  ρ_a = air density, g/m^3
  M_l = average molar mass of oil, g/mol
  P_0 = total pressure, Pa
  ρ_l = oil density, g/m^3
  M_a = molar mass of air, 28.8 g/mol

As noted earlier, vertical eddy diffusion is highly dependent on velocity and geometry of the body of water and is inhibited by thermal stratification in lakes and salinity gradients in estuaries. Resistance to the vertical transport of chemical in the bulk of the water column may then contribute to overall resistance to volatilization. In this case, terms of the form $1/k_e$ may be added to the expression for $1/k$ in Equation 14.9, where k_e is the coefficient of mass transfer due to eddy diffusion in a layer of the water column. For deep bodies of water, the coefficient of mass transfer due to eddy diffusion may be roughly estimated from its relationship to aquatic turbulent diffusivity (Thomas, 1990),

$$k_e = 1.3 \frac{D_e}{Z}$$
(14.31)

* Oxygen reaeration and CO_2 exchange are markedly inhibited by oil films thicker than about 10 μm (Downing and Truesdale, 1955; Broecker et al., 1978).

where k_e = eddy diffusion mass-transfer coefficient, m/s
   D_e = aquatic turbulent diffusivity, m²/s
   Z = depth, m

Typical values of vertical turbulent diffusivity in lakes and oceans were compiled by Schwarzenbach et al. (1993). In the mixed layer of oceans and lakes, D_e varies from 10^{-5} m²/s during calm conditions to 1 m²/s during storms. In deep ocean water, D_e varies from 10^{-4} to 10^{-2} m²/s, while in deep lake water, it ranges from 10^{-7} to 10^{-5} m²/s.

REFERENCES

Baker, J. E. and Eisenreich., S. J., 1990. Concentrations and fluxes of polycyclic aromatic hydrocarbons and polychlorinated biphenyls across the air–water interface of Lake Superior. *Environ. Sci. Technol.*, **24**, 342–352.

Broecker, H. C., Petermann, J., and Seims, W., 1978. The influence of wind on CO₂ — exchange in a wind–wave tunnel, including the effects of monolayers. *J. Mar. Res.*, **36**, 595–610.

Broecker, W. S. and Peng, T. H., 1984. Gas measurements in natural systems. In *Gas Transfer at Water Surfaces*. Brutsaert, W. H., and Jirka, G. H., Eds. D. Reidel Publishing, Dordrecht, Holland, pp. 479–493.

Cirpka, O., Reichert, P., Wanner, O., Muller, S. R., and Schwarzenbach, R. P., 1993. Gas exchange at river cascades: field experiments and model calculations. *Environ. Sci. Technol.*, **27**, 2086–2097.

Cohen, Y., Cocchio, W., and Mackay, D., 1978. Laboratory study of liquid–phase controlled volatilization rates in presence of wind waves. *Environ. Sci. Technol.*, **12**, 553–558.

Danckwerts, P. V., 1951. Significance of liquid–film coefficients in gas adsorption. *Ind. Eng. Chem.*, **43**, 1460–1467.

Dickey, T. D., Hartman, B., Hammond, D., and Hurst, E., 1984. A laboratory technique for investigating the relationship between gas transfer and fluid turbulence. In *Gas Transfer at Water Surfaces*. Brutsaert, W. H. and Jirka, G. H., Eds. D. Reidel Publishing, Dordrecht, Holland, pp. 93–100.

Dobbins, W. E., 1964. BOD and oxygen relationships in streams. *J. Sanit. Eng. Div. ASCE*, **90**, 53–78.

Downing, A.L. and Truesdale, G.A., 1955. Some factors affecting the rate of solution of oxygen in water. *J. Appl. Chem.*, **5**, 570–581.

Duran, A. P. and Hemond, H. F., 1984. Dichlorodifluoromethane (Freon–12) as a tracer for nitrous oxide release from a nitrogen-enriched river. In *Gas Transfer at Water Surfaces*. Brutsaert, W. and Jirka, G. H., Eds. D. Reidel Publishing, Dordrecht, Holland, pp. 421–429.

Elmore, H. L. and West, W. F., 1961. Effect of water temperature on stream reaeration. *J. Sanit. Eng. Div., Amer. Soc. Civ. Eng.*, **87**, 59–71.

Emerson, S., 1975. Gas exchange in small Canadian shield lakes. *Limnol. Oceanogr.*, **20**, 754–761.

Frexes, P., Jirka, G. H., and Brutsaert, W., 1984. Examination of recent field data on stream reaeration. *J. Env. Eng. ASCE*, **110**, 1179–1183.

GCA Corp., 1985. *Emissions Data and Model Review for Wastewater Treatment Operations*. Contract No. 68-01-6871, Assignment 49. Draft Technical Note. Prepared for the U.S. Environmental Protection Agency, Washington, D.C.

Hunt, J. C. R., 1984. Turbulence structure and turbulent diffusion near gas–liquid interfaces. In *Gas Transfer at Water Surfaces*. Brutsaert, W. and Jirka, G. H., Eds., D. Reidel Publishing, Dordrecht, Holland, pp. 67–82.

Hwang, S. T., 1982. Toxic emissions from land disposal facilities. *Environ. Prog.*, 1, 46–52.

Jahne, B., Huber, W., Dutzi, A., Wais, T., and Ilmberger, J., 1984. Wind/wave tunnel experiment on the Schmidt number — and wave field dependence of air/water gas exchange. In *Gas Transfer at Water Surfaces*. Brutsaert, W. and Jirka, G. H., Eds. D. Reidel Publishing, Dordrecht, Holland, pp. 303–309.

Jury, W.A., Spencer, W.F., and Farmer, W.J., 1983. Behavior and assessment model for trace organics in soil. I model description. *J. Environ. Qual.*, 12, 558–564.

Ledwell, J.J., 1984. The variation of the gas transfer coefficient with molecular diffusivity. In *Gas Transfer at Water Surfaces*. Brutsaert, W. and Jirka, G. H., Eds. D. Reidel Publishing, Dordrecht, Holland, pp. 293–302.

Liss, P. S. and Slater, P. G., 1974. Flux of gases across the air–sea interface. *Nature*, 247, 181–184.

Liss, P. S. and Merlivat, L., 1986. Air–sea gas exchange: introduction and synthesis. In *The Role of Air–Sea Exchange in Geochemical Cycling*. Volume 185. Buat-Menard, P., Ed. NATO ASI Series C, D. Reidel Publishing, Dordrecht, Holland, pp. 113 -127.

Lunney, P. D., Springer, C., and Thibodeaux, L. J., 1985. Liquid-phase mass transfer coefficients for surface impoundments. *Environ. Prog.*, 4, 203–211.

Mackay, D., 1981. Environmental and laboratory rates of volatilization of toxic chemicals from water. In *Hazard Assessment of Chemicals: Current Developments*. Volume 1. Saxena, J. and Fischer, F., Eds. Academic Press, New York, pp. 303–322.

Mackay, D. and Leinonen, P. J., 1975. The rate of evaporation of low solubility contaminants from water bodies. *Environ. Sci. Technol.*, 9, 1178–1180.

Mackay, D. and Yeun, A., 1983. Mass transfer coefficient correlations for volatilization of organic solutes from water. *Environ. Sci. Technol.*, 17, 211–217.

McLachlan, M., Mackay, D., and Jones, P. H., 1990. A conceptual model of organic chemical volatilization at waterfalls. *Environ. Sci. Technol.*, 24, 252–257.

Neely, W. B., Blau, G. E., and Alfrey, T., Jr., 1976. Mathematical models predict concentration–time profiles resulting from chemical spill in a river. *Environ. Sci. Technol.*, 10, 72–76.

O'Conner, D. J. and Dobbins, W. E., 1958. Mechanism of reaeration in natural streams. *Trans. Amer. Soc. Civ. Eng.*, 123, 641–666.

Rathbun, R. E., 1990. Prediction of stream volatilization coefficients. *J. Environ. Eng.*, 116, 615–631.

Rathbun, R. E. and Tai, D. Y., 1982. Volatilization of organic compounds from streams. *J. Environ. Eng. Div. ASCE*, 108, 973–989.

Schwarzenbach, R. P., Molnar-Kubica, E., Giger, W., and Wakeham, S. G., 1979. Distribution, residence time, and fluxes of tetrachloroethylene and 1,4-dichlorobenzene in Lake Zurich, Switzerland. *Environ. Sci. Technol.*, 13, 1367–1373.

Schwarzenbach, R. P., Gschwend, P. M., and Imboden, D. M., 1993. The Gas–Liquid Interface: Air–Water Exchange. Chapter 10. In *Environmental Organic Chemistry*. John Wiley & Sons, New York.

Smith, J. H., Bomberger, D. C., Jr., and Haynes, D. L., 1980. Prediction of the volatilization rates of high-volatility chemicals from natural water bodies. *Environ. Sci. Technol.*, 14, 1332–1337.

Smith, J. H., Bomberger, D. C., and Haynes, D. L., 1981. Volatilization rates of intermediate and low volatility chemicals from water. *Environ. Sci. Technol.*, 10, 281–289.

Spencer, W.F., Cliath, M.M., Jury, W.A., and Zhang, L.-Z., 1988. Volatilization of organic chemicals from soil as related to their Henry's law constants. *J. Environ. Qual.,* **17**, 504–509.

St. John, J. P., Gallagher, T. W., and Paquin, P. R., 1993. The sensitivity of the dissolved oxygen balance to predictive reaeration equations. In *Gas Transfer at Water Surfaces.* Brutsaert, W. and Jirka, G. H., Eds. D. Reidel Publishing, Dordrecht, Holland, pp. 577–588.

Suntio, L. R., Shiu, W. Y., Mackay, D., Sieber J. N., and Glotfelty, D., 1987. Critical review of Henry's law constants for pesticides. *Rev. Environ. Contam. Toxicol.,* **103**, 3–59.

Thomann, R. V. and Mueller, J. A., 1987. *Principles of Surface Water Quality Modeling and Control.* Chapters 6 and 8. Harper & Row, New York.

Thomas, R. G., 1990. Volatilization from Water. Chapter 15. In *Handbook of Chemical Property Estimation Methods. Behavior of Organic Compounds.* Lyman, W. J., Reehl, W. F., and Rosenblatt, D. H., Eds. American Chemical Society, Washington, D.C.

Torgersen, T., Mathieu, G., Hesslein, R. H., and Broeker, W. S., 1982. *J. Geophys. Res.,* **87**, 546–556.

Upstill-Goddard, R. C., Watson, A. J., Liss, P. S., and Liddicoat, M. I., 1990. *Tellus,* **42B**, 364–377.

U.S. EPA, 1989. *Hazardous Waste Treatment, Storage, and Disposal Facilities (TSDF) — Air Emission Models.* Chapter 4. Publication No. EPA-450/3-87-026. November 1989. U.S. Environmental Protection Agency, Office of Air and Radiation, Office of Air Quality Planning and Standards, Research Triangle Park, NC.

Wanninkhof, R., Ledwell, J. R., and Broecker, W. S., 1985. Gas exchange–wind speed relation measured with sulfur hexafluoride on a lake. *Science,* **227**, 1224–1226.

Wanninkhof, R., Ledwell, J. R., and Broecker, W. S., 1987. Gas exchange on Mono and Crowley Lake, California. *J. Geophys. Res.,* **92**, 14567–14580.

Watson, A. J., Upstill-Goddard, R. C., and Liss, P. S., 1991. Air–sea gas exchange in rough and stormy seas measured by a dual-tracer technique. *Nature,* **349**, 145–147.

Westat Corp., 1984. *National Survey of Hazardous Waste Generators and TSDFs Regulated Under RCRA in 1981.* Contract No. 68-01-6861, April 1984. Prepared for the U.S. Environmental Protection Agency, Washington, D.C.

Whitman, W.G., 1923. The two-film theory of gas absorption. *Chem. Metal. Eng.,* **29**, 146–148.

Wilcock, R. J., 1984. Reaction studies of some New Zealand rivers using methyl chloride as a gas tracer. In *Gas Transfer at Water Surfaces.* Brutsaert, W. and Jirka, G. H., Eds. D. Reidel Publishing, Dordrecht, Holland, pp. 413–420.

APPENDIX 1

Fundamental and Defined Constants

Constant	Symbol	Value
Gas constant	R	8.31451 J K^{-1} mol^{-1}
		8.2058×10^{-2} l atm K^{-1} mol^{-1}
		62.364 l torr K^{-1} mol^{-1}
		1.9872 cal K^{-1} mol^{-1}
Ideal gas molar volume, 273.15 K, 1 atm		22.414 l mol^{-1}
Avogadro constant	N_A, L	6.02214×10^{23} mol^{-1}
Ice point of air-saturated water	0°C	273.15 K
Boltzmann constant	k, k_B	1.38066×10^{-23} J K^{-1}
Faraday constant	$\mathscr{F}$	9.64853×10^{4} C mol^{-1}
(RT/$\mathscr{F}$)ln 10, 298.15 K		59.16 mV
Plank constant	h	6.62608×10^{-34} J s
Speed of light in vacuum	c	2.99792458×10^{8} m s^{-1}
Atomic mass unit	amu	1.66054×10^{-27} kg
Elementary charge	e	1.602177×10^{-19} C
Vacuum permeability	μ_0	$4\pi \times 10^{-7}$ N A^{-2}
Vacuum permittivity	ε_0	$8.8541878 \times 10^{-12}$ F m^{-1}
Standard gravitational acceleration	g	9.80665 m s^{-2}

APPENDIX 2

Units of Measure

A2.1 THE INTERNATIONAL SYSTEM OF UNITS

Measurements and estimates of physical and chemical quantities are always reported as products of a numerical value and a unit of measure. The SI system of units, *Le Système International d'Unites*, was proposed to promote uniformity and reduce confusion in reporting technical information. It is now used globally and should be used to report chemical property estimates.

There are seven base SI units from which all other SI units are derived. The base units are listed in Table A2.1.

The base SI units can be multiplied, divided, and canceled just like algebraic quantities in order to produce derived SI units, to simplify complex mathematical relationships, and to perform dimensional analyses. No conversion factors are needed. Some useful derived units are listed in Table A2.2.

In modeling chemical distribution and fate, the units of area, volume, and concentration are of particular importance. The SI unit of area is the m^2. The hectare (ha) equals 10^4 m^2 and is sometimes more convenient to use. Also, the SI unit of volume is the m^3, but the liter (L), which is 10^{-3} m^3, is firmly entrenched by tradition, and its use is almost universal.

Many different units are used to report chemical density and concentration. Density is best reported in units of kg/m^3. The preferred units of concentration are mol/m^3, kg/m^3, and g/m^3, but mol/L and g/L are widely used. Units of mass-per-volume concentration of solute in water, such as parts per million (ppm), parts per billion (ppb), and parts per trillion (ppt), are not recommended. In using these, the solution density is assumed to be $1000 kg/m^3$, which is rarely the case. It is better to report mass-per-volume concentration in units of mg/L, μg/L, and ng/L.

Units of volume-per-volume concentration of vapor in air, such as parts per thousand, parts per million, parts per billion, and parts per trillion, are commonly used. This can cause confusion, however. The parts per billion unit is particularly troublesome. One billion is variously defined as 10^9 in North America and 10^{12} in Europe. Also, the abbreviation of parts per thousand, ppt, is easily confused with

313

Table A2.1 The Base SI Units

Quantity	Unit	Symbol
Length	Meter (m)	l
Mass	Kilogram (kg)	m
Time	Second (s)	t
Electric current	Ampere (A)	l
Temperature	Kelvin (K)	T
Quantity	Mole (mol)	n
Luminous intensity	Candela (cd)	l_v

Table A2.2 Some Derived SI Units

Derived Quantity	Name	Derived Unit
Force	Newton (N)	$kg\ m\ s^{-2}$
Energy, work, heat	Joule (J)	$kg\ m^2\ s^{-2}$, N m, Pa m^3
Power	Watt (W)	$kg\ m^2\ s^{-3}$, J s^{-1}
Pressure, stress	Pascal (Pa)	$kg\ m^{-1}\ s^{-2}$, N m^{-2}
Molar mass[a]		$kg\ mol^{-1}$
Electric charge	Coulomb (C)	A s
Electric potential	Volt (V)	$kg\ m^2\ s^{-3}\ A^{-1}$, J C^{-1}
Electric resistance	Ohm (Ω)	$kg\ m^2\ s^{-3}\ A^{-2}$, V A^{-1}
Electric capacitance	Farad (F)	$kg^{-1}\ m^{-2}\ s^4\ A^2$, C V^{-1}
Electric conductance	Siemens (S)	$kg^{-1}\ m^{-2}\ s^3\ A^2$, Ω^{-1}
Magnetic flux density	Tesla (T)	$kg\ s^{-2}\ A^{-1}$, V s m^{-2}

[a] The mass of 1 mol (6.022×10^{23} units) of substance is called the molar mass and is often given in g/mol. The unit was formerly called the molecular weight. It is less commonly called the molecular mass.

Table A2.3 SI Decimal Prefixes

Prefix	Factor	Name	Prefix	Factor	Name
d	10^{-1}	Deci[a]	da	10	Deca[a]
c	10^{-2}	Centi[a]	h	10^2	Hecto[a]
m	10^{-3}	Milli	k	10^3	Kilo
μ	10^{-6}	Micro	M	10^6	Mega
n	10^{-9}	Nano	G	10^9	Giga
p	10^{-12}	Pico	T	10^{12}	Tera
f	10^{-15}	Femto	p	10^{15}	Peta
a	10^{-18}	Atto	E	10^{18}	Exa

[a] The prefixes deka, hecto, deci, and centi are used only with units of length, area, and volume.

that of parts per trillion. Accordingly, units of volume-per-volume concentration should be clearly defined when they are used.

The SI system employs metric units and is a decimal system. Both base and derived units may be modified with a prefix that denotes a power of ten. The decimal prefixes are listed in Table A2.3.

Mathematical operations apply to both the prefix and the unit it modifies. This is illustrated in the example below.

Table A2.4 Some Non-SI Units and Factors for Conversion to SI Units

Quantity	Unit	Symbol	SI Equivalent
Length	Ångstrom	Å	10^{-10} m
	Inch	in	0.0254 m
	Foot	ft	0.3048 m
	Mile	mi	1609.344 m
Volume	Liter	L	0.001 m³, 1 dm³
Mass	Atomic mass unit	amu	1.66054×10^{-27} kg
	Pound	lb	0.45359 kg
	Metric tonne	t	10^3 kg
Force	Dyne	dyn	10^{-5} N
Pressure	Bar		10^5 Pa
	Atmosphere	atm	1.01325×10^5 Pa
	Mm of mercury	torr mmHg	133.322 Pa
	Pound per square inch	psi	6.89473×10^3 Pa
Energy	Erg	erg	10^{-7} J
	Calorie	cal_{th}	4.184 J
	British thermal unit	BTU	1055 J
	Electronvolt	eV	1.602178×10^{-19} J
Viscosity	Poise	P	0.1 kg m⁻¹ s⁻¹
Dipole moment	Debye	D	3.335641×10^{-30} C m
Magnetic flux density	Gauss	G	10^{-4} T

Example

Convert one cubic centimeter to cubic meters.

$$1 \ cm^3 = \left(1 \times 10^{-2} \ m\right)^3 = 1 \times 10^{-6} \ m^3$$

A2.2 NON-SI UNITS

Non-SI units are frequently encountered in reports, and many are still widely used. For instance, units of time such as minutes (min), hours (hr), days (d), and years (yr) are often more convenient to use than the second. Some other non-SI units that are employed for chemical property estimates are listed in Table A2.4 along with factors for converting them to SI units. Many of the conversion factors are based on values of the fundamental constants listed in Appendix 1 and must be revised as the fundamental constants are revised.

The degree Celsius (°C) is often used as the unit of temperature. If temperature on the Kelvin scale is denoted by T and temperature on the Celsius scale is denoted by θ, the two scales are related by the equation:

$$\frac{T}{K} = \frac{\theta}{°C} + 273.15 \qquad\qquad (A2.1)$$

In the U.S., it is popular to report ambient temperature on the Fahrenheit scale (°F). If temperature on the Fahrenheit scale is denoted by ϕ, the Fahrenheit and Celsius scales are related by the equation

$$\frac{\theta}{°C} = \left(\frac{5}{9}\right)\frac{\phi}{°F} - 32 \qquad\qquad (A2.2)$$

A2.3 DIMENSIONAL ANALYSIS

When estimating chemical properties, it is useful to perform a dimensional analysis in order to verify that the equations and conversion factors are correct. The following example of dimensional analysis relates one unit of pressure, the atmosphere, to another unit in common use, the bar.

Example

Find the relationship between the standard atmosphere and the bar.

A 760-mm column of mercury exerts a pressure of 1 atm at its base. The relationship between pressure, column height, and fluid density of mercury is pressure = (column height)(column density)(g), where g is the gravitational acceleration.

$$1 \text{ atm} = (0.7600 \text{ m})\left(13.5951 \times 10^3 \text{ kg/m}^3\right)\left(9.80665 \text{ m/s}^2\right)$$

$$= 1.01325 \times 10^5 \left(\text{kg}\cdot\text{m/s}^2\right)\left(1/\text{m}^2\right)$$

$$= 1.01325 \times 10^5 \text{ N/m}^2$$

$$= 1.01325 \times 10^5 \text{ Pa}$$

$$= 1.01325 \text{ bar}$$

REFERENCES

Homann, K. H., 1987. Abbreviated list of Quantities, Units and Symbols in Physical Chemistry. International Union of Pure and Applied Chemistry, Physical Chemistry Division, Commission on Physicochemical Symbols, Terminology and Units. Seacourt Press, Oxford, England.

Lide, D. R., Ed., 1993. *CRC Handbook of Physics and Chemistry.* 74th ed. CRC Press, Boca Raton, FL.

Uncertainty of the Chemical Property Estimate

Information of unknown reliability is as useless as no information at all. Accordingly, the probable error in a property estimate is of critical interest and should be reported correctly. This section presents statistical concepts and techniques used to evaluate and report uncertainty in chemical property estimates. Uncertainty in regression model estimates is discussed in Appendix 4. Calculus is reviewed in Appendix 7. Additional information on error analysis can be found in books by Bennet and Franklin (1954), by Dixon and Massey, Jr. (1969), by Davies and Goldsmith (1972), and by Berthouex and Brown (1994).

A3.1 ACCURACY AND PRECISION

The total uncertainty in the magnitude of a quantity, q, is determined by the accuracy and precision of the data used to derive the value. What do the terms accuracy and precision mean?

Accuracy is *a measure of the degree to which a quantity approximates its true or accepted value*, q_{true}. If μ is the arithmetic mean value of q, the absolute error is $\mu - q_{true}$ and the relative error is $(\mu - q_{true})/q_{true}$. Uncertainty is increased and accuracy is reduced by determinate errors (bias) such as method errors and personal mistakes. Accuracy is evaluated with standard samples and literature values, but since q_{true} is usually unknown, it is hard to gauge with any certainty.

Precision is *a measure of the reproducibility of a result as indicated by the spread of the estimated values of q about the mean value*. In making measurements, uncertainty is increased and precision is reduced by indeterminate (random) error due to instrument noise and random fluctuations in environmental conditions. Precision can be estimated with statistical methods when sufficient data are available and when a normal (Gaussian) distribution of individual values of q is observed.

The **population standard deviation**, σ, is *the statistic that describes the spread of the distribution of all possible estimates of a single value of q*. The familiar

Gaussian (normal) probability curve is described by the distribution function, G(q), of individual values of q, q_i, produced by random deviations from the mean value μ. The distribution function is

$$G(q) = \left(\frac{1}{\sigma\sqrt{2\pi}} \right) e^{\frac{-(q_i-\mu)^2}{2\sigma^2}} \qquad (A3.1)$$

In a Gaussian distribution, 68.3% of the individual q_i lies within $\pm1\sigma$ of the mean, 95.5% lies within $\pm2\sigma$ of the mean, and 99.7% lies within $\pm3\sigma$ of the mean.

A3.2 SIGNIFICANT FIGURES AND ROUNDING ESTIMATED VALUES

The number of **significant figures** in a measured or derived result, the number of digits in the reported value, indicates the precision of the result. Therefore, the least significant digit of the result should be of the same order of magnitude as σ, the uncertainty in the result. This may require rounding the estimated value of the chemical property to the correct number of digits.

In scientific notation, all digits are significant except leading zeros. Trailing zeros to the right of a decimal point are always significant regardless of notation. For example, the values 2.270 and 0.02270 both have four significant figures. If scientific notation is not used, trailing zeros to the left of a decimal point may not be significant. For example, it isn't clear whether the value 5000 has one, two, three, or four significant figures. In proper scientific notation, the number 5000 might denote either 5.000×10^3, 5.00×10^3, 5.0×10^3, or 5×10^3. Such needless uncertainty should be avoided.

The following rules for rounding to the correct number of digits are intended to avoid introducing bias in reporting results. If the left-most digits to be dropped are greater than 5, round up to the correct number of significant digits. If they are less than 5, round down. If the digits to be dropped are exactly 5, round to an even number by either dropping the digits or rounding up.

Example

Round the following estimates of log K_{ow} for benzene at 25°C to three significant figures.

Round 2.2751 to 2.28

2.2743 to 2.27

2.2650 to 2.26

2.2750 to 2.28

When multiplying or dividing quantities, the uncertainty of the calculated result is as great as the uncertainty of the least certain input value. The calculated result should have the same number of significant figures as the input value with the least number of significant figures.

Example

Calculate the value of trichloroethylene's air–water partition coefficient from its Henry's law constant at 25°C, 1.18×10^4 Pa m^3 mol^{-1}.
The relationship between the air–water partition coefficient, K_{AW}, and Henry's law constant, H, is

$$K_{AW} = H/R \cdot T$$

where R is the gas constant and T(K) is the temperature.

$$K_{AW} = \left(1.18 \times 10^4 \ \text{Pa m}^3 \ \text{mol}^{-1}\right) / \left(8.3145 \ \text{m}^3 \ \text{Pa mol}^{-1} \ \text{K}^{-1}\right) (298.15 \ \text{K})$$

$$= 0.478$$

When adding or subtracting quantities, the uncertainty of the calculated result is as great as the uncertainty of the input value with the least number of decimal places. Therefore, the derived quantity should have as many decimal places as the input value with the least number of decimal places.

Example

Estimate the solubility of 2-chloroethyl ether in water at 25°C from its octanol–water partition coefficient. For 2-chloroethyl ether, $K_{ow} = 1.12$. Solubility, S, and octanol–water partition coefficient, K_{OW}, are related by

$$\log(1/S) = 1.214 \log K_{OW} - 0.850$$

$$= 1.214 \ (1.12) - 0.850$$

$$= 1.36 - 0.850$$

$$= 0.51$$

A3.3 MEASURES OF PRECISION

In the absence of bias, the uncertainty in a quantity is indicated by reporting various measures of precision. Some statistics which are useful measures of precision are described below.

The **sample standard deviation**, s, is an estimate of the population standard deviation, σ. The sample and population standard deviations are assumed to be identical when $N = 20$.

$$s = \sqrt{\frac{\sum_{i=1}^{N}(q_i - \mu)^2}{f}} \tag{A3.2}$$

where N = the number of q_i
   f = degrees of freedom
   f = N for 20 or more values
    = N − 1 for less than 20 values

As explained above, the standard deviation indicates the probability that an individual q_i lies near the true mean, μ. The **standard error of the mean**, s_M, indicates the probability that the arithmetic average, q_{avg}, lies near the true mean. It is related to the sample standard deviation by

$$s_M = \frac{s}{N^{1/2}} \tag{A3.3}$$

Example

Four measurements of the solubility of chloroethyl ether produced values of 0.309, 0.301, 0.408, and 0.291 mol/l. Calculate the mean value and standard deviation.

S_i (mol/L)	$(S_i - \mu)^2$ (mol²/L²)
0.309	$(-0.018)^2 = 0.32 \times 10^{-3}$
0.301	$(-0.026)^2 = 0.68 \times 10^{-3}$
0.408	$(+0.081)^2 = 6.6 \times 10^{-3}$
0.291	$(-0.036)^2 = 1.3 \times 10^{-3}$
$\sum_{i=1}^{N} S_i = 1.309$	$\sum_{i=1}^{N}(S_i - \mu)^2 = 8.9 \times 10^{-3}$

$$S_{avg} = \left(\sum_{i=1}^{N} S_i\right)\Big/N = 1.309/4 \text{ mol/L} = 0.327 \text{ mol/L}$$

$$s = \left(\sum_{i=1}^{N}(S_i - \mu)^2\Big/f\right)^{1/2} = (8.857 \times 10^{-3})^{1/2}\Big/3 \text{ mol/L} = 0.031 \text{ mol/L}$$

$$S = 0.327 \pm 0.031 \text{ mol/L}$$

$$s_M = s/N^{1/2} = 0.031/(4)^{1/2} \text{ mol/L} = 0.016 \text{ mol/L}$$

Note that s and s_M have the same units as the quantity q. Also, s_M is smaller than s, implying that q_{avg} is a better estimate than any q_i of the true mean μ.

Measured values of a chemical property are usually accumulated over time using several series of samples, and estimates of μ and σ calculated with the pooled data set will be better than those from a single subset. The sample standard deviation for pooled data is

$$s = \sqrt{\frac{\sum_{j=i}^{M} \sum_{i=1}^{N} \left(q_i - \mu_j \right)^2}{N-M}} \qquad (A3.4)$$

Example

Four series of measurements of the saturation vapor pressure of 2-chlorobiphenyl at 25°C produced the following results. Calculate the mean and standard deviation of the pooled data.

Saturation Vapor Pressure of 2-Chlorobiphenyl at 298.15 K

Subset	P_i (Pa)	P_{avg}	$\Sigma\,(P_i - P_{avg})^2$
1	1.53, 1.55, 1.52, 1.44, 1.60, 1.51	1.53	0.0139
2	1.64, 1.53, 1.57, 1.59, 1.61	1.59	0.0069
3	1.84, 1.72, 1.80, 1.75, 1.84	1.79	0.0116
4	1.38, 1.45, 1.47, 1.53, 1.60	1.49	0.0278
Total	33.47	Total	0.0602

$$P_{avg} = 33.47/21 = 1.59$$

$$\sum_{j=1}^{M} \sum_{i=1}^{N} \left(P_i - P_{avg} \right)^2 = 0.0602$$

$$f = 21 - 4 = 17$$

$$s = \left(0.0602/17 \right)^{1/2} = 0.0595$$

$$P_{sat} = 1.59 \pm 0.060$$

The **variance**, var, is equal to s^2. The total variance is equal to the sum of individual variances of the independent sources of uncertainty. This is discussed in more detail in the section below on propagation-of-error calculations.

The **coefficient of variation**, CV, is the percent standard deviation relative to the arithmetic mean, μ. It is used to compare the uncertainty in different methods of estimating q.

$$CV = \left(\frac{s}{\mu} \right) \times 100 \qquad (A3.5)$$

Example

Two independent sets of measurements produced two values for the aqueous solubility of DDT, $2.00 \pm 0.50 \times 10^1$ µg/L and $1.00 \pm 0.25 \times 10^3$ µg/L. Compare the variability of the two sets of measurements.

$$CV = (s/\mu) \times 100$$

$$CV_1 = (0.50/2.00) \times 100 = 0.25\%$$

$$CV_2 = (0.25/1.00) \times 100 = 0.25\%$$

The precision of the two sets of measurements is the same.

The arithmetic average, q_{avg}, approximates the true mean of all possible measurements, μ. The true mean lies, at a specified level of certainty, within a confidence interval defined by confidence limits about q_{avg}. The **confidence limits**, CL, are calculated as follows:

If $N \geq 20$,

$$CL \text{ for } \mu = q_{avg} \pm z\sigma/N^{1/2} \tag{A3.6}$$

where the value of z employed depends upon the confidence level or probability that the actual deviation, $q_{avg} - \mu$, is less than $z\sigma$. Values of z for various confidence levels are given in Table A3.1.

If $N < 20$,

$$CL \text{ for } \mu = q_{avg} \pm ts/N^{1/2} \tag{A3.7}$$

where the value of t employed depends upon the confidence level or probability that the actual deviation, $q_{avg} - \mu$, is less than $t\sigma$. Values of t for various confidence levels are given in Table A3.2.

Table A3.1 Confidence Interval Parameter z Vs. Confidence Level

Confidence Level, %	z
90	1.64
95	1.96
99	2.58
99.7	3.00
99.9	3.29

Table A3.2 Confidence Interval Parameter t Vs. Confidence Level

	Confidence Interval, %			
f	90	95	99	99.9
1	6.31	12.7	63.7	637
2	2.92	4.30	9.92	31.6
3	2.35	3.18	5.84	12.9
4	2.13	2.78	4.60	8.60
5	2.02	2.57	4.03	6.86
6	1.94	2.45	3.71	5.96
10	1.81	2.23	3.17	4.59
20	1.64	1.96	2.58	3.2

Example

Twenty-five measurements of the pKa of p-cresol produced an average value of 10.26 ± 0.13. Calculate the 95% confidence limits of the true mean pKa.
From Table A3.1, $z = 1.96$ at the 95% confidence level.

$$\text{CL for } \mu = q_{avg} \pm z\sigma/N^{1/2}$$

$$= 10.26 \pm (1.96)0.13/25^{1/2} \text{ pka units}$$

$$= 10.26 \pm 0.05 \text{ pKa units}$$

It is 95% certain that the true mean pKa of p-cresol lies between $10.26 + 0.05$ and $10.26 - 0.05$ pKa units.

Example

Three measurements of the melting point of phthalic anhydride produced an average value of $131.0 \pm 0.5°C$. Calculate the 99% confidence limits of the true mean melting point.
From Table A3.2, $t = 9.92$ at the 99% confidence level when $f = 3 - 1 - 2$.

$$\text{CL for } \mu = q_{avg} \pm ts/N^{1/2}$$

$$= 131.0 \pm (9.92)(0.5)/3^{1/2}°C$$

$$= 131.0 \pm 3°C$$

It is 99% probable that the true mean melting point of phthalic anhydride lies within the range of $131 + 3$ to $131 - 3°C$.

A.3.4 PROPAGATION OF ERROR

Uncertainty in the true value of the input parameters, x, y, z, ... propagates through model calculations and contributes to the uncertainty of an estimated value of property q. How is the resulting estimate error calculated? In this section, we develop a general relationship describing the propagation of uncertainty in a mathematical model. Then, we derive some simple formulas for calculating the probable error in common types of chemical property estimates.

Our objective is to calculate the standard deviation, σ_q, of an estimate of chemical property q. If q is a function of the input parameters x, y, z, ...,

$$q = q(x,y,z, \ldots)$$

the uncertainty in the ith measurement of q, dq_i, due to the uncertainties dx_i, dy_i, dz_i, ... in the input parameters is

$$dq_i = q(dx_i, \ dy_i, \ dz_i, \ldots)$$

$$= q(q_i - \mu)$$

In order to calculate σ_q, we must calculate $(dq_i)^2$ and sum it from $i = 1$ to $i = N$. The relationship between dq and dx, dy, dz, ... is

$$dq = \left[\left(\frac{\delta q}{\delta x}\right)_{y,z,\ldots} dx + \left(\frac{\delta q}{\delta y}\right)_{x,z,\ldots} dy + \left(\frac{\delta q}{\delta z}\right)_{x,y,\ldots} dz + \ldots\right]$$

and the relationship for $(dq)^2$ is

$$dq^2 = \left[\left(\frac{\delta q}{\delta x}\right)_{y,z,\ldots} dx + \left(\frac{\delta q}{\delta y}\right)_{x,z,\ldots} dy + \left(\frac{\delta q}{\delta z}\right)_{x,y,\ldots} dz + \ldots\right]^2$$

When the right-hand side of this equation is expanded, two types of terms arise, squared terms like

$$\left[\left(\frac{\delta q}{\delta x}\right)dx\right]^2, \left[\left(\frac{\delta q}{\delta y}\right)dy\right]^2, \left[\left(\frac{\delta q}{\delta z}\right)dz\right]^2, \ldots$$

and cross-terms like

$$\left(\frac{\delta q}{\delta x}\right)dx\left(\frac{\delta q}{\delta y}\right)dy, \left(\frac{\delta q}{\delta x}\right)dx\left(\frac{\delta q}{\delta z}\right)dz, \left(\frac{\delta q}{\delta y}\right)dy\left(\frac{\delta q}{\delta z}\right)dz, \ldots$$

The squared terms are always positive. Since the dx, dy, dz, ... can be positive or negative, the cross-terms can be positive or negative. Furthermore, dx, dy, dz, ... are random uncertainties, and the cross-terms will nearly cancel when N is large. The result is

$$\sum_{i=1}^{N} dq_i^2 = \left(\frac{\delta q}{\delta x}\right)^2 \sum_{i=1}^{N} dx_i^2 + \left(\frac{\delta q}{\delta y}\right)^2 \sum_{i=1}^{N} dy_i^2 + \left(\frac{dq}{dz}\right)^2 \sum_{i=1}^{N} dz_i^2 + \dots$$

Dividing this relationship by N,

$$\frac{\sum_{i=1}^{N} dq_i^2}{N} = \left(\frac{\delta q}{\delta x}\right)^2 \frac{\sum_{i=1}^{N} dx_i^2}{N} + \left(\frac{\delta q}{\delta y}\right)^2 \frac{\sum_{i=1}^{N} dy_i^2}{N} + \left(\frac{dq}{dz}\right)^2 \frac{\sum_{i=1}^{N} dz_i^2}{N} + \dots$$

But

$$\frac{(dq_i)^2}{N} = \frac{(q_i - \mu)^2}{N} = \sigma_q^2, \quad \frac{(dx_i)^2}{N} = \sigma_x^2, \dots$$

and substituting these terms, we obtain Equation A3.8, a general relationship for the propagated error in a calculated result:

$$\sigma_q^2 = \left(\frac{\delta q}{\delta x}\right)^2 \sigma_x^2 + \left(\frac{\delta q}{\delta y}\right)^2 \sigma_y^2 + \left(\frac{\delta q}{\delta z}\right)^2 \sigma_z^2 + \dots \qquad (A3.8)$$

We will use Equation A3.8 to calculate σ_q for common types of chemical property calculations.

A3.4.1 Calculations Involving Addition and Subtraction

The model is

$$q = x + y - z$$

and

$$\frac{\delta q}{\delta x} = \frac{\delta q}{\delta y} = -\frac{\delta q}{\delta z} = 1$$

Equation A3.8 shows that the variance of q is

$$s_q^2 = \left(\frac{\delta q}{\delta x}\right)^2 s_x^2 + \left(\frac{\delta q}{\delta y}\right)^2 s_y^2 + \left(\frac{\delta q}{\delta z}\right)^2 s_z^2$$

Substituting 1 for $\delta q/\delta x$, $\delta q/\delta y$, and -1 for $\delta q/\delta z$,

$$s_q^2 = s_x^2 + s_y^2 + s_z^2 \tag{A3.9}$$

Equation A3.9 reveals that the model variance is the sum of variances of the model's parameters, and

$$s_q = \sqrt{s_x^2 + s_y^2 + s_z^2} \tag{A3.10}$$

Example

The measured log K_{OW} of benzene is 2.13 ± 0.02 log units. Log K_{OW} for chlorobenzene is estimated using the fragment constants $f(H) = +0.23 \pm 0.12$ for hydrogen atoms and $f(Cl) = +0.94 \pm 0.12$ for chlorine atoms.

$$\log\ K_{OW}\ (C_6H_5Cl) = \log\ K_{OW}\ (C_6H_6) - f(H) + f(Cl)$$

$$= 2.13 - 0.23 + 0.94$$

$$= 2.84\ \text{log units}$$

Calculate the standard deviation of the log $P(C_6H_5Cl)$ estimate.

$$s_{\log KOW}^2 = \left(s_{\log KOW(C_6H_6)}^2 + s_{f(H)}^2 + s_{f(Cl)}^2 \right)^{1/2}$$

$$= \left[(0.02)^2 + (0.12)^2 + (0.12)^2 \right]^{1/2}$$

$$= 0.2\ \text{log units}$$

$$\log\ K_{OW}\ (C_6H_5Cl) = 2.1 \pm 0.2\ \text{log units}$$

3.4.2   Calculations Involving Multiplication and Division

The model is

$$q = x \cdot y / z$$

Applying Equation A3.8,

$$s_q^2 = \left(\frac{y}{z}\right)^2 s_x^2 + \left(\frac{x}{z}\right)^2 s_y^2 + \left(-\frac{xy}{z^2}\right)^2 s_z^2$$

Dividing this equation by q²,

$$\left(\frac{S_q}{q}\right)^2 = \left(\frac{S_x}{x}\right)^2 + \left(\frac{S_y}{y}\right)^2 + \left(\frac{S_z}{z}\right)^2 \tag{A3.11}$$

Equation A3.11 shows that the relative variance of the model is the sum of relative variances of model's parameters, and

$$S_q = q\sqrt{\left(\frac{S_x}{x}\right)^2 + \left(\frac{S_y}{y}\right)^2 + \left(\frac{S_z}{z}\right)^2} \tag{A3.12}$$

Example

The saturation vapor pressure of 2,4-D at 25°C is $5.6 \pm 0.2 \times 10^{-5}$ Pa. Its solubility at 25°C is $4.03 \pm 0.05 \times 10^{-3}$ mol/l. Henry's law constant for 2,4-D is estimated as

$$H = P_{sat}/S$$

$$= \left(5.6 \times 10^{-5}\ Pa\right)/\left(4.03 \times 10^{-3}\ mol/l\right)$$

$$= 1.39 \times 10^{-2}\ Pa\ mol/l = 1.39 \times 10^{-5}\ Pa\ mol/m^3$$

Calculate the standard deviation of H.
Using Equation A3.12,

$$s_H = H\sqrt{\left(\frac{s_{P_s}}{P_s}\right)^2 + \left(\frac{s_S}{S}\right)^2}$$

$$= \left(1.39 \times 10^{-5}\right)\sqrt{\left(\frac{0.2}{5.6}\right)^2 + \left(\frac{0.05}{4.03}\right)^2}\ \frac{Pa \cdot mol}{m^3}$$

$$H = 1.39 \pm 0.06 \times 10^{-5}\ Pa\ mol/m^3$$

3.4.3   Calculations Involving Logarithms and Antilogarithms

The model is

$$q = \log x$$

$$= 0.434\ \ln x$$

and

$$\left(\frac{\delta q}{\delta x}\right) = 0.434\frac{\delta \ln x}{\delta x} = \frac{0.434}{x}$$

Applying Equation A3.8,

$$s_q^2 = \left[0.434\frac{\delta(\ln x)}{\delta x}\right]^2 s_x^2 \qquad (A3.13)$$

Substituting $1/x$ for $\delta \ln x/\delta x$,

$$s_q = 0.434\frac{s_x}{x} \qquad (A3.14)$$

The model standard deviation equals the relative deviation of the model parameter.

Example

The hydrogen ion concentration of a water sample is $1.59 \pm 0.03 \times 10^{-4}$ M. Calculate the sample's pH.

$$pH = -\log\left[H^+\right]*$$

$$= -\log\left(1.59 \times 10^{-4}\right)$$

$$= 3.80$$

$$S_{pH} = 0.434\, s_{[H+]}\left/\left[H^+\right]\right.$$

$$= 0.434\,(0.03)/1.59$$

$$= 0.008$$

$$pH = 3.80 \pm 0.01$$

3.4.4   Calculations Involving Exponentials

The model is

$$q = x^y$$

* The argument of a logarithm must be dimensionless, of course. In fact, pH = $-\log a_m$ (H^+), where a_m (H^+) is the hydrogen ion activity calculated on the molality scale, a dimensionless quantity. a_m (H^+) = γ (H^+)m(H^+)/m^0, where γ (H^+) is the hydrogen ion activity coefficient, m(H^+) is the hydrogen ion molality, and m^0 is unit molality (= 1 mol/kg) introduced to make the argument of the logarithm dimensionless.

and

$$\left(\frac{\delta q}{\delta x}\right) = \frac{\delta(x^y)}{\delta x} = yx^{y-1}$$

Applying Equation A3.8,

$$s_q^2 = \left[\frac{\delta(x^y)}{\delta x}\right]^2 s_x^2 = \left(yx^{y-1}\right)^2 s_x^2$$

and

$$\left(\frac{s_q}{q}\right)^2 = \left(\frac{yx^{y-1}}{x^y}\right)^2 s_x^2 = \left(\frac{y}{x}\right)^2 s_x^2$$

or

$$\frac{s_q}{q} = y\frac{s_x}{x} \qquad (A3.15)$$

The relative standard deviation of the calculated result is the relative standard deviation of the input parameter multiplied by the exponent.

Example

The organic–carbon sorption coefficient, K_{OC}, of DDT is $2.43 \pm 1.03 \times 10^5$. Estimate DDT's bioconcentration factor, BCF.

Using the relationship,

$$BCF = 0.0264 \ K_{OC}^{1.119}$$

$$= 0.0264 \ (243,000)^{1.119}$$

$$= 2.81 \times 10^4$$

$$s_{BCF}/BCF = 0.434 \ s_{Koc}/K_{OC}$$

$$= 0.434 \ (1.03/2.43)$$

$$= 0.184$$

$$s_{BCF} = 0.184 \ (28,100)$$

$$= 5.17 \times 10^3$$

$$BCF = 2.81 \pm 0.52 \times 10^4$$

Example

The parachor of water at 20°C is calculated from its molar volume, $V = 18.0479 \pm 0.0001$ cm$_3$/mol, and its surface tension, $\sigma = 72.75 \pm 0.05 \times 10^{-3}$ N/m:

$$P = \overline{V}\sigma^{\frac{1}{4}}$$

$$P(H_2O) = \left(18.0479 \times 10^{-6} \text{ m}^3/\text{mol}\right)\left(72.75 \times 10^{-3} \text{ kg/s}^2\right)^{1/4}$$

$$= 9.373 \times 10^{-6} \text{ m}^3 \text{ kg}^{1/4}/\text{mol s}^{1/2}$$

Calculate the standard deviation of the parachor estimate.

$$s_P = P\sqrt{\left(\frac{s_V}{V}\right)^2 + \left(\frac{1}{4}\frac{s_\sigma}{\sigma}\right)^2}$$

$$= 9.373 \times 10^{-6}\sqrt{\left(\frac{0.0001}{18.0479}\right)^2 + \left(\frac{1}{4}\frac{0.05}{72.75}\right)^2}\frac{m^3 \cdot kg^{\frac{1}{4}}}{mol \cdot s^{\frac{1}{2}}}$$

$$P = 9.373 \pm 0.002 \times 10^{-6} \text{ m}^3 \text{ kg}^{1\cdot4}/\text{mol s}^{1/2}$$

REFERENCES

Bennet, C. A. and Franklin, N. L., 1954. *Statistical Analysis in Chemistry and the Chemical Industry.* John Wiley & Sons, New York.

Berthouex, P. M. and Brown, L. C., 1994. *Statistics for Environmental Engineers.* Lewis Publishers, Boca Raton, FL.

Davies, O. L. and Goldsmith, P. L., Eds., 1972. *Statistical Methods in Research and Production.* Fourth Edition. Longman, New York.

Dixon, W. J. and Massey, F. J., Jr., 1969. *Introduction to Statistical Analysis.* Third Edition. McGraw-Hill, New York.

APPENDIX **4**

Empirical Modeling

A4.1 INTRODUCTION

How is an empirical model derived from experimental data? If the independent variable is not subject to significant error, a standard least-squares regression analysis is the usual method. This will be the case, for instance, when deriving a model of the correlation between a chemical property and a computable molecular descriptor.

The least-squares method of regression analysis is outlined below. Also, a procedure is decribed for estimating the predictive error of the resulting model. Simple regression analysis routines are available as part of all statistical software and most spreadsheet programs for digital computers. Many scientific hand calculators offer regression analysis routines as well. Since access to these programs is nearly universal, detailed instruction for performing the analysis by hand is probably of limited interest. However, the error analysis is not routinely performed by available software, and the following discussion of predictive error should be useful. The subject is described in more detail by Draper and Smith (1981).

In practice, a least-squares analysis is used if the measurement variance of the independent variable is less than a tenth of the average scatter of this variable about its mean value in the data set (Daniel et al., 1971). (Be aware, however, that simple regression analysis ignores error in the independent variable, and the usual calculation procedures underestimate model errors.) A regression analysis may not be appropriate if both independent and dependent variables are subject to significant experimental error, as might occur if both describe chemical properties. There is no agreed-upon way to proceed in this case (Weisberg, 1985).

Wolberg (1967) describes an iterative regression method for use when both the dependent variable, q, and the independent variable, x, are subject to substantial measurement error. The method requires that errors in q and x be independent, normally distributed, and have constant variance. An initial guess is made of the magnitude of the regression parameters, and convergence to the true values is obtained by a process of successive approximation. To do this properly requires experience and care, and the procedure is time-consuming. On the other hand,

Bartlett describes a simple procedure which quickly produces a linear model (Bartlett, 1949). Bartlett's three-group procedure is described below.

The vocabulary of data analysis is not used consistently, and we need to define our terms. Independent variables are called **predictor variables**. The dependent variable is called the **response variable**. A set of chemicals used to derive a model is called either a **training set** or a **learning set**. An independent set of chemicals used to test or validate the model is variously called a **validation set**, an **evaluation set**, a **test set**, or a **prediction set**. The terms 'validation set' and 'evaluation set' always describe a set of chemicals for which property values are used to evaluate the model but not to derive it. The terms 'test set' and 'prediction set' are sometimes used specifically to describe a set of chemicals for which the property values are not known.

With the methods described here, a good fit between the data and the resulting model can be obtained even if there is no theoretical basis for the model. So long as the model is not extrapolated beyond the limits of the training set, the model will be satisfactory for estimation purposes.

Calculus is central to the discussion of empirical modeling. A review of differential and integral calculus is presented in Appendix 7.

A4.2 LEAST-SQUARES REGRESSION METHOD

Suppose that a linear relationship exists between quantities q and x. The relationship can be modeled as:

$$q = \beta_0 + \beta_1 x + e \qquad\qquad (A4.1)$$

where β_0, β_1 = estimates of the true intercept and slope, respectively, of the regression curve

e   = residual error due to model inaccuracy

Assume that (1) the values of the predictor variable, x, have negligible error, (2) for any value of x, the values of the response variable, q, exhibit a Gaussian (normal) distribution about a mean value of zero, (3) the q-value errors are not correlated, and (4) the variance of q is constant across the range of x values. Then, a least-squares regression analysis will produce the "best fit" linear relationship between q and x, one that minimizes the sum of squared residual differences between data points and model predictions:

$$\sum_{i=1}^{N}\left(e_i\right)^2 = \sum_{i=1}^{N}\left(q_i - \underline{q_i}\right)^2$$

where q_i is the measured value of q when $x = x_i$ and $\underline{q_i}$ is the value of q predicted by Equation A4.1 when $x = x_i$.

A least-squares analysis involves calculating the following quantities:

The **corrected sum of cross-products,**

$$S_{xq} = \sum_{i=1}^{N} x_i q_i - \frac{\sum_{i=1}^{N} x_i \sum_{i=1}^{N} q_i}{N} \qquad (A4.2)$$

where N is the number of samples in the training set.
The **corrected sum of squares for the x_i,**

$$S_{xx} = \sum_{i=1}^{N} x_i^2 - \frac{\left(\sum_{i=1}^{N} x_i\right)^2}{N} \qquad (A4.3)$$

The **sample average of x,**

$$\bar{x} = \frac{\sum_{i=1}^{N} x_i}{N} \qquad (A4.4)$$

The **sample average of q,**

$$\bar{q} = \frac{\sum_{i=1}^{N} q_i}{N} \qquad (A4.5)$$

The **corrected sum of squares of the q_i,**

$$S_{qq} = \sum_{i=1}^{N} q_i^2 - \frac{\left(\sum_{i=1}^{N} q_i\right)^2}{N} \qquad (A4.6)$$

The **correlation coefficient,** a measure of the degree of association of q and x,

$$r = \frac{S_{xq}}{\sqrt{S_{xx} S_{qq}}} \qquad (A4.7)$$

If q and x are perfectly correlated, $r = +1$. If q and x are perfectly anticorrelated, $r = -1$. If q is not at all correlated with x, $r = 0$.
The **slope** of the regression curve,

$$\beta_1 = \frac{S_{xq}}{S_{xx}} \qquad (A4.8)$$

The **q-intercept** of the regression curve,

$$\beta_0 = \bar{q} - \beta_1 \bar{x} \qquad (A4.9)$$

The **x-intercept** of the regression curve,

$$x_{q=0} = \frac{-\beta_0}{\beta_1} \qquad (A4.10)$$

As a last step in the procedure, it is a good idea to plot the regression formula along with the data points. This ensures, among other things, that the regression coefficients are not determined primarily by the presence of an outlier. It is also useful to plot the residuals, e_i, as a function of the predictor variable. If curvature is noted in the scatter plot or a trend is noted in the plot of residuals, a change of variable may be required in order to fit a straight line. However, when changing variables, the relative importance of the calibration data is also changed, and a weighted least-squares analysis may be necessary.

Example

Derive a least-squares relationship between the molar volume, V_m, and the octanol–water partition coefficient, K_{ow}, of the chlorinated alkanes.

1. We transform the data from K_{OW} to log K_{OW} to obtain a linear correlation with V_M. Values of V_M and log K_{OW} are given in Table A4.1 for some chlorinated alkanes.
2. Calculate the values of $x_i^2, q_i^2,$ and $x_i q_i$. These and their sums are shown in Table A4.2.

Table A4.1 V_M and Log K_{OW} of Some Chlorinated Alkanes at 25°C

Compound	V_M (cm³/mol)	log K_{OW}
Methyl Chloride	50.5	0.91
Dichloromethane	71.0	1.25
Carbon tetrachloride	113.0	2.62
Chloroethane	72.7	1.43
1,1-Dichloroethane	93.6	1.79
1,1,1-Trichloroethane	115.0	2.47
1,1,2,2-Tetrachloroethane	135.0	2.39
Pentachloroethane	156.0	2.89
Hexachloroethane	177.0	3.93
1,2-Dichloropropane	103.3	1.99
1,2,3-Trichloropropane	129.3	2.63
1-Chlorobutane	117.1	2.39

From Mackay, D., Shiu, W.Y., and Ma, K.C., 1993. *Illustrated Handbook of Physical-Chemical Properties and Environmental Fate for Organic Chemicals, Volume III, Volatile Organic Chemicals.* Lewis Publishers, Boca Raton, FL, pp. 616–624.

Table A4.2 Chloroalkane Model Regression Data

V_M, x_i	log K_{OW}, q_i	x_i^2	q_i^2	$x_i q_i$
50.5	0.91	2550	0.8281	45.95
71.0	1.25	5041	1.5625	88.75
113.0	2.62	12769	6.8644	296.06
72.7	1.43	5285	2.0449	103.96
93.6	1.79	8761	3.2041	167.54
115.0	2.47	13225	6.1009	284.05
135.0	2.39	18225	5.7121	322.65
156.0	2.89	24336	8.3521	450.84
177.0	3.93	31329	15.4449	695.61
103.3	1.99	10671	3.9601	205.567
129.3	2.63	16718	6.9169	340.059
117.1	2.39	13712	5.7121	279.869
Σ 1333.5	Σ 26.69	Σ 162623	Σ 66.7031	Σ 3280.915

3. The corrected sum of cross-products of V_M and log K_{OW}, Equation A4.2, is calculated.

$$S_{xq} = \sum_{i=1}^{N} x_i q_i - \frac{\sum_{i=1}^{N} x_i \sum_{i=1}^{N} q_i}{N}$$

$$= 3280.915 - (1333.5)(26.69)/12$$

$$= 314.98875$$

4. The corrected sum of squares of x, Equation A4.3, is calculated.

$$S_{xx} = \sum_{i=1}^{N} x_i^2 - \frac{\left(\sum_{i=1}^{N} x_i\right)^2}{N}$$

$$= 162623.3 - (1333.5)^2/12$$

$$- 14438.103$$

5. The arithmetic mean of x, Equation A4.4, is calculated.

$$\bar{x} = \frac{\sum_{i=1}^{N} x_i}{N}$$

$$= 1333.5/12$$

$$= 111.125 = 111.1$$

6. The arithmetic mean of q, Equation A4.5, is calculated.

$$\bar{q} = \frac{\sum_{i=1}^{N} q_i}{N}$$

$$= 26.69/12$$

$$= 2.224 = 2.22$$

7. The corrected sum of squares of q, Equation A4.6, is calculated.

$$S_{qq} = \sum_{i=1}^{N} q_i^2 - \frac{\left(\sum_{i=1}^{N} q_i\right)^2}{N}$$

$$= 66.7031 - (26.69)^2/12$$

$$= 7.34009$$

8. The correlation coefficient, Equation A4.7, is calculated.

$$r = \frac{S_{xq}}{\sqrt{S_{xx}S_{qq}}}$$

$$= 314.9888/\left[(14438.1)(7.34009)\right]^{1/2}$$

$$= 0.967585$$

9. The slope of the regression curve, Equation A4.8, is calculated.

$$\beta_1 = \frac{S_{xq}}{S_{xx}}$$

$$= 314.9888/14438.1$$

$$= 0.0218 \ \text{mol}/\text{cm}^3$$

10. The q-intercept of the regression curve, Equation A4.9, is calculated.

$$\beta_0 = \bar{q} - \beta_1\bar{x}$$

$$= 2.224 - (0.0218)(111)$$

$$= -0.20$$

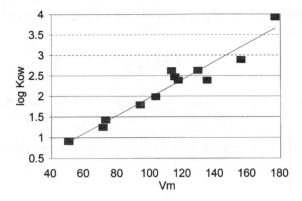

Figure A4.1 Plot of log octanol–water partition coefficient, log K_{OW}, against molar volume, V_M (cm³/mol), for chlorinated alkanes at 25˚C.

11. The x-intercept of the regression curve, Equation A4.10, is calculated.

$$x_{q=0} = \frac{-\beta_0}{\beta_1}$$

$$= -(-0.20)/0.0218$$

$$= 9.2$$

The least-squares regression formula is

$$\log K_{OW} = -0.20 + 0.0218 V_M$$

The regression model is plotted along with the data points in Figure A4.1. There are no apparent problems with the fit, and the model should be reliable if used in the range of log K_{OW} values between 1 and 4 and V_M values between 50 and 180 cm³/mol.

Uncertainty in the model parameters is evaluated with the following quantities:

The **standard deviation about the regression**, the standard deviation of q when deviations are measured from the regression curve,

$$s_R = \sqrt{\frac{S_{qq} - \beta_1^2 S_{xx}}{N-2}} \tag{A4.11}$$

The **standard deviation of the slope**,

$$s_\beta = \sqrt{\frac{s_R^2}{S_{xx}}} \tag{A4.12}$$

The **confidence limit of the slope** is estimated using the confidence interval parameter t from Table A3.2.

$$CL = \beta_1 \pm ts_\beta \qquad\qquad (A4.13)$$

The number of degrees of freedom, f, is set equal to $N - 2$, since one degree of freedom is lost in calculating β_0 and another is lost in calculating β_1.

The **coefficient of determination**, a measure of the proportion of variance explained by the model,

$$R^2 = \frac{\sum_{i=1}^{N}(q_i - \bar{q})^2}{\sum_{i=1}^{N}(q_i - \bar{q})^2} \qquad\qquad (A4.14)$$

where the q_i are values of q predicted by the model. When $R^2 = 1$, there is a perfect fit between the model and the data. When $R^2 = 0$, there is a complete lack of fit.

In estimating uncertainty in the model parameters, we ignore measurement error in the predictor variable, x. If the measurement variance of x is significant compared with the range of x values in the training set (more than one tenth of the average scatter of x values in the set), our calculations will significantly underestimate the model error.

Example

Evaluate the uncertainty in the chloroalkane regression model derived in the previous example.

Calculate the standard deviation about the regression, the standard deviation of the slope, the 95% confidence interval of the slope, and the coefficient of determination of the chloroalkane regression model.

1. The standard deviation about the regression, Equation A4.11, is

$$s_R = \sqrt{\frac{S_{qq} - \beta_1^2 S_{xx}}{N-2}}$$

$$= \left[(7.340092 - (0.0218)^2 (14438.1) (12-2)\right]^{1/2}$$

$$= 0.216366$$

$$= 0.22$$

2. The standard deviation of the slope, Equation A4.12, is

Table A4.3 Predictions of the Chloroalkane Regression Model

V_M, x_i	log K_{OW}, q_i	log K_{OW}, q_i	$(q_i - \bar{q})^2$	$(q_i - \bar{q})^2$
50.5	0.91	0.90	1.7270	1.7493
71.0	1.25	1.35	0.9490	0.7663
113.0	2.62	2.27	0.1567	0.0017
72.7	1.43	1.39	0.6307	0.7027
93.6	1.79	1.84	0.1885	0.1462
115.0	2.47	2.31	0.0604	0.0072
135.0	2.39	2.75	0.0275	0.2713
156.0	2.89	3.20	0.4433	0.9585
177.0	3.93	3.66	2.9099	2.0654
103.3	1.99	2.05	0.0548	0.0291
129.3	2.63	2.62	0.1647	0.1572
117.1	2.39	2.35	0.0275	0.0170
			Σ 7.3401	Σ 6.8720

$$s_\beta = \sqrt{\frac{s_R^2}{S_{xx}}}$$

$$= \left[(0.216366)^2 \; 14438.1 \right]^{1/2}$$

$$= 0.0018$$

3. The confidence limit of the slope, Equation A4.13, is

$$CL = \beta_1 \pm t s_\beta$$

$$= 0.0218 \pm (2.23)(0.0018)$$

$$= 0.0218 \pm 0.0040$$

4. Predicted values of K_{OW} for the chlorinated alkanes are compared with measured values in Table A4.3. The arithmetic mean value of q, $-\bar{q}$, was found above to be 2.22. The coefficient of determination, Equation A4.14, is

$$R^2 = \frac{\sum_{i=1}^{N} (q_i - \bar{q})^2}{\sum_{i=1}^{N} (q_i - \bar{q})^2}$$

$$= 6.8719/7.3401$$

$$= 0.93622$$

The regression model will be used to estimate the magnitude of the response variable for a given value of the predictor variable:

$$q_i = \beta_0 + \beta_1 x_i$$

Assuming that the value of the predictor variable is obtained under experimental conditions similar to those used to obtain the training set data and that the value lies within the range of x in the training set, the **estimated standard error of the prediction** is

$$se(q_i) = s_r \sqrt{1 + \frac{1}{N} + \frac{(x_i - \bar{x})^2}{S_{xx}}} \qquad (A4.15)$$

If repeated measurements are made of the magnitude of q when $x = x_i$, the true mean value of q_i would lie with probability $1 - \alpha$ within a **predictive interval** around q_i, an interval determined using the confidence interval parameter $t(1 - \alpha, N - 2)$ given in Table A3.2. The predictive interval is

$$q_i \pm t(1 - \alpha, N - 2) \cdot se(q_i)$$

Equation A4.15 shows that the error of the prediction increases with increasing difference between x_i and $\bar{x}$. Our estimate will be most accurate near the center of the model's range. Also, there is no way to tell if the relationship between q and x is linear outside of the range of data used to derive the regression model. Therefore, estimates of the magnitude of q made by extrapolating beyond the limits of the training set are not reliable.

Example

Estimate log K_{OW} of chloropropane using the chloroalkane model derived above. For chloropropane, $V_M = 87.5$ cm³/mol.

1. The regression model predicts:

$$\log K_{OW} = -0.20 + 0.0218 V_M$$
$$= -0.20 + 0.0218(87.5)$$
$$= 1.71$$

2. The estimated standard error of the prediction, Equation A4.15, is

$$se(q_i) = s_r \sqrt{1 + \frac{1}{N} + \frac{(x_i - \bar{x})^2}{S_{xx}}}$$

The values are substituted from Table A4.3 above.

$$se\left(\underline{q_i}\right) = 0.216\sqrt{1 + \frac{1}{12} + \frac{(87.5 - 111.125)^2}{14438}}$$

$$= 0.229$$

3. The 95% predictive interval is determined using t(0.95,10).

$$q = \underline{q_i} \pm t(0.95,10) \cdot se\left(q_i\right)$$

$$\log K_{OW} = 1.71 \pm 2.23(0.229)$$

$$= 1.71 \pm 0.51$$

A4.3 BARTLETT'S THREE-GROUP METHOD

Bartlett's three-group method is used to derive a linear relationship between two chemical properties when both are subject to significant measurement error (Bartlett, 1949; Draper and Smith, 1981). The N calibration measurements are divided into three nonoverlapping groups, one with a high, one with a median, and one with a low range of x values. The number of data points in the extreme groups should be made equal and as close to N/3 as possible. The estimated slope of the regression line, β_1, is given by the slope of the line joining the mean coordinates of the two extreme groups. The final model is derived by requiring that the fitted line pass through the midpoint of the data set, $(\bar{q}, \bar{x})$.

Example

Derive a relationship between the solubility and octanol–water partition coefficient of the chlorinated alkanes.

1. Data on the solubility, S, and octanol–water partition coefficient, K_{OW}, of 12 chlorinated alkanes were used to prepare Table A4.4. The data are sorted by magnitude of K_{OW} and divided into three equal, nonoverlapping groups. Each group has four data points.
2. The mean coordinates of the first and third groups are $(\bar{q}_1, \bar{x}_1) = (-1.9681, 1.35)$ and $(\bar{q}_3, \bar{x}_3) = (-0.3857, 3.02)$, respectively. The estimated slope of the fitted line, β_1, is given by $(\bar{q}_3 - \bar{q}_1)/(\bar{x}_3 - \bar{x}_1) = (-0.3857 - (-1.9681)/(3.02 - 1.35) = 0.948$.
3. The q intercept, β_0, is found as follows. The fitted line must pass through the point $(\bar{q}, \bar{x}) = (-1.1606, 2.22)$, and $q = \beta_0 + \beta_1 x$. Then $-1.1606 = \beta_0 + 0.948 (2.22)$, and $\beta_0 = -3.27$. The relationship between S and K_{OW} is

$$\log (1/S) = -3.27 + 0.948 \ \log K_{OW}$$

Table A4.4 Solubility and K_{OW} of Some Chlorinated Alkanes at 25°C

Compound	Log 1/S, q	Log K_{OW}, x	Group Average
Methyl chloride	−2.0157	0.91	$^-q_1 = -1.9681$
Dichloromethane	−2.1915	1.25	
Chloroethane	−1.9462	1.43	$^-x_1 = 1.35$
1,1-Dichloroethane	−1.6828	1.79	
1,2-Dichloropropane	−1.3941	1.99	$^-q_2 = -1.606$
1-Chlorobutane	−0.8222	2.39	
1,1,2,2-Tetrachloroethane	−1.2467	2.39	$^-x_2 = 2.22$
1,1,1-Trichloroethane	−1.0496	2.47	
Carbon Tetrachloride	−0.7160	2.62	$^-q_3 = -0.3857$
1,2,3-Trichloropropane	−1.1092	2.63	
Pentachloroethane	−0.3927	2.89	$^-x_3 = 3.02$
Hexachloroethane	0.6753	3.93	
Training set average	−1.1606	2.22	

From Mackay, D., Shiu, W.Y., and Ma, K.C., 1993. *Illustrated Handbook of Physical-Chemical Properties and Environmental Fate for Organic Chemicals, Volume III, Volatile Organic Chemicals.* Lewis Publishers, Boca Raton, FL, pp. 616–624.

The error in measuring log K_{OW} is small compared to the range of log K_{OW} in the chloroalkane training set, and Bartlett's three-group method should produce results similar to the results of a regression analysis. A least-squares analysis of the chloroalkane data produces

$$\log(1/S) = -3.28 + 0.952 \ \log \ K_{OW}$$

The two methods produce nearly identical results.

The confidence interval of the true slope of the fitted curve is found in Bartlett's method by solving the quadratic equation that defines t (Bartlett, 1949):

$$\left(\overline{x_3} - \overline{x_1}\right)^2 \left(\beta_1 - \beta\right)^2 \frac{N}{6} = t^2 \left(s_{qq} - 2\beta s_{xq} + \beta^2 s_{xx}\right) \tag{A4.16}$$

where β is the true slope, t is the confidence interval parameter for $f = N - 2$ from Table A3.2, and

$$s_{qq} = \frac{1}{N-2}\left[\Sigma_1 \left(q_i - \overline{q_1}\right)^2 + \Sigma_2 \left(q_i - \overline{q_2}\right)^2 + \Sigma_3 \left(q_i - \overline{q_3}\right)^2\right] \tag{A4.17}$$

$$s_{xq} = \frac{1}{N-2}\left[\Sigma_1 \left(x_i - \overline{x_1}\right)\left(q_i - \overline{q_1}\right) + \Sigma_2 \left(x_i - \overline{x_2}\right)\left(q_i - \overline{q_2}\right)\right.$$
$$\left. + \Sigma_3 \left(x_i - \overline{x_3}\right)\left(q_i - \overline{q_3}\right)\right] \tag{A4.18}$$

$$s_{xx} = \frac{1}{N-2}\left[\Sigma_1\left(x_i - \overline{x_1}\right)^2 + \Sigma_2\left(x_i - \overline{x_2}\right)^2 + \Sigma_3\left(x_i - \overline{x_3}\right)^2\right] \quad (A4.19)$$

Σ_1, Σ_2, and Σ_3 in Equations A4.17 to A4.19 are sums of values restricted to the first, second, and third groups of data, respectively.

Equation A4.16 is quadratic in β. It can be rearranged to

$$a\beta^2 + b\beta + c = 0 \quad (A4.20)$$

where a, b, and c are constants. Solutions to Equation A4.20 are given by the quadratic formula:

$$\beta = \frac{-b \pm \sqrt{b^2 - 4ac}}{2a} \quad (A4.21)$$

Example

Estimate the 95% confidence interval of the slope of the relationship between log (1/S) and log K_{OW} derived in the last example. Calculate s_{qq}, s_{xq}, and s_{xx} with Equations A4.17 to A4.19. The group sums are given in Table A4.5.

Table A4.5 Group Sums for the Chloroalkane Data Set

q	x	q − q₁	(q − q₁)²	x − x₁	(x − x₁)²	(x − x₁)(q − q₁)
−2.0157	0.91	−0.0567	0.0032	−0.4350	0.1892	0.02246
−2.1915	1.25	−0.2325	0.0540	−0.0950	0.0090	0.0221
−1.9462	1.43	0.0129	0.0002	0.0850	0.0072	0.0011
−1.6828	1.79	0.2763	0.0763	0.4450	0.1980	0.1229
			Σ_1 0.1337		Σ_1 0.4035	Σ_1 0.1707

q	x	q − q₂	(q − q₂)²	x − x₂	(x − x₂)²	(x − x₂)(q − q₂)
−1.3941	1.99	−0.2365	0.0559	−0.2342	0.0548	0.0554
−0.8222	2.39	0.3354	0.1125	0.1658	0.0275	0.0556
−1.2467	2.39	−0.0891	0.0079	0.1658	0.0275	−0.0148
−1.0496	2.47	0.1080	0.0117	0.2458	0.0604	0.0266
			Σ_2 0.1880		Σ_2 0.1703	Σ_2 0.1228

q	x	q − q₃	(q − q₃)²	x − x₃	(x − x₃)²	(x − x₃)(q − q₃)
−0.7160	2.62	−0.3304	0.1091	−0.3975	0.1580	0.1313
−1.1092	2.63	−0.7236	0.5235	−0.3875	0.1501	0.2804
−0.3927	2.89	−0.0071	0.0001	−0.1275	0.0163	0.0009
0.6753	3.93	1.0610	1.1256	0.9125	0.8327	0.9681
			Σ_3 1.0758		Σ_3 1.1571	Σ_3 1.3807

1. Calculate s_{qq} using Equation A4.17.

$$s_{qq} = \frac{1}{N-2}\left[\Sigma_1\left(q_i - \overline{q_1}\right)^2 + \Sigma_2\left(q_i - \overline{q_2}\right)^2 + \Sigma_3\left(q_i - \overline{q_3}\right)^2\right]$$

$$s_{qq} = \frac{1}{10}[0.13372 + 0.18803 + 1.7583]$$

2. Calculate s_{xq} using Equation A4.18.

$$s_{xq} = \frac{1}{N-2}\left[\Sigma_1\left(x_i - \overline{x_3}\right)\left(q_i - \overline{q_1}\right) + \Sigma_2\left(x_i - \overline{x_2}\right)\left(q_i - \overline{q_2}\right) + \Sigma_3\left(x_i - \overline{x_3}\right)\left(q_i - \overline{q_3}\right)\right]$$

$$s_{xq} = \frac{1}{10}[0.1707 + 0.12278 + 1.3807]$$

3. Calculate s_{xx} using Equation A4.19.

$$s_{xx} = \frac{1}{N-2}\left[\Sigma_1\left(x_i - \overline{x_1}\right)^2 + \Sigma_2\left(x_i - \overline{x_2}\right)^2 + \Sigma_3\left(x_i - \overline{x_3}\right)^2\right]$$

$$s_{xx} = \frac{1}{10}[0.4035 + 0.1703 + 1.1571]$$

4. Equation A4.16 with $t = 2.23$ for 10 degrees of freedom is

$$\left(\overline{x_3} - \overline{x_1}\right)^2\left(\beta_1 - \beta\right)^2\frac{N}{6} = t^2\left(s_{qq} - 2\beta s_{xq} + \beta^2 s_{xx}\right)$$

$$(3.0175 - 1.340)^2\,(0.9475 - \beta)^2\,\frac{12}{6} = (2.23)^2\left[0.2080 - 2(0.1674)\beta + 0.1731\beta^2\right]$$

or

$$4.7329\beta^2 - 8.9342\beta + 3.9869 = 0$$

Using the quadratic formula, Equation A4.21, to solve for β,

$$\beta = \frac{-b \pm \sqrt{g^2 - 4ac}}{2a}$$

$$\beta = \frac{-(-1.7938) \pm \sqrt{(1.7938)^2 - 4(0.9519)(0.8014)}}{2(0.9519)}$$

$$= 0.942 \pm 0.214$$

The three-group model's 95% confidence interval estimate is wider than the estimated 95% confidence interval of a simple least-square model because measurement error in K_{OW} is accounted for in Bartlett's method but ignored in the regression analysis.

A4.4 MULTIVARIATE REGRESSION

Regression analysis can be used to derive models involving K predictor variables:

$$q_i = \beta_0 + \beta_1 x_{1i} + \beta_2 x_{2i} + \ldots + \beta_K x_{Ki} + e_i \qquad (A4.22)$$

where the individual values of the response variable, q_i, are taken from a Gaussian (normal) distribution with mean $\beta_0 + \beta_1 x_{1i} + \beta_2 X_{2i} + \ldots$ and variance σ^2. The maximum likelihood estimates of $\beta_0, \beta_1, \beta_2, \ldots$ are obtained by minimizing the sum of square residual differences

$$\sum \left(q_i - \beta_0 - \beta_1 x_{1i} - \beta_2 x_{2i} - \ldots \right)^2 \qquad (A4.23)$$

Setting the derivatives of the sum of squares with respect to $\beta_0, \beta_1, \beta_2, \ldots$ equal to zero produces a set of equations that can be used to estimate the regression parameters.

Regression models are usually reported along with statistics describing the number of chemicals in the training set, the correlation of variables, and the predictive error. For multiple regression models, the value of the **F ratio**, $F(v_1, v_2)$, where $v_1 = K$ and $v_2 = N - K - 1$, is usually reported, also. It is

$$F(v_1, v_2) = \frac{N - K - 1}{K} \frac{R^2}{1 - R^2} \qquad (A4.24)$$

The statistic is the ratio of the variance explained by the model to the residual variance. It is compared with standard F values that are tabulated for various levels of significance. If the calculated value is larger than the tabulated value, the correlation is significant

The **robustness** or stability of a model is demonstrated with a cross-validation procedure in which successive subsets of samples are left out of the training set and the resulting models are compared. If the model is stable, the cross-validation procedure will not markedly alter the regression and correlation coefficients.

REFERENCES

Bartlett, M.S., 1949. Fitting a straight line when both variables are subject to error. *Biometrics*, **September**, 207–212.

Daniel, C., Wood, F.S., and Gorman, J.W., 1971. *Fitting Equations to Data. Computer Analysis of Multifactor Data for Scientists and Engineers.* John Wiley & Sons, New York, p. 32.

Draper, N.R. and Smith, H., 1981. *Applied Regression Analysis*. Second Edition. John Wiley & Sons, New York, pp. 1–55.

Mackay, D., Shiu, W.Y., and Ma, K.C., 1993. *Illustrated Handbook of Physical-Chemical Properties and Environmental Fate for Organic Chemicals, Volume III, Volatile Organic Chemicals*. Lewis Publishers, Boca Raton, FL, pp. 616–624.

Weisberg, S., 1985. *Applied Linear Regression*, John Wiley & Sons, New York, pp. 76–78.

Wolberg, J.R., 1967. *Prediction Analysis*. D. Van Nostrand, Princeton, NJ.

The Molecular Connectivity Index

A5.1 INTRODUCTION

A **computable molecular descriptor** is a numerical function that encodes a topological feature* of molecular structure such as the number of chlorine atoms, chain branching, ring structure, the substitution pattern of heteroatoms, etc. Many different descriptors have been correlated with chemical properties. The molecular connectivity indices developed by Kier and Hall (1976, 1986) are widely used because they are easy to calculate, they encode structural features that control additive and constitutive chemical properties, and there is no restriction on the type of chemical structure to which they apply. This appendix describes procedures for calculating simple and valence molecular connectivity indices. The zero- and first-order indices are easy to calculate by hand. Because of the difficulty in identifying all of the components, the second- and higher-order indices are much harder to calculate. It is safer to calculate these with a computer application program. Application programs are available from Hall (1996) and Sabljic (1991).

A5.2 SIMPLE CONNECTIVITY INDICES

The procedure for calculating simple molecular connectivity indices starts with the molecular skeleton or **hydrogen-suppressed graph** of the molecule. The graph shows all atoms except hydrogen and all covalently bonded connections except to hydrogen. (Hydrogen is shown attached to heteroatoms to help identify the molecular fragments.) Intramolecular interactions such as hydrogen bonding are not shown. For example, the hydrogen-suppressed graph of isopentane is shown in Figure A5.1.

---

* Molecular *topology* is the molecular structure as defined by the number and type of individual atoms in a molecule and the bonds or connections between them. Molecular *topography* is the three-dimensional geometric characteristics of a molecule such as size, shape, volume, and surface area as determined by molecular topology. Most chemical properties depend to some degree on the detailed arrangement of atoms in a molecule, which is to say that they are constitutive properties.

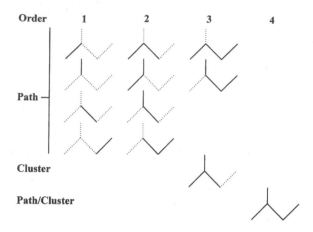

Figure A5.1 All of the distinct subgraphs (dark lines) of isopentane. (After Kier, L.B. and Hall, L.H., *Molecular Connectivity in Chemistry and Drug Research.* Academic Press, New York. With permission.)

Each atom in the graph is assigned a simple connectivity value or **delta value**, δ, equal to the number of adjacent atoms other than hydrogen to which it is connected. In formal terms, $\delta = \sigma - h$, where σ is the number of σ electrons in the atom's valence shell and h is the number of hydrogens to which the atom is bonded. For example, the δ of carbon is 4 in >C<, 3 in >CH–, 2 in –CH$_2$–, and 1 in –CH$_3$. When calculating simple connectivity values, heteroatoms are treated just like carbon atoms; adjacent hydrogen atoms are not counted.

The zero-, first-, and second-order connectivity indices encode properties associated with the size of a molecule such as surface area and volume. The **zero-order molecular connectivity index** describes the atoms in a molecular structure. It is the sum of $\delta^{-0.5}$ values for all atoms in the skeleton.

$$\cdot\ ^{0}\chi = \sum_{atoms}\left(\delta_i\right)^{-0.5} \tag{A5.1}$$

A **subgraph** is a fragment of the molecular structure. The subgraphs of isopentane are shown in Figure A5.1. The simplest subgraph consists of one chemical bond and is called a first-order subgraph. For each pair of atoms i and j that are linked by chemical bonds, a product $\delta_i\delta_j$ can be calculated. The simple **first-order molecular connectivity index** or simple **chi-one index** is the sum of the values of $(\delta_i\delta_j)^{-0.5}$ for all first-order subgraphs (σ-bonds) in the skeleton:

$$\cdot\ ^{1}\chi = \sum_{bonds}\left(\delta_i\delta_j\right)^{-0.5} \tag{A5.2}$$

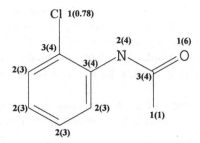

Figure A5.2 The hydrogen-suppressed graph of 2-chloroacetanilide. δ values are shown next to each atom. δ^v values are given in parentheses.

Example

Calculate the simple first-order molecular connectivity index of 2-chloro-acetanilide.

1. The hydrogen-suppressed graph of 2-chloroacetanilide is shown in Figure A5.2. Each atom is labeled with its simple δ value.
2. The first-order connectivity index is calculated using Equation A6.2.

$$^1\chi = \Sigma_{bonds}\left(\delta_i\delta_j\right)^{-0.5}$$

$$= (4)\,(2\cdot 3)^{-0.5} + (3)\,(3\cdot 1)^{-0.5} + (3)\,(2\cdot 2)^{-0.5} + (3\cdot 3)^{-0.5}$$

$$= 5.198$$

The value of δ indicates the amount of branching occurring at an atom in a molecule, and so the first-order connectivity index is inversely correlated with molecular branching. As a result, this index correlates well with molecular surface area.

Figure A5.1 shows how to divide the molecular skeleton into fragments (subgraphs) of different sizes. For all skeletal fragments in the molecular structure in which three atoms are linked by two bonds, the product $\delta_i\delta_j\delta_k$ can be calculated. The **second-order molecular connectivity index** is the sum of values of $(\delta_i\delta_j\delta_k)^{-0.5}$ for all three-atom fragments in the molecular skeleton:

$$\cdot\,^2\chi = \sum_{fragments}\left(\delta_i\delta_j\delta_k\right)^{-0.5} \tag{A5.3}$$

The $^1\chi$ and $^2\chi$ indices both encode information about the number of atoms in a molecule and about the amount of branching in a molecule. The $^1\chi$ index is more sensitive to the number of atoms than $^2\chi$. The $^2\chi$ index conveys more information about branching than $^1\chi$.

Higher-order connectivity indices, $^{m-1}\chi_P$, are calculated by summing the values of $(\delta_i \delta_j \ldots \delta_m)_{-0.5}$ for all fragments in the molecular structure in which m atoms are linked by $m - 1$ contiguous chemical bonds. The subscript P indicates that the index is calculated using path subgraphs, fragments in which the bonding between atoms is contiguous. Figure A5.1 illustrates the first-, second-, and third-order path subgraphs of isopentane.

In addition, a molecule can be broken into cluster fragments (subscript C), path/cluster fragments (PC), and chain fragments (subscript CH). A cluster is a star-shaped fragment with three or more peripheral atoms connected to a central atom. Cluster fragments are third or higher order. The third-order cluster subgraph of isopentane is shown in Figure A5.1.

A path/cluster fragment contains at least one additional atom connected to a nonperipheral atom. Path/cluster fragments are fourth or higher order. The path/cluster subgraph of isopentane is shown in Figure A5.1.

A chain fragment consists of a closed ring of atoms in the skeleton. Chain fragments describe the type of rings present in a molecule and are fifth order or higher. The $^5\chi$ and $^6\chi$ indices denote a cyclopentane ring and a cyclohexane ring, respectively.

The higher-order indices encode information about various structural details of a molecule. The $^3\chi_p$ index encodes information about a molecule's three-dimensional conformation. The $^4\chi_{pc}$ index encodes information such as the degree of substitution and the proximity of substituents. The $^5\chi$ and $^6\chi$ indices are descriptors of ring structures and also encode information about the amount of substitution on the ring.

Example

Calculate the simple molecular connectivity indices of isopentane up to fourth order.

1. The hydrogen-suppressed graph of isopentane is shown in Figure A5.1. The delta values for $CH_3CH(CH_3)CH_2CH_3$ are $\delta_1 = 1$, $\delta_2 = 3$, $\delta_3 = 1$, $\delta_4 = 2$, and $\delta_5 = 1$. The zero-order connectivity index is calculated using Equation A5.1:

$$^0\chi_P = \Sigma_{atoms} \left(\delta_i\right)^{-0.5}$$

$$= (3)\,(1)^{-0.5} + (1)\,(2)^{-0.5} + (1)\,(3)^{-0.5}$$

$$= 4.284$$

2. The first-order connectivity index is calculated using Equation A5.2:

$$^1\chi_P = \Sigma_{bonds} \left(\delta_i \delta_j\right)^{-0.5}$$

$$= (2)\,(1\cdot3)^{-0.5} + (1)\,(2\cdot3)^{-0.5} + (1)\,(2\cdot1)^{-0.5}$$

$$= 2.270$$

3. The second-order connectivity index is calculated using Equation A5.3:

$$^2\chi_P = \Sigma_{fragments} \left(\delta_i \delta_j \delta_k\right)^{-0.5}$$

$$= (3)\,(1 \cdot 3 \cdot 2)^{-0.5} + (1)\,(1 \cdot 3 \cdot 1)^{-0.5}$$

$$= 1.802$$

4. The third-order connectivity indices are

$$^3\chi_P = (2)\,(1 \cdot 3 \cdot 2 \cdot 1)^{-0.5}$$

$$= 0.816$$

$$^3\chi_c = (1 \cdot 1 \cdot 2 \cdot 3)^{-0.5}$$

$$= 0.408$$

5. The fourth-order connectivity indices are

$$^4\chi_P = 0$$

$$^4\chi_{PC} = (1 \cdot 1 \cdot 2 \cdot 3 \cdot 1)^{-0.5}$$

$$= 0.408$$

A5.3 VALENCE CONNECTIVITY INDICES

Simple molecular connectivity indices encode no information about unsaturation and the presence of heteroatoms in a molecule. This information is conveyed by the valence connectivity indices. The **valence delta**, δ^v, is calculated by counting the number of bonds to an atom in the skeleton:

$$\delta^v = \frac{Z^v - h}{z - z^v - 1} \tag{A5.4}$$

where Z^v = number of valence electrons on the atom
       h = number of suppressed hydrogen atoms bonded to atom
       Z = atomic number of atom

For the atoms of the second period of the periodic table, this formula becomes $\delta^v = Z^v - h$. The difference between the valence delta value and the simple delta value, $\delta^v - \delta$, equals the number of π-bonds and lone-pair electrons on an atom. Valence delta values for various atoms are given in Table A5.1.

Table A5.1 Valence Delta Values of Some Atoms

Group	δ^v	Group	δ^v
>N–	5	>C<	4
N in –NH–	4	C in >CH–	3
N in –NH$_2$	3	C in –CH$_2$–	2
–O–	6	–S–	0.67
O in –OH	5	–SH	0.56
–F[a]	7	S in –S–S–	0.89
–Cl[b]	0.78	S in –SO–	1.33
–Br[c]	0.26	S in –SO$_2$–	2.67
–I[d]	0.16	P in >PO–	2.22

[a] The δ^v value of F has been reported earlier as –20 (Kier and Hall, 1976) and as –7 (Kier, 1980) and the subgraph term for a C–F bond subtracted in calculating valence indices for fluorine-containing molecules.
[b] The δ^v value of Cl has been reported earlier as 0.690 (Kier and Hall, 1976) and as 0.7 (Kier, 1980).
[c] The δ^v value of Br has been reported earlier as 0.254 (Kier and Hall, 1976) and as 0.25 (Kier, 1980).
[d] The δ^v value of I has been reported earlier as 0.085 (Kier and Hall, 1976) and as 0.15 (Kier, 1980).

From Kier, L.B. and Hall, L.H., 1986. *Molecular Connectivity in Structure-Activity Analysis.* John Wiley & Sons, New York.

The rules for calculating δ^v values have changed over time. For example, the δ^v value of Cl has been given earlier as 0.690 (Kier and Hall, 1976) and as 0.7 (Kier, 1980). You should be aware that some workers still use the older values, while others use the current value for Cl of 0.78 in developing quantitative structure–property relationships. It is important to verify which convention was used by the method's author, particularly when working with halogenated hydrocarbons.

The bonding of sulfur and phosphorous atoms in sulfites, sulfates, and phosphates involves coordinate covalent bonding and expanded valence shells. In this case, the appropriate valence delta values cannot be calculated according to simple rules. Instead, empirical values are used. The empirical values are listed in Table A5.1.

Valence connectivity indices take account of the hybrid state of atoms and the type of chemical bonding between atoms. They are calculated by substituting δ^v values in the formulas given above for the simple connectivity indices. For instance, the **first-order valence connectivity index** is

$$\cdot {}^1\chi^v = \sum_{bonds} \left(\delta_i^v \delta_j^v \right)^{-0.5} \tag{A5.5}$$

Example

Calculate the first-order valence connectivity index of 2-chloroacetanilide.

1. The hydrogen-suppressed graph of 2-chloroacetanilide is shown in Figure A5.2. Each atom is labeled with its δ^v value in parenthesis.
2. The first-order valence connectivity index is calculated using Equation A5.5.

$$^1\chi^v = \Sigma_{bonds}\left(\delta_i^v\delta_j^v\right)^{-0.5}$$

$$= (3)\,(4\cdot4)^{-0.5} + (3)\,(3\cdot3)^{-0.5} + (2)\,(3\cdot4)^{-0.5} + (4\cdot6)^{-0.5} + (1\cdot4)^{-0.5} + (0.78\cdot4)^{-0.5}$$

$$= 3.597$$

Example

Calculate the third-order valence connectivity index of trifluoroacetic acid, CF$_3$COOH.

1. The δ^v values are assigned to the atoms in the hydrogen-suppressed graph. Values of δ^v are listed in Table A5.1.

Atom	δ^v
Carbon	4
Fluorine	7
Oxygen (=O)	6
Oxygen (−OH)	5

2. The third-order valence connectivity indices are calculated as follows:

$$^3\chi_P = (3)\,(7\cdot4\cdot4\cdot6)^{-0.5} + (3)\,(7\cdot4\cdot4\cdot5)^{-0.5}$$

$$= 0.243$$

$$^3\chi_C = (7\cdot7\cdot7\cdot4)^{-0.5} + (3)\,(7\cdot7\cdot4\cdot4)^{-0.5} + (1)\,(4\cdot4\cdot6\cdot5)^{-0.5}$$

$$= 0.180$$

REFERENCES

Hall, L. H., 1996. Molconn-X. Department of Chemistry, Eastern Nazarene College, Quincy, MA 02170.

Kier, L.B., 1980. Molecular Connectivity as a Descriptor of Structure for SAR Analysis. In *Physical Chemical Properties of Drugs*. Yalkowsky, S.H., Sinkula, A.A., and Valvani, S.C., Eds. Marcel Decker, New York, pp. 277–320.

Kier, L.B. and Hall, L.H., 1976. *Molecular Connectivity in Chemistry and Drug Research*. Academic Press, New York.

Kier, L.B. and Hall, L.H., 1986. *Molecular Connectivity in Structure-Activity Analysis*. John Wiley & Sons, New York.

Sabljic, A., 1991. GRAPH III. Apple Center Ljubljana, Parmova 41, 61000 Ljubljana, Slovenia, Yugoslavia.

APPENDIX 6

Special Functions

A6.1 THE ERROR FUNCTION

Most people are familiar with the **Gaussian or normal distribution** function:

$$y(x) = \frac{1}{\sqrt{2\pi}} e^{-\frac{x^2}{2}} \tag{A6.1}$$

where y(x) is the probability per unit interval of observing a deviation x from the mean value of a set of measurements. The area under the normal curve is related to the probability of making a measurement that falls within a specified interval. The area is given by the **error function**:

$$erf(x) = \frac{2}{\sqrt{\pi}} \int_0^x e^{-x^2} \, dt \tag{A6.2}$$

Evaluation methods and tables of values of erf(x) are given in standard mathematics handbooks such as Beyer (1991). Various graphical and numerical methods are summarized by Dence (1975).

The error function can be expressed in closed form only for x = 0 and x = 1: erf(0) = 0 and erf(∞) = 1. Otherwise, *for small values of x*, it is evaluated with the series expansion:

$$erf(x) = \frac{2}{\sqrt{\pi}} \sum_{n=0}^{\infty} \frac{(-1)^n x^{2n+1}}{n!(2n+1)} \tag{A6.3}$$

The value of the first m terms of the expansion approximates, within an order of magnitude, the series' asymptotic value. So for most purposes, it is adequate to retain only the first few terms of the series (Thomas, 1990).

$$erf(x) = \frac{2}{\sqrt{\pi}} \left(x - \frac{x^3}{3} + \frac{1}{2!}\frac{x^5}{5} - \frac{1}{3!}\frac{x^7}{7} - \frac{1}{4!}\frac{x^9}{9} + ... \right) \qquad \text{(A6.4)}$$

For large values of x, the error function is expressed in terms of a reciprocal power series:

$$erf(x) = \cong \frac{e^{-x^2}}{x\sqrt{\pi}} \left[1 + \sum_{n=1}^{m} (-1)^n \frac{(2n-1)!\, x^{-2n}}{(n-1)!\, 2^{2n-1}} \right] \qquad \text{(A6.5)}$$

The series diverges when x is positive, but, as above, the value of the first few terms is roughly equal to the series' asymptotic value.

The **complementary error function**, erfc(x) = 1 − erf(x), is

$$erfc(x) = \frac{2}{\sqrt{\pi}} \int_{x}^{\infty} e^{-t^2}\, dt \qquad \text{(A6.6)}$$

REFERENCES

Beyer, W. H., 1991. *CRC Standard Mathematical Tables and Formulae.* 29th Edition. CRC Press, Boca Raton, FL.

Dence, J. B., 1975. *Mathematical Techniques in Chemistry.* John Wiley & Sons, New York.

Thomas, R. G., 1990. Volatilization from Soil. Chapter 16. In *Handbook of Chemical Property Estimation Methods.* Lyman, W.J., Reehl, W.F., and Rosenblatt, D.H., Eds. American Chemical Society, Washington, D.C.

The Calculus

The fundamental ideas and vocabulary of differential and integral calculus are reviewed in this appendix. Familiarity with calculus will help you to understand the discussion of the dynamic behavior of chemicals in this book and the review of propagation of error in Appendix 3. Many good presentations of the subject are in print. See the texts by Dence (1975) and by Mortimer (1981), for instance. Tables of derivatives and integrals are also available (Beyer, 1991).

A7.1 DIFFERENTIAL CALCULUS

Suppose that the variable y is a function of the variable x, y = y(x). This means that a specific value of y(x) is associated with each value of x. The variable y(x) is called the **dependent variable**, and x, the **argument** of y(x), is the **independent variable**. We are concerned only with **single-valued** functions of x: y(x) has only one value associated with each value of x. For example, the normal probability (Gaussian) curve has the form:

$$y(x) = \frac{1}{\sqrt{2\pi}} e^{-\frac{1}{2}x^2} \tag{A7.1}$$

where y(x) is the probability per unit interval of observing a deviation of magnitude x from the mean value of a measurement. When x = 3.00, y(x) = $1/(2{\cdot}\pi)^{1/2}e - 3.00^2/2$ = 0.9987.

The statement "as x approaches the value p, the limit of y(x) equals q" means that the difference y(x) − q grows smaller as the value of x approaches p. Even if y(x) cannot be evaluated at x = p, the difference y(x) − q can be made as small as desired by choosing an appropriate value of x close to p. In mathematical shorthand, the statement is

$$\lim_{x \to p} y(x) = q \tag{A7.2}$$

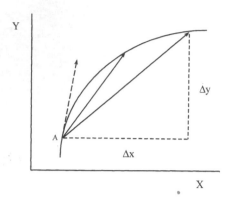

Figure A7.1 The quantity $\Delta y/\Delta x$ is the approximate slope of the curve at point A. As $\Delta x \to 0$, $\Delta y/\Delta x$ approaches m, the slope of the tangent to the curve at point A.

In what follows, the function must be continuous: $y(x)$ must be defined throughout the range of x, and the limit must equal q whether x approaches it by increasing or decreasing in value.

A function that has the form $y(x) = mx + b$ is a **linear function** of x, where m is the **slope** of the graph of y vs. x and $b = y(0)$ is the **intercept** of the graph with the y axis. Given the coordinates of two points on the graph, (y_1,x_1) and (y_2,x_2), the slope equals $(y_2 - y_1)/(x_2 - x_1) = \Delta y/\Delta x$, where $\Delta y = y_2 - y_1$, and $\Delta x = x_2 - x_1$. If $y(x)$ is not a linear function, the slope of the graph at a point is approximately equal to $\Delta y/\Delta x$. As shown in Figure A7.1, the smaller the value of Δx, the closer $\Delta y/\Delta x$ is to the value of m. In the limit as $\Delta x \to 0$,

$$m = \lim_{\Delta x \to 0} \frac{\Delta y}{\Delta x} = \frac{dy}{dx} \tag{A7.3}$$

The limit dy/dx in Equation A7.3 is called the **derivative** of y with respect to x. By definition, it is the slope of the graph of $y(x)$ or the gradient of $y(x)$ along x at point (x,y). See Figure A7.1. The quantities dy and dx are called **differentials**.

The derivatives of algebraic functions are easily found. For example, the derivative of $y = ax^2$ is obtained as follows:

$$
\begin{aligned}
\Delta y &= y_2 - y_1 & &\therefore y = ax^2 \\
&= ax_2^2 - ax_1^2 & &\therefore x_2 = x_1 + \Delta x \\
&= a(x_1 + \Delta x)^2 - ax_1^2 \\
&= a(x_1^2 + 2x_1\Delta x + \Delta x^2) - ax_1^2 \\
&= 2ax_1\Delta x + \Delta x^2
\end{aligned}
\tag{A7.4}
$$

Table A7.1 Ordinary Derivatives of Functions $y(x)$[a]

$y(x)$	dy/dx
a	0
ax^n	nax^{n-1}
e^{ax}	ae^{ax}
$\ln(ax)$	x^{-1}
$\sin(ax)$	$a\cos(ax)$
$\cos(ax)$	$-a\sin(ax)$

[a] a and n are constants.

$$\Delta y/\Delta x = 2ax_1 + \Delta x$$

$$\frac{dy}{dx} = \lim_{\Delta x \to 0} \frac{\Delta y}{\Delta x} = 2ax$$

The derivatives of some simple functions are listed in Table A7.1.

The derivative of any function found in this book can be obtained using the information in Table A7.1 along with the following rules involving two functions of x, $y(x)$ and $z(x)$, and the constant a:

$$\frac{d(y+a)}{dx} = \frac{dy}{dx} \tag{A7.5}$$

$$\frac{d(ay)}{dx} = a\frac{dy}{dx} \tag{A7.6}$$

$$\frac{d(y+z)}{dx} = \frac{dy}{dx} + \frac{dz}{dx} \tag{A7.7}$$

$$\frac{d(yz)}{dx} = y\frac{dz}{dx} + z\frac{dy}{dx} \tag{A7.8}$$

$$\frac{d(y/z)}{dx} = \frac{x\left(\dfrac{dy}{dx}\right) - y\left(\dfrac{dz}{dx}\right)}{z^2} \tag{A7.9}$$

If y is a **composite function** of x, that is, if $y(z)$ and $z(x)$,

$$\frac{dy}{dx} = \frac{dy}{dz}\frac{dz}{dx} \tag{A7.10}$$

The **second derivative** of y,

$$\frac{d^2y}{dx^2} = \frac{d(dy/dx)}{dx} \tag{A7.11}$$

is a measure of the curvature of the graph of y(x).

To this point, we were concerned only with functions of one independent variable, but we often deal with functions of two or more independent variables. In the next few paragraphs, we review procedures for obtaining derivatives of multivariate functions like f = f(x,y,z). The **partial derivative** of f(x,y,z) with respect to x, $(\delta f/\delta x)_{y,z}$, is defined as the ordinary derivative of f(x,y,z) derived as if y and z are constants:

$$\left(\frac{\delta f}{\delta x}\right)_{y,z} = \lim_{\Delta x \to 0} \frac{f(x+\Delta x, y, z) - f(x, y, z)}{\Delta x} \tag{A7.12}$$

that is, $(\delta f/\delta x)_{y,z} = df_{y,z}/dx_{y,z}$. When there is no danger of confusion, the subscripts can be dropped, so that $(\delta f/\delta x)_{y,z}$ is usually written $(\delta f/\delta x)$. For example, functions of the type $c(t,z) = e^{at}\sin(bz)$ are solutions to one-dimensional diffusion equations. Using the ordinary derivatives given in Table A7.1, we see that one partial derivative of the function is $(\delta f/\delta y)_z = ae^{ay}\sin(bz)$. The other is $(\delta f/\delta z)_y = be^{yz}\cos(bz)$.

The function f(x,y,z) describes a three-dimensional surface. Imagine that the surface is composed of a set of parallel lines placed one next to another. The partial differential $(\delta f/\delta x)_{y,z}$ is the slope of the line in the f(x),x plane that is tangent to the surface at point (x,y,z). It is the gradient of f(x,y,z) in the x direction.

The derivative of any function f(x,y,z) that appears in this book can be derived with the information given in Table A7.1 and the following rules:

$$df = \left(\frac{\delta f}{\delta x}\right)_{y,z} dx + \left(\frac{\delta f}{\delta y}\right)_{x,z} dy + \left(\frac{\delta f}{\delta z}\right)_{x,y} dz \tag{A7.13}$$

$$\left(\frac{\delta x}{\delta y}\right)_z = \frac{1}{(\delta y/\delta x)_z} \tag{A7.14}$$

$$\left(\frac{\delta x}{\delta y}\right)_z \left(\frac{\delta y}{\delta z}\right)_x \left(\frac{\delta z}{\delta x}\right)_y = -1 \tag{A7.15}$$

Notice that in Equation A7.15 the differentials δx_z, δy_x, ... don't cancel because a different variable is held constant in each derivative. This means that the three gradients in Equation A7.15 are along three different directions.

$$\left(\frac{\delta f}{\delta z}\right)_y = \left(\frac{\delta f}{\delta x}\right)_y \left(\frac{\delta x}{\delta z}\right)_y \tag{A7.16}$$

Table A7.2 Integrals of Simple Functions f(x)[a]

f(x)	$\int f(x)dx$
a	$ax + C$
ax^n	$(n+1)^{-1}ax^{n+1} + C$
e^{ax}	$a^{-1}e^{ax} + C$
ln (ax)	$ax^{-1} + C$
sin (ax)	$-a^{-1}\cos(ax) + C$
cos (ax)	$a^{-1}\sin(ax) + C$

[a] a and n are constants. C is an arbitrary constant.

In Equation A7.16, all derivatives are for constant y.

Second-order partial derivatives are quantities such as $\delta^2 f/\delta x^2$ and $\delta^2 f/\delta x\delta y$. Such terms are found in diffusion models, for example. For all multivariate functions described in this book, the order of partial differentiation is immaterial, so that $\delta^2 f/\delta x\delta y = \delta^2 f/\delta y\delta x$. Consider the term $\delta(D\delta c/\delta z)/\delta z$. Its meaning depends on the nature of D. If $D = D(z)$ is a function of z, then we use Equation A7.8 to show that $\delta[D(z)\delta c/\delta z]/\delta z = [\delta D(z)/\delta z][\delta c/\delta z] + D(z)[\delta^2 c/\delta z^2]$. On the other hand, if D is not a function of z, then $\delta(D\delta c/\delta z)/\delta z = D(\delta^2 c/\delta z^2)$.

A7.2 INTEGRAL CALCULUS

Integration is the process of finding a function y(x) when its derivative f(x) = dy/dx is known. The solution to this problem is given by the **indefinite integral** of f(x),

$$y(x) = \int f(x)\,dx \qquad (A7.17)$$

For example, if dy/dx = 1/x, we use the information in Table A7.1 to find the general solution y(x) = ∫(1/x)dx = ln x + C, where C is an arbitrary constant added so that the result is a general solution to the problem. Integrals of some simple functions are listed in Table A7.2.

The difference y(b) − y(a), where a and b are two values of x, is given by the **definite integral** of f(x) between the limits a and b:

$$y(b) - y(a) = \int_a^b f(x)\,dx \qquad (A7.18)$$

Here, the integration constant C cancels itself in the result.

The definite integral is a number equal in value to the area between the graph of f(x) and the x axis. The area is estimated as follows. The interval from a to b is

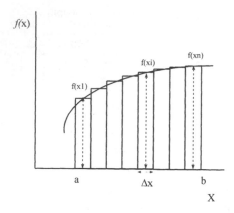

Figure A7.2 The approximate area under the curve between points a and b is the sum of the areas of the n segments, $\Sigma_i f(x_i) \cdot \Delta x$. The value of the sum approaches the exact area as $\Delta x \to 0$.

divided into n segments each with height $f(x_i)$, width Δx, and area $f(x_i)\Delta x$. The total area under $f(x)$ between points a and b is approximately equal to the sum of the areas of the segments. As $\Delta x \to 0$, the value of the sum approaches the value of the total area. The limit of the sum as $\Delta x \to 0$ is defined as the definite integral of $f(x)$,

$$\int_a^b f(x)\,dx = \lim_{\Delta x \to 0} \sum_{i=1}^n f(x_i)\Delta x \qquad (A7.19)$$

The value of the definite integral exactly equals the area under the graph of $f(x)$ between points a and b. This is illustrated in Figure A7.2.

The following integration rules are useful to know:

$$\int_a^b f(x)\,dx = -\int_b^a f(x)\,dx \qquad (A7.20)$$

$$\int_a^c f(x)\,dx = \int_a^b f(x)\,dx + \int_b^c f(x)\,dx \qquad (A7.21)$$

Multiple integrals must sometimes be solved when dealing with multivariate functions. The order in which multiple integrals are expanded is as follows:

$$\int_a^b \int_c^d \int_t^s f(x)f(y)\,f(z)\,dxdydz = \int_t^s f(x)\,dx \int_c^d f(y)\,dy \int_a^b f(z)\,dz \qquad (A7.22)$$

The individual integrals are then solved using the information given in Table A7.2. For example,

$$\int_{0}^{2\pi}\int_{0}^{\pi}\int_{0}^{\infty} r^2 \sin\theta\, dr\, d\theta\, d\phi = \int_{0}^{2\pi} d\phi \int_{0}^{\pi} \sin\theta\, d\theta \int_{0}^{a} r^2\, dr = 1 \cdot 1 \cdot \frac{a^3}{3}$$

REFERENCES

Beyer, W. H., 1991. *Standard Mathematical Tables and Formulae.* 29th Edition. CRC Press, Boca Raton, FL.

Dence, J. B., 1975. *Mathematical Techniques in Chemistry.* John Wiley & Sons, New York.

Mortimer, R. G., 1981. *Mathematics for Physical Chemistry.* Macmillan, New York.

Software and Data Sources
for Chemical Property Estimation

A8.1 APPLICATION SOFTWARE WITH WHICH TO ESTIMATE
SUITES OF CHEMICAL PROPERTIES

Computer programs used to estimate various chemical properties are listed.

Table A8.1 Software to Estimate Chemical Partitioning

Property	QPPR Method	QSPR Method
Melting point	CHEMEST	
Boiling point		CHEMEST, DSOC
Activity coefficient		PC-UNIFAC 4.0
Aqueous solubility	DTEST, CHEMEST, WSKOWWIN	CHEMEST, DSOC, MIXTURES 1.1
Vapor pressure	DTEST, CHEMEST, DSOC, MPBPVP	MIXTURES 1.1
K_{AW}	DTEST, CHEMEST	CHEMEST, DSOC, HENRY
K_{OW}	DTEST, CHEMEST	CLOGP, DSOC, LOGKOW
K_{OC}	DTEST, CHEMEST, DSOC	PCKOC
Bioconcentration factor	DTEST, CHEMEST	
Liquid viscosity	DSOC	
Surface tension	DSOC	
Parachor	DSOC	
Liquid density	DSOC	
K_A		CHEMEST

ADAPT: The program computes an array of molecular descriptors. Jurs, P.C., 1986. *Computer Software Applications in Chemistry.* John Wiley & Sons, New York.

CHEMEST: Lyman, W. J. and Potts, R. G., 1987. *CHEMEST: User's Guide – A Program for Chemical Property Estimation*, Version 2.1. A. D. Little, Cambridge,

MA; Boethling, R. S., Campbell, S. E., Lynch, D. G., and LaVeck, G. D., 1988. Validation of CHEMEST, an on-line system for the estimation of chemical properties. *Ecotoxicol. Environ. Saf.,* **15**, 21–30; Lynch, D.G., Tirado, N.F., Boethling, R.S., Huse, G.R., and Thom, G.C., 1991. Performance of on-line chemical property estimation methods with TSCA premanufacture notice chemicals. *Sci. Total Environ.,* **109/110**, 643–648.

E4CHEM DTEST: Bruggemann, R. and Munzer, B., 1990. A Graph Theoretical Method to Estimate Substance Data. In *Software-Developments in Chemistry, 4.* Gasteiger, J., Ed. Springer-Verlag, Berlin, pp. 85-96.

DSOC: Kohlenbrander, J.P., Drefahl, A., and Reinhard, M., 1995. *DSOC User's Guide.* Stanford Bookstore, Stanford, CA 94305.

MIXTURES 1.1 bri, P.O. Box 7834, Atlanta, GA 30357-0834.

PCCHEM: U.S. Environmental Protection Agency, Washington, D.C.

POLLY 2.3. The program computes a wide array of molecular descriptors. Basak, S.C., Harris, D.K., and Magnuson, V.R. Natural Resources Institute, University of Minnesota, Duluth, MN 55811.

A8.2 APPLICATION SOFTWARE TO ESTIMATE INDIVIDUAL PROPERTIES

ARVOMOL/CONTOUR: Pacios, L.F., 1994. ARVOMOL/CONTOUR: Molecular surface areas and volumes on personal computers. *Comput. Chem.,* **18**, 377–385.

CLOGP: Leo, A. J. and Weininger, D., 1984. *CLOGP, Version 3.2 — User Reference Manual.* Medicinal Chemistry Project. Pomona College, Claremont, CA.

Coe, D.A., 1995. Henry's law constant program. *J. Chem. Inf. Comput. Sci.,* **35**, 168–169.

HENRY: Howard, P.H., 1992. *Henry's Law Constant Program.* CRC Press/Lewis Publishers, Boca Raton, FL.

LOGKOW: Meylan, W.M. and Howard, P.H., 1995. Atom/fragment contribution method for estimating octanol/water partition coefficients. *J. Pharm. Sci.,* **84**, 83–92.

MOLCONN-Z: A program to calculate molecular connectivity indices. Hall Associates, 2 Davis Street, Quincy, MA 02170.

MPBPVP: A program to estimate vapor pressure from boiling point. Syracuse Research Corporation, Merrill Lane, Syracuse, NY 13210.

PCKOC: Howard, P.H., 1992. *PCKOC Soil/Sediment Adsorption Constant Program.* CRC Press/Lewis Publishers, Boca Raton, FL.

PC-UNIFAC 4.0 bri, P.O. Box 7834, Atlanta, GA 30357-0834.

WSKOWWIN: Meylan, W.M. et al., 1996. Improved method for estimating water solubility from octanol/water partition coefficient. *Environ. Toxicol. Chem.,* **15**, 100–106.

A8.3 APPLICATION SOFTWARE TO ESTIMATE CHEMICAL FATE

Hamrick, K.J., Kollig, H.P., and Bartell, B.A., 1992. Computerized extrapolation of hydrolysis rate data. *J. Chem. Inf. Comput. Sci.,* **32**, 511–514.

Klopman, G., Dimayuga, M., and Talafous, J., 1994. META 1. A program for the evaluation of metabolic transformation of chemicals. *J. Chem. Inf. Comput. Sci.,* **34**, 1320–1325.

Howard, P.H., 1992. *Hydrolysis Rate Program.* CRC Press/Lewis Publishers, Boca Raton, FL.

Howard, P.H. and Meylan, W.M., 1992. Biodegradation Probability Program. CRC Press/Lewis Publishers, Boca Raton, FL.

Nelson, R.N., 1995. HYDRO. *J. Chem. Inf. Comput. Sci.,* **35**, 327.

A8.4 SOME DATA SOURCES

(Additional sources are given in the introduction to individual chapters.)

Aldrich Handbook of Fine Chemicals. 1996. Aldrich Chemical Company, Milwaukee, WI (for boiling point, melting point).

Boublik, T., Fried, V., and Hala, E., 1984. *The Vapor Pressures of Pure Substances: Selected Values of the Temperature Dependence of the Vapor Pressures of Some Pure Substances in the Normal and Low Pressure Region. Vol. 17.* Elsevier, Amsterdam.

Dean, J.R., 1966. *Lange's Handbook of Chemistry.* McGraw-Hill, New York (primarily for density, boiling point, melting point).

Graselli, J., Ed., 1990. *Handbook of Data of Organic Compounds.* CRC Press, Boca Raton, FL.

Hansch, C. and Leo, A., 1995. STARLIST database of Log K_{OW} values. MEDCHEM Project, Pomona College, Claremont, CA.

Howard, P.H., 1992. *Database of SMILES Notations.* CRC Press/Lewis Publishers, Boca Raton, FL.

Howard, P.H. and Meylan, W.M., 1997. *Handbook of Physical Properties of Organic Chemicals.* CRC Press/Lewis Publishers, Boca Raton, FL (C_W^S, K_{OW}, P^S, pKa, K_{AW}).

Lide, D.R., Ed., 1996. *Basic Laboratory and Industrial Chemicals. A CRC Quick Reference Handbook.* CRC Press, Boca Raton, FL (for boiling point, melting point, and thermodynamic properties).

Lide, D.R., Ed., 1996. *CRC Handbook of Chemistry and Physics.* CRC Press, Boca Raton, FL (primarily for density, boiling point, and melting point).

Mackay, D., Shiu, W.Y., and Ma, K.C., 1992. *Illustrated Handbook of Physical-Chemical Properties and Environmental Fate for Organic Chemicals.* Volume 1. Monoaromatic Hydrocarbons, Chlorobenzenes, and PCBs. Lewis Publishers, Boca Raton, FL (T_m, T_b, V_M, C_W^S, K_{OW}, P^S, K_{AW}).

Mackay, D., Shiu, W.Y., and Ma, K.C., 1992. *Illustrated Handbook of Physical-Chemical Properties and Environmental Fate for Organic Chemicals*. Volume 2. Polynuclear Aromatic Hydrocarbons, Polychlorinated Dioxins, and Dibenzo-furans. Lewis Publishers, Boca Raton, FL (T_m, T_b, V_M, C_W^S, K_{OW}, P^S, K_{AW}).

Mackay, D., Shiu, W.Y., and Ma, K.C., 1993. *Illustrated Handbook of Physical-Chemical Properties and Environmental Fate for Organic Chemicals*. Volume 3. Volatile Organic Chemicals. Lewis Publishers, Boca Raton, FL (T_m, T_b, V_M, C_W^S, K_{OW}, P^S, K_{AW}).

Mackay, D., Shiu, W.Y., and Ma, K.C., 1995. *Illustrated Handbook of Physical-Chemical Properties and Environmental Fate for Organic Chemicals*. Volume 4. Oxygen, Nitrogen, and Sulfur Containing Compounds. CRC Press/Lewis Publishers, Boca Raton, FL (T_m, T_b, V_M, C_W^S, K_{OW}, P^S, K_{AW}).

Miller, T.M., Boiten, J.W., Ott, M.A., and Noordik, J.H., 1994. Organic reaction database translation from REACCS to ORAC. *J. Chem. Inf. Comput. Sci.,* **34,** 653–660.

Sangster, J., 1993. LOGKOW DATABANK. Sangster Research Laboratories, Montreal, Quebec, Canada (log K_{OW} data on disk).

Syracuse Research Corp., 1997. *PHYSPROP Database*. Syracuse Research Corporation, Merrill Lane, Syracuse, NY; Bloch, D.E., 1995. Review of PHYSPROP database (Version 1.0). *J. Chem. Inf. Comput. Sci.,* **35,** 328 (C_W^S, K_{OW}, P^S, pKa, K_{AW}).

U.S. Department of Commerce, 1995. *DIPPR Data Compilation of Pure Compound Properties, 1995. Version 10.0.* Project sponsored by the Design Institute for Physical Property Data. American Institute of Chemical Engineers, NIST Standard Reference Database 11, U.S. Department of Commerce, Gaithersburg, MD 20899 (melting point, boiling point, solubility, density, vapor pressure, viscosity, surface tension, and thermodynamic properties).

Verschueren, K., 1983. *Handbook of Environmental Data on Organic Chemicals,* Second Edition. Van Nostrand Reinhold, New York.

Windholz, M., Ed., 1986. *The Merck Index*. Merck & Co, Rahway, NJ.

Yalkowsky, S.H. and Dannenfelser, R.M., 1992. Arizona DATABASE, Version 5. College of Pharmacy, University of Arizona, Tucson (AQUASOL aqueous solubility data base).

A8.5 ON-LINE DATA BASES

Beilstein OnLine. Springer-Verlag Electronic Information Services, New York, NY.

Dialog Information Services. Palo Alto, CA. Provides data on suites of physical properties through Beilstein Online and other sources.

ECDIN, Environmental Chemicals Data and Information Network. Commission of the EC, JRC Ispra (Italy); Penning, W., Roi, R., and Boni, M., 1990. ECDIN — the European Data Bank on Environmental Chemicals. *Toxicol. Environ. Chem.,* **25,** 251–264.

STN International. Columbus, OH. Provides data on suites of physical properties through Design Institute for Physical Property Data, Beilstein Online, and other sources.

Voigt, K., Benz, J., and Mucke, W., 1989. Databanks of data sources for environmental chemicals. *Toxicol. Environ. Chem.*, **23**, 243–250.

Voigt, K., Pepping, T., Kotchetova, E., and Mucke, W., 1992. Testing of online databases in the information system for environmental chemicals with a test set of 68 chemicals. *Chemosphere*, **24**, 857–866.

A8.6 INTERNET RESOURCES

American Chemical Society, Washington, D.C. (http://www.ChemCenter.org) offers all on-line resources of the society.

Biobyte Corp., Claremont, CA. (http://www.biobyte.com/~clogp/) offers ClogP and QSAR regression software.

Biosym-Molecular Simulations Inc., San Diego, CA. (http://www.msi.com/weblab) offers computational chemistry application programs for molecular properties.

CambridgeSoft Corp., Cambridge, MA. (http://www.chemfinder.camsoft.com) offers a database of chemical properties plus links to other chemical information resources on the internet.

CaChe Scientific, Beaverton, OR. (http://www.ig.com) offers molecular modeling, computational chemistry, and QSAR software.

Daylight Chemical Information Systems, Irvine, CA. (http://www.daylight.com) offers a variety of chemical databases.

MDL Information Systems, San Leandro, CA and Current Science Group, London. (http://www.ChemWeb.com) offers a data base of chemical properties.

Syracuse Research Corporation, Syracuse, NY. (http://esc.syrres.com) offers the PHYSPROP database and software to estimate vapor pressure (MPBPVP), log K_{OW} (LOGKOW), aqueous solubility (WSKOWWIN), Henry's law constant (HENRY), and other chemical properties.

Tripos Inc., St. Louis, MO. (http://www.tripos.com) offers molecular modeling and computational chemistry software.

Wavefunction Inc., Irvine, CA. (http://www.wavefun.com) offers molecular modeling and computational chemistry software.

Index

Atmospheric pollutants, plant uptake of, 234
Atomic diffusion volumes, 249, 251
Atomic mass unit, 311, 315
Atrazine, 94, 191, 198, 202, 224, 281
AUTOCHEM, 83, 187
Avogadro constant, 311
Azinphos-methyl, 184, 188
Azobenzene, 92

B

Bahnick and Doucette's method, 173–177
Barometric pumping, 5, 271, 273–274
Bartlett's three-group procedure, 332, 341–345
Bendiocarb, 191
Benefin (benfluralin), 188
Benomyl, 190
Bensulide, 188
Benz(a)anthracene, 180
Benzaldehyde, 89
Benzamide, 93, 184, 196, 198
Benzene
　air–water partitioning of, 277, 301
　aqueous solubility of, 84, 114–116
　density of, 40
　diffusivity of, 255
　octanol-water partitioning of, 143, 159, 160, 318, 326
　organic carbon partitioning of, 277
　partial pressure of, 65–66
　soil sorption of, 173, 196, 202
　volatilization of, 277–278, 290, 298, 301–302
Benzenediol, 89
Benzidine, 185
Benzo(b)fluoranthene, 85
Benzo(j)fluoranthene, 85
Benzo(k)fluoranthene, 85
Benzo(a)fluorene, 85
Benzo(b)fluorene, 85
Benzo(b)furan, 224
Benzoic acid, 32, 90, 184, 198
Benzoic acid butyl ester, 185
Benzoic acid ethyl ester, 185, 198
Benzoic acid methyl ester, 185, 198
Benzoic acid phenyl ester, 185, 198
Benzo(b)naphtho(2,3-d)thiophene, 224
Benzonitrile, 92
Benzophenone, 89
Benzo(a)pyrene, 85, 180, 217, 224, 243
Benzo(e)pyrene, 85
Benzo(f)quinoline, 180
Benzo(b)thiophene, 180, 224
Benzyl alcohol, 255
Benzylamine, 92
Benzyl bromide, 150
Benzyl butyl phthalate, 91
o-t-Benzyl carbamate, 93
BHC (benzene hexachloride), 180, 202. *See also* Lindane

Bibenzyl, 85
Bifenox, 191
Bifenthrin, 188
Bioaccumulation factors (BAF), 212, 229–237
Bioconcentration factors (BCF), 211–237
　definition of, 211
　estimation of, 214–232
　overview of, 4, 212–214
　partitioning and, 5
　sensitivity of, 226
　software for, 365
　tabulated values of, 217–225
Biphenyl, 85, 180
2,2'-Biquinoline, 203
Bis(2-chloroethyl)ether, 217
2,2'-Bithiophene, 224
BLOGP, 159
BMPC, 185
Boiling point, 5, 14, 20, 23–30, 57, 365
Boltzmann constant, 311
Boundary layers, 268–270, 286–287, 293
BPMC, 217
Brachydanio rerio, 217–218, 220–225
Branching factors, 149–150
Bromacil, 166, 175, 203, 217
Bromacil acid, 191
Bromobenzene, 86, 172
1-Bromobutane, 86
Bromochlorobenzene, 86
Bromochloromethane, 85
1-Bromo-3-chloropropane, 86
Bromodichloromethane, 85
Bromoethane, 85
5-Bromoindole, 217
1-Bromo-3-methylbutane, 86
1-Bromo-2-methylpropane, 86
Bromonaphthalene, 37–38, 88
4-Bromophenol, 45, 91, 173, 180, 198, 217
Bromophos, 95, 217
Bromopropane, 86
Bromoxynil octanoate ester, 188
Butane, 255
Butanethiol, 94
Butanol, 88, 184, 255, 298
2-Butoxyethanol, 91, 217
Butralin, 175, 190
Butyl acetate, 90
1-Butylamine, 185
Butylate, 191
Butylbenzene, 84, 172, 196
Butyl benzyl phthalate, 185, 217
Butyl carbamate, 93
t-Butyl isopropyl ether, 217
t-Butyl methyl ether, 217
Butylphenol, 89, 217
t-Butylphenyldiphenyl phosphate, 224
Butyraldehyde, 89
Butyramide, 255
Butyranilide, 185
Butyric acid, 89, 255

C

M

Oxalic acid, 89
Oxamyl, 192
Oxycarboxin, 189
Oxydemeton-methyl, 192
Oxyfluorfen, 189
Oxygen mass transfer, 293, 295–299
Oxythioquinox (Quinomethionate), 192

P

PAHs. *See* Polycyclic aromatic hydrocarbons
Parachor, 47–53, 330, 365
Parathion (ethyl parathion), 95, 189, 198, 202
Partial vapor pressure, 55, 65. *See also* Vapor
    pressure
Particle transport, 205
Partition coefficients, 3–5, 13. *See also* Air-water
    partition coefficient; Octanol-water partition
    coefficient; Organic carbon-water partition
    coefficient; Organic matter-water partition
    coefficient; Soil-water partition coefficient
PCBs. *See* Polychlorinated biphenyls
PCCHEM, 83, 187, 366
PCGEMS, 83, 187
PCKOC, 365–366
PCNB, 192
Peat soils, 166
Pebulate, 95, 189
Pelargonaldehyde, 89
Pendimethalin, 192
Pentachloroaniline, 221, 225
Pentachlorobenzene
    aqueous solubility of, 80, 87, 99, 114
    bioconcentration of, 216, 221, 225
    octanol-water partitioning of, 160, 198, 221,
        225
    organic carbon partitioning of, 176
    soil sorption of, 172
1,1,3,4,4-Pentachloro-1,2-butadiene, 86
Pentachloroethane, 85, 221, 334, 342
Pentachloronitrobenzene, 94, 221
Pentachlorophenol
    aqueous solubility of, 91
    bioconcentration of, 221, 237
    cosolvent effects on, 206
    octanol-water partitioning of, 160, 196, 198,
        221
    organic carbon partitioning of, 175, 194–195,
        197, 200, 205
    soil sorption of, 200
Pentachlorophenyl, 172
1,4-Pentadiene, 84
Pentamethylbenzene, 84
Pentane, 84, 255
Pentanol, 88, 184, 298
2-Pentanone, 298
Pentene, 84
Pentylbenzene, 84
1-Pentyne, 84
Percolation, 5
Permethrin, 186, 189, 221

Peroxides, 28
Pesticides
    air-water partitioning of, 126, 133
    bioconcentration of, 215, 231, 233, 235–236
    diffusivity of, 259
    organic carbon partitioning of, 174, 179, 181,
        191–192, 201, 204
    soil sorption of, 167–168
    transport of, 205, 265, 269, 281–283
pH
    bioconcentration and, 228
    calculation of, 328
    octanol-water partitioning and, 161
    soil sorption and, 161, 170, 195–200, 205
Phenanthrene, 85, 172, 202
sec-Phenethyl alcohol, 184
Phenmedipharm, 190
Phenobarbital, 93
Phenol
    aqueous solubility of, 89, 97
    bioconcentration of, 221
    boiling point of, 26–27
    melting point of, 31
    octanol-water partitioning of, 199, 221
    organic carbon partitioning of, 180
    vapor pressure of, 63–64
Phenolphthalein, 91
Phenothiazin, 94
2-Phenoxyethanol, 91
Phenthoate, 221
Phenylacetic acid, 90, 184, 198
Phenylalanine, 93, 256
3-Phenyl-1-cyclohexylurea, 186
3-Phenyl-1-cyclopentylurea, 175
2-Phenyldodecane, 221
Phenylenediamine, 92
1-Phenylethanol, 89
Phenylmethanol, 89
N-Phenyl-2-naphthylamine, 221
p-Phenylphenol, 89
Phenyl salicylate, 91
Phenylthiourea, 94
Phenylurea, 175, 186
Phenytoin, 93
Phorate, 95, 189, 198
Phosalone, 192
Phosmet, 189, 221
Phosphamidon, 190
Phthalamide, 93
Phthalic acid, 90, 184
Phthalic anhydride, 323
Phthalimide, 93
Phthalonitrile, 92
Picloram, 175
Pimelic acid, 256
Pimephales promelas, 217, 219–225
Pinuophalespromelas, 217
Piperalin, 189
Piperophos, 186, 189
Pirimicarb, 186, 191
Pirimiphos-methyl, 192